Pirate Radio Stations

Tuning in to Underground Broadcasts in the Air and Online

Andrew Yoder

BRAD!
HOPE YOU HAVE FUN
LISTENING TO THE
PIRATES!

McGraw-Hill

New York   Chicago   San Francisco
Lisbon   London   Madrid   Mexico City
Milan   New Dehli   San Juan   Seoul
Singapore   Sydney   Toronto

Cataloging-in-Publication Data is on file with the Library of Congress.

McGraw-Hill

A Division of The McGraw-Hill Companies

1 2 3 4 5 6 7 8 9 0   DOC/DOC   0 9 8 7 6 5 4 3 2 1

P/N 0-07-137564-3
Part of ISBN 0-07-137563-5

The sponsoring editor for this book was Scott Grillo, the editing supervisor was Daina Penikas, and the production supervisor was Sherri Souffrance. It was set in Bancroft Book through the services of Cabinet Communications (Editing, Design, and Production).

Printed and bound by R.R. Donnelley & Sons Company.

This book is printed on recycled, acid-free paper containing a minimum of 50% recycled, de-inked fiber.

McGraw-Hill books are available at special quantity discounts to use as premiums and sales promotions, or for use in corporate training programs. For more information, please write to the Director of Special Sales, McGraw-Hill, 2 Penn Plaza, New York, NY 10121. Or contact your local bookstore.

Dedication

This book is dedicated to all of the pirate stations who take risks to be creative, funny, or informative--to provide good radio. You make radio listening not only fun, but worthwhile.

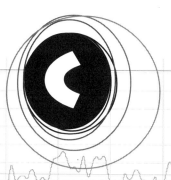

Contents

Contents

Contents

Acknowledgments

Thanks to the following hobbyists (in no particular order): Jon Anderson, editor of the pirate radio section on About.com; John Cruzan of the Free Radio Network; Jonny of SRS; Dave Valko; David Hodgson; I.O. Snopes; Beerus Maximus; Pete Costello; Frederick Moe; Niel Wolfish, Greg Majewski, and Harold Frodge of *Free Radio Weekly*; Hans-Joachim Koch; Martin Schoech of SRS-Germany; John Sedlacek; Ed Cummings; Ron Kocher; John Calabro and Perry Cavalieri of Radio Free New York; Allan Weiner, Tim Smith, and Scott Becker of WBCQ; Captain Fred of the AMPB; Pete triDish of the Prometheus Radio Project; Sig Benson and Dale Kinsinger; Alex Draper; John Arthur of *The ACE*; George Zeller of *Monitoring Times* and *The ACE*; Harry Helms; Janice Laws; Craig and Kim Harkins; Pat Murphy of the Free Radio Network; Alan Handelman; Steve Coletti; Axel Rose; Bill Finn; Jerry Coatsworth; Phil Schoenthal; Dawn and Steve Foehner; Art Johnson; Rob Keeney; Terry Provance; Guy Connor; Bill Martin; Mike Townsend; Christopher Maxwell of the Radio Free Richmond project; Martin Van der Ven; Mike Brand; Kris Field; Robert Whyley; Zacharias Liangas; Robert Gregory; I. O. Snopes; Andrew Howlett; Dave Valko; Jim Kay; Scott Hepler; Dr. David Benson; Dave Homan; Bill Taylor; Wes Harris; Paul Johnson; Johnny-John Baker; Aaron Bittner; David McCandless; Paul Sanders; JohnSedlacek; Al Fansome; Jesse Walker; Kai Salvesen, Nick Catford; Patrick Willfur; Charles Collins; Bob Gardner; and John S. Platt

Thanks to the following stations (in no particular order): Captain Disturbio of WVDA, KMUD, Blackbeard of Jolly Roger International, Radio Cochiguaz, Andino Relay Service, Radio Blandengue, Fearless Fred of Radio Garbanzo, Radio Animal of WKND, W.D.C.D., Radio Dr. Tim; Joe Stalin and Pigmeat Martin of Voice of the Angry Bastard; Captain Ron; Radio Black Arrow, Radio Foxfire; Radio Borderhunter; Radio Torenvalk; Voice of the Netherlands; Radio Nova International

(UK); Mike Radio; Crazy Wave Radio; Britain's Better Music Station; Union Radio; Radio Astoria; A.J. Michaels of Action Radio; Allan Maxwell of KIPM; Owsley of Up Against the Wall Radio; Ground Zero Radio; Radio Neptune; Peter Worth of Anteater Radio; Bill O. Rights of Radio Free Speech; Dr. A.G. Bell of the NFFR; Mike Martin and Scott Blixt of the Voice of the Voyager; He-Man Radio; Pirate Rambo of CSIC; Moe Howard of WMOE; V-Man at Radio Free Santa Cruz; E. H. Pirate Relay Service; James Beebop Brown of Solid Rock Radio; Pat Edison of Radio Kaleidoscope; Radio Aquarius; Frank Carson of Radio Nova International (Netherlands); Master Control; Captain Ganja of Radio Free Euphoria; James Brownyard of WHYP; Radio East Coast Holland; Radio Doomsday; Captain Eddy of Radio Airplane; Phil Muzik of KNBS; and Radio Black Power.

Thanks to Kellie Hagan, Sally Craley, Jessie McCleary, Amanda Rudisill, and Crystal Clifton for reinvigorating my sense of wonder and appreciation for the whole book-production process.

Thanks to Scott Grillo of McGraw-Hill for first believing in this project and then being unbelievably patient with me as I slowly assembled this third edition. Thanks also to Daina Penikas, of McGraw-Hill, for the great help with the editing and layout.

Special thanks to Richard and Judy Yoder; Corbin and Bryn Yoder; Angie, Edward, and Molly Piwonka. Ultra-special thanks to Yvonne for tolerating wires, radios, late-night phone calls, even later-night DX sessions, piles of cassettes, and stories about pirate radio for all of these years.

My apologies to those I missed!

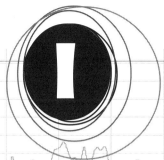

Introduction

About the Third Edition

In many ways, it's hard to believe that I wrote the first edition of *Pirate Radio Stations* just over 13 years ago. When I wrote that edition, pirate radio in the United States and Canada was at its lowest level in activity since the pirates began broadcasting en masse around 1976. Also, the stations were very disconnected in terms of interpersonal communications. Sure, a few different pirate operators knew each other, but most were fairly solitary figures—an anonymous hobbyist airing an occasional program.

Around the time that the book was published, more radio hobbyists began working together. Some of the earliest little pirate gatherings (that I know of) occurred in the late 1980s. The annual Shortwave Listener (SWL) Winterfest, which has a pirate radio talk and numerous pirate fans in attendance, started in 1987. Radio Newyork International, which broadcast from a ship off the coast of Long Island in 1987 and 1988, also helped inspire (with its broadcasts and the massive national attention that it received) a number of shortwave pirates and a new wave of political "microbroadcasters" and "low-power FM" stations.

All of this was fuel for the pirate radio explosion in the 1990s, which I feel was caused by (or at least significantly aided by) the Internet. The Internet has made it possible for information to be spread very quickly, such as shortwave radio loggings listed on *The ACE* page while the broadcasts are still on the air. Or test schedules being passed along by e-mail or posted on the Free Radio Network, hobbybroadcasting.com, or alt.radio.pirate. Also, ICQ and chat rooms have made it possible for pirates and their listeners to communicate in real time, sometimes while a broadcast is still occurring.

Since the dawn of computers, radio hobbyists have complained about this evil pied piper, who was luring the once-happy children of the world away into a hellish digital stupor. Sure, computers do take some potential radio hobbyists away because our time in this world is limited, not just in length, but by what we can squeeze into our daily schedule. However, the communications potential of the Internet for radio applications is tremendous. You can buy your shortwave receiver or transmitter online, read receiver reviews on the Radio Nederland web page, and e-mail the station or another listener. The advantages of listening for pirates in the 21st century are vast compared to 1988.

But where the Internet has had its greatest effect is with the FM pirates, who have exploded from essentially nothing at the writing of the first edition to a real force that has hit the headlines of nearly every major newspaper in the United States. Information about the FM stations were included in the second edition of the book, but the Internet was still in its toddler stage when that edition was written in 1995. The FM stations are primarily operated and programmed by those with no prior technical or radio experience. Because of this trend, the Internet has served as both disseminator and educator for the masses. Prior to the late 1990s, those with an interest in community radio would have either never discovered pirate radio or would not have known where to start.

Since this time, the synonymous terms FM pirate radio and microcasting have sparked radio rallies, protests, and even some legalization. Because of the efforts of FM station operators and listeners, the FCC has instituted a new class of FM radio license, which allows noncommercial, nonprofit community radio stations to operate with either 10 or 100 watts (depending on the class of service). This new, legal service falls under the term LPFM.

This book is about pirate radio, not licensed radio, so although it doesn't include some information concerning the LPFM movement, I'm not heavily covering the rules, regulations, and procedures for LPFM.

Another adjacent topic to pirate radio is netcasting. Netcasting or webcasting is broadcasting an audio signal via the Internet. Although I've heard some people mistakenly refer to it as pirate radio, it is neither pirate nor radio in the true sense of either words. However, a chapter on netcasting is included because a number of pirates simulcast or offer old programs on the Internet. By the radio hobbyist definition, hopping on a web page and clicking the mouse doesn't mean that you have "heard the station" and you don't deserve a QSL for it. However, it is worth checking out if you have an interest in simply hearing the programming on some of these stations.

This book is dedicated to pirate radio, a topic that I have been fascinated with for two decades. My goals are to accurately provide a history of this topic, document the activity of present-day stations, and provide information so that readers can get a taste of underground radio or even make a hobby of listening and QSLing stations.

Contact

I don't do equipment recommendations and I don't have the resources to respond to all of the questions about what receiver to buy. However, if you would like to drop me a line concerning this book or your personal experiences with pirate radio, feel free to contact me. Some of the material in this book was received from pirate radio enthusiasts who wrote in after reading a previous edition of *Pirate Radio Stations.*

Andrew Yoder
P.O. Box 642
Mont Alto, PA 17237 USA
info@hobbybroadcasting.com
http://www.hobbybroadcasting.com

Because of the volume of mail that I've been receiving, I can't guarantee a response, but I will try.

the History of Pirate Radio

1900-1975

Every hobby or profession, no matter how outwardly mundane, has an underground component that borders on the bizarre. For example, professional baseball has seen Germany Schaeffer steal first base and the 1919 Chicago White Sox, which were the first team to be caught fixing a game. Aviation experts whisper about top-secret flights and such antics as those of Mathias Rust, the teen pilot who flew a plane through the Soviet air-defense system in the mid-1980s. The print media has thousands of small, underground publishing operations. Similarly, there is an esoteric underground "personality" to radio broadcasting: pirate radio.

Straggly, irregular branches in the family trees of various activities rarely destroy the attractiveness of the foliage. The Chicago White Sox scandal did raise questions about the integrity of the sport itself, but baseball has grown since then and outlived many attacks on its character (even bouncing back from the 1994 season-ending strike). Further, the banned players from this "Black Sox" team, such as "Shoeless" Joe Jackson, Ed Cicotte, and Bucky Weaver, achieved an almost cult-type following and have been the focus of several movies. Although the freaks and outlaws among the enthusiasts of a particular activity might not always be welcomed at first, they sometimes create a nostalgia for themselves and their antics that continues for decades after their disappearance.

In addition to creating a web of stories that will be passed through many generations, these offbeats usually strengthen the fiber of their particular hobby. Satirical newspapers and magazines, for instance, have given readers many laughs and sometimes provide inspiration to initiate changes that eliminate the cause of the mockery. Underground activities have often provided better services to the public than the establishment, with competition forcing the latter to work harder. This has been true for pirate radio. Unlicensed offshore European pirate stations, for example, succeeded in forcing government broadcasters to diversify if they were to compete. In fact, governments often hired announcers from these unlicensed radio stations to provide what the listening public wanted.

Today, radio broadcasting has become more nationalized and uniform due to music services and talk shows (Rush Limbaugh, Dr. Laura, *etc.*) delivered by satellite. The end result is that radio has become sterile and dull in most markets. In fact, many local AM stations do nothing but relay programming delivered by satellite and are completely automated and unattended operations; not only are there no local announcers or locally produced programming, sometimes no one is even present at the station while it is broadcasting! Pirate radio, on the other hand, is performed by people who truly love the medium and are not inspired by greed to broadcast. That alone is inspirational.

What Is a Pirate?

When a radio station hits the airwaves without a license for any reason, it falls into one of two categories: pirate or clandestine. Pirates (also known as *free radio stations*) broadcast information and music because the operators want to be radio personalities or because they feel that an alternative to commercial radio needs to be presented.

These stations have broadcast to the general public on nearly every part of the radio spectrum, including the longwave (below the bottom of the AM band, from 530 kHz on down), AM (also known as *medium wave*), shortwave, FM, television, and even the satellite bands.

It is important to remember that pirate stations broadcast directly to the public. They are not to be confused with amateurs (ham radio operators) or CBers. Hams and CBers are not broadcasting because their transmissions are two-way, point-to-point communications, and are not intended for the general public. Some CBers and hams, however, use fake callsigns and deliberately jam other stations with Morse code, music, or by "swishing" the transmitter's frequency control (VFO) back and forth across a frequency (producing an annoying "swishing" sound). These transmissions are both annoying and illegal. Even if the station is playing music, they are also not pirate broadcasts because the intent is to ruin someone else's signal, not broadcast to the general public.

To add to the confusion, clandestine (or guerrilla) stations also get thrown into the illegal transmitting melting pot. Clandestines are politically motivated radio stations that are usually operated by an opposing government or as the voice of a revolutionary group. They usually beam their broadcasts toward countries offering little political freedom. They typically advocate (and sometimes help organize) the overthrow of these countries' governments. Although some pirates might be politically motivated or outspoken, the matter of violence and the lack of a large supporting organization separate the pirates from the clandestines. Political pirates and American stations that border on the clandestine are covered more thoroughly in Chaper 9.

Although many of the world's countries have been besieged internally with clandestine broadcasters at one time or another, the United States and Canada have rarely been targetted. The racist, ultra-right-wing Voice of Tomorrow, which operated on 1616 kHz (in the days when the AM broadcast band ended at 1610 kHz), 6240, 7410, and 15040 kHz (the latter three are shortwave frequencies) between 1983 an 1991, was one exception–a pirate that sometimes was considered to be a clandestine as well. A few other stations have been similarly labeled. The Menomonee Warrior's Station operated in 1975 and supported Indian rights during a time of civil unrest in Wisconsin. Several of the better-known stations in the FM "microbroadcasting" movement (such as Berkeley Liberation Radio, Steal This Radio, and Black Liberation Radio) have been considered as clandestines by some, except that most of the media coverage has been from the traditional mass media (who would not know how to differentiate between pirates and clandestines) instead of the shortwave-listening hobby press.

Likewise, few foreign clandestine broadcasts have ever been aimed at the United States from outside its boundaries, but this country has been the location for a handful of stations advocating the overthrow of various Central American countries. The Federal Communications Commission (FCC) seized two transmitters of La Voz del CID (Cuba Independiente y Democratica) in the summer of 1982 and the transmitter of La Voz del Alpha 66 early the following year in south Florida. Some of these clandestine stations, which beam programming to Cuba, are still audible legally, via the commercial shortwave transmitters of WRMI (Radio Miami International). The operators have discovered that it is less expensive to buy airtime on a "transmitter for hire" than to be occasionally raided and fined by the FCC. These stations are entirely separate from the free radio movement and are closely linked with democratic and right-wing revolutionary groups. Aside from the few North American stations, other clandestine stations from around the world are audible via shortwave (Fig. 1-1).

One type of station that falls into a gap between definitions is the Europrivate. These broadcasters operate from Ireland and Italy, often with high power and regular schedules. Both countries have passed bills allowing their citizens to broadcast within government standards. Because of these regulations, unlicensed transmissions are usually ignored. With this in mind, many commercialized, independent, unlicensed stations operate openly, daily or weekly, from Italy and Ireland. The privates, along with the extralegal offshore stations (unlicensed, but still legally able to broadcast from international waters), form a group of what we might call "unlicensed but not illegal" stations. These have typically been, by far, the easiest unlicensed

And finally I want to tell you that we are not used to such kinds of
Radio DX activities which we are receiving from you. We want to know
more about them. Will you please inform us more. Can we have any kind
of relations with them. (Relation on developing radio broadcasting activities. We want to work together.

We hope to hear from you once again.

With best wishes

Fre Tesfamichael
Director, VORT

Fig. 1-1 *The end of a three-page QSL letter from Fre Tesfamichael, director of the clandestine station, Voice of the Revolution of Tigray (central eastern Africa).*

European stations to hear in North America. Laser Hot Hits, Reflections
Europe, Radio Caroline, Radio Dublin International, Radio Fax, and Dun
Laoghaire Local Radio are a few of those heard in the 1990s. Pirates from
Europe, South America, and Oceania are covered more thoroughly in
Chapter 7.

Early Pirate Broadcasting

Pirate broadcasting in the United States is usually considered to be a recent
phenomena because few of today's radio listeners can remember hearing
any of these stations before 1966. Although most of the loosely organized
free radio movement has occurred since 1976, a large group of stations
operated before the 1940s. These were nearly always regional commercial
operations with low power on frequencies in or near the AM broadcast
band. Most of these stations existed under the regulations adopted by the
Federal Radio Commission (a precursor to the FCC) and were later forced
out by a crackdown resulting from the passage of the Communications
Act of 1934.

It's fascinating to see just how similar the earliest broadcasts were
to modern-day pirate radio. Although some articles credit the forerunner of
KDKA in Pittsburgh as making the first broadcasts during the 1920 presidential campaign, most texts actually credit Reginald Fessenden with the
first test broadcasts 14 years earlier. Fessenden was an important pioneer in
the field of radio, primarily from a technical perspective. Formerly an

engineer for Edison laboratories, Fessenden is credited with inventing the continuous wave (CW) mode used for present-day Morse code transmissions and for being the first to superimpose (modulate) audio on continuous wave radio signals.

Although he had been experimenting with telephony as early as 1902, his first official test occurred on Christmas Eve 1906. This test was much more complex than Alexander Graham Bell's initial telephone experiments; no "Come here, Watson, I need you" from Reginald. Instead, he made the first broadcast, which consisted of Fessenden reading from the Bible, playing "Oh Holy Night" on the violin, and playing a phonograph or Edison cylinder recording of a woman singing. Obviously, the few radio operators who were tuning in were stunned to hear any type of voices, let alone a variety program. Keeping up with the irregular holiday schedule that is still popular with unlicensed stations nearly 100 years later, Fessenden returned with another test broadcast on New Year's Eve 1906. The more-famous radio pioneer, Lee De Forest, began broadcasting in 1907, although little is known about these transmissions.

Numerous part-time experimental radio stations were operated between approximately 1912 and 1917 (Fig. 1-2). These stations, along with all American private two-way amateur radio stations, were closed from the

Fig. 1-2 *A photo of amateur radio station 4BQ from Rome, Georgia, in 1920. He was heard as far away as North Dakota on 1500 kHz (then used for amateur radio). Although it's unknown whether 4BQ was used for broadcasting, the owner did relay early radio broadcasts to friends via telephone.*

U.S. participation in World War I until 1920, when only government stations were allowed on the air. One of the first post-war broadcasting experiments was founded by Westinghouse employee and amateur radio operator, Frank Conrad. His station began as a hobby news operation, broadcasting the results of the 1920 Presidential election. Conrad didn't know what to do when he ran out of things to say, so he began playing records between news reports. The community was thrilled with the broadcasting. The operator continued transmitting every evening after returning from work at Westinghouse. Soon, Westinghouse paid him to broadcast every day and the station that would later be known as *KDKA* became more than just a hobby.

Within several years, hundreds of broadcasting stations sprang up across the country in much the same manner as KDKA. Many were started by ham radio operators for fun and were funded by companies; others started when a company would ask a ham to construct a station for them. After a few years of this type of operating, the AM broadcast band began to resemble the present-day Citizens Band, with stations signing on and off, and changing frequency as often as they pleased. Also, some stations were funded by companies that aired deceptive claims in describing their products (especially true for the "snake-oil salesmen" who sold various "medicinal" products). At this time, little could be done to punish stations for causing willful interference or deliberately misinforming their listeners.

Along these lines, one of my favorite radio quotes comes from Aimee Semple McPherson, a popular radio evangelist, who owned a radio station in Los Angeles. The broadcaster was closed down by the FRC because the signal was drifting across the band. The following is an excerpt that McPherson sent to then-Secretary Hoover: "Please order your minions of Satan to leave my station alone. You cannot expect the Almighty to abide by your wavelength nonsense. When I offer my prayers to him I must fit into his wave reception. Open this station at once."

Obviously, broadcasting as a whole had more than a few problems that were awaiting solutions. Even though the Federal Radio Commission (FRC) had been formed to control the airwaves, it had little power to enforce its rules and many stations deliberately disobeyed it. So, in the early 1930s, the government began constructing a new agency to be built around a piece of legislation later known as the *Communications Act of 1934*.

Although now considered to be completely out of date by its critics, the Communications Act of 1934 is still law. The new government agency, the Federal Communications Commission (FCC), proved to be not only a durable regulatory body but also to be more powerful than the defunct Federal Radio Commission. Immediately after the FCC was given authority and funding, it set about finding and dismantling unlicensed broadcast stations.

Early radio pirates differed from those after World War II in that they were often commercial or public service stations without licenses. Few operated as pirates do today; most were low powered, daily stations that sounded much like legal, licensed broadcasters. When the FCC hunted such broadcasters in the early years, the operators often claimed their use of low power kept the signal within the state. Because no interstate boundaries were being crossed by the signals, they argued, no government agency had the right to interfere with their broadcasting. Although this argument had worked in the past with the FRC, those operators that attempted to use it against the new FCC they found themselves being fined and forced off the air.

When the FCC first plodded off in its quest to eliminate unlicensed radio stations, it seemed that it would be nearly impossible to completely eliminate them, considering the large number of low powered pirates and the lack of sophisticated directionfinding equipment. But because most of these stations operated in the open, frequently announcing their location, address, or telephone number. By 1937, few pirates remained on the air. Because of the tightened control of the airwaves during World War II, combined with the loss of radio operators (and manpower in general) to the branches of the military, the original group of unlicensed radio stations had nearly evaporated by the end of the war in 1945.

Unlicensed radio of the 1920s and 1930s failed to receive the fame and notoriety given to stations of the present era. Today's pirates capture media attention for their underground exploits and daring challenges to the system to "uphold their guaranteed freedom of speech." The earlier stations were sometimes shunned by both media and listeners. They were thought to be outlaws or con artists because they claimed to be legal, but merely lacked a license. DXers (radio hobbyists who competitively listen for distant or unusual stations, so named for DX, the early radiotelegraph abbreviation for "distance") and regular radio listeners alike seemed to think that the lack of a license somehow defrauded them. In contrast, popular opinion today seems to be that pirate radio is special because the stations don't have a license. Given this turn of public opinion over the past 65 years, you can only wonder how it will change in the next 60 years.

A great example of how early listeners felt defrauded by pirate stations are the hoax stations that occasionally appeared during the golden age of DXing in the 1920s and early 1930s. These days, it's hard to imagine, but literally thousands of DXers were actively hunting for new stations regularly. It was a common occurrence for listeners to have a small article written up in the local newspaper when they heard or verified their first

European AM station. In fact, one newspaper account from February 1926 stated that more than 500 people in Omaha, Nebraska, alone reported hearing an English AM station that was discovered to be a pirate hoax.

Unlike the long-forgotten hoax stations, one of these early pirates managed to become a legend in the shortwave listening hobby. This station, WUMS, survived federal attacks for decades as a true radio pirate. WUMS (for *We're Unknown Mystery Station* or *We're Unlicensed Marine Station*) was actively operated by David Thomas from 1925 to 1948. It continued operating on an infrequent basis for many more years. According to excellent articles by Tom Kneitel in *Popular Communications* and John Santosuosso in the monthly publication of the Newark News Radio Club (NNRC) about the station, Thomas originally operated WUMS as an emergency broadcaster from his father's ferry on the Ohio River during flood conditions. Programming consisted of music, news, weather reports, and information on when and where the ferry would deliver emergency supplies.

WUMS' true identity and location were kept secret throughout the 1920s to avoid the FRC. Despite only broadcasting with a few watts on 1560 and 2004 kHz, WUMS must have become widely popular because the FRC began looking into its operation. In spite of attempts to end his broadcasting hobby, Thomas always slid past FRC (and later, FCC) charges by telling them of the thousands of dollars (and maybe lives) that had been saved because of his activities. Even though the FRC/FCC desperately wanted to finally close WUMS, these arguments always sent them away to look for another approach to the situation.

In addition to spanning many decades with his unlicensed broadcasting, Thomas stepped nearly 50 years ahead of his time by offering special DX tests (a broadcast scheduled for a time when conditions for distant reception are optimum) and other programming directed toward shortwave listeners (SWLs). These tests were coordinated with several DX clubs (clubs for DXers and others interested in reception of distant radio stations) and listeners on the WUMS mailing list. Thomas further assisted listeners by promptly verifying reports sent to him. However, the reports had to be "100% correct" in the most literal sense of that term. All WUMS reception reports were read and considered for verification, but the qualifications were probably the most stringent of any radio station ever. To receive a QSL (a card or letter verifying that the listener actually heard the station, see Chapter 11 for more information on QSLing), the listener had to copy the title of every musical selection and precisely transcribe different sets of letters sent in Morse code. The report had to be mailed within 24 hours of the broadcast. If any details were missed or if the report was mailed late, a "cannot verify" card (Fig. 1-3) would be sent to the listener. Out of the hundreds of reports mailed to WUMS over the years, only 30 QSL

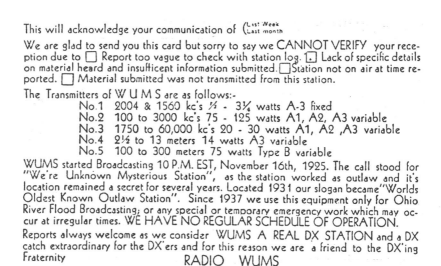

This will acknowledge your communication of (Last Week / Last month

We are glad to send you this card but sorry to say we CANNOT VERIFY your reception due to ☐ Report too vague to check with station log. ☐ Lack of specific details on material heard and insufficent information submitted. ☐ Station not on air at time reported. ☐ Material submitted was not transmitted from this station.

The Transmitters of W U M S are as follows:-

    No.1 2004 & 1560 kc's ½ - 3¼ watts A-3 fixed
    No.2 100 to 3000 kc's 75 - 125 watts A1, A2, A3 variable
    No.3 1750 to 60,000 kc's 20 - 30 watts A1, A2 ,A3 variable
    No.4 2½ to 13 meters 14 watts A3 variable
    No.5 100 to 300 meters 75 watts Type B variable

WUMS started Broadcasting 10 P.M. EST, November 16th, 1925. The call stood for "We're Unknown Mysterious Station", as the station worked as outlaw and it's location remained a secret for several years. Located 1931 our slogan became "Worlds Oldest Known Outlaw Station". Since 1937 we use this equipment only for Ohio River Flood Broadcasting; or any special or temporary emergency work which may occur at irregular times. WE HAVE NO REGULAR SCHEDULE OF OPERATION.

Reports always welcome as we consider WUMS A REAL DX STATION and a DX catch extraordinary for the DX'ers and for this reason we are a friend to the DX'ing Fraternity   RADIO WUMS

Fig. 1-3 *One of the later "cannot verify" cards from WUMS. Notice the pre-1937 history noted on the card.*

cards were ever issued.

    Of course, the FCC continued to hassle WUMS over the question of unlicensed broadcasting throughout the 1930s and into the 1940s. They didn't seem to mind the emergency flood broadcasts, but the DX tests had to go. Thomas claimed that he quit airing DX tests in 1937, but he actually hadn't. Although the FCC knew this, they could never quite catch WUMS in the act. Instead, they resorted to threats with the hope that Thomas would "get his day."

    As it turned out, WUMS did get its day, although it wasn't for several more years. In the midst of a disagreement over a particular procedure in February 1948, his adversaries in the shortwave-listening community informed the FCC when the next WUMS DX test would occur. The agents waited for the broadcast to begin and within minutes, the station was closed. Amazingly, the tests were heard as far away as China. Later, Thomas was convicted in a lower court for operating an unlicensed broadcasting station, but he appealed and somehow wriggled free again in the higher federal court. Thomas later said, "It has been said the FCC spent about $10,000 prosecuting WUMS and that it was the only station that *ever* beat them." Thomas might have made it to safety, but WUMS essentially disappeared after this point.

    Thomas virtually ended his broadcasts in 1948, but he did occasionally return from retirement for a flood program or a DX test for many more years. A former columnist for *Popular Communications*, Harry Helms, vividly remembers a visit from Thomas in 1970 during which Thomas claimed to have recently aired a DX test and said that he received a correct reception report from a well-known DXer in New York. Helms was also struck by the fact that Thomas' car had Ohio license plates reading "WUMS," even though he was a legal resident of Florida at the time.

Thomas was entirely different from the other pirates in his era. Perhaps the greatest tribute to David Thomas is that the WUMS equipment was requested by the Ohio Historical Society and by the Smithsonian Institution. Unfortunately, some of the equipment was stolen and its current whereabouts are unknown.

Pirate Radio During World War II

World War II signaled the end of the early era of unlicensed broadcasting. With the war paranoia (some of which was justified) in full swing, the FCC was expanded and better funded in order to prevent Axis spy and clandestine radio stations from operating within the country.

Because of the massive war, clandestine and spy stations were active from the United States, but the FCC quickly "DFed" (located with direction-finding equipment) and raided these operations. World War II made a great impact on pirate radio in the sense that the FCC was maintained at a higher level than it was before the war. The war propaganda of the day also helped establish the popular notion that unlicensed radio was a criminal activity and not a right.

One interesting bootlegger hoax was covered by Tom Kneitel in an article in *Popular Communications*. According to this article, the mystique of covert spy radio in the United States intrigued more than just the FCC. For some time, American troops and other two-way shortwave radio stations in the United States were taunted with often-sarcastic comments from a German-accented male who frequently ended transmissions with "Heil Hitler!" After months of activity, the FCC raided the station. Instead of finding a German spy, they found an Iowa college student with a penchant for mischief. Stranger yet, this station wasn't even the "real" one; the operator merely copied someone else's hoax and the origininator of the prank was still at large!

Post-World War II Piracy

World War II sealed the fate of the pirate broadcasting scene for several years. The rest of the 1940s and 1950s proved to be a graveyard for unlicensed broadcasters. Even when the few pirates did broadcast, the information was not publicized or otherwise recorded for posterity. Aside from the rarely heard WUMS, most pirates were young radio experimenters–either those with a love for the technical aspects of radio or kids with dreams of becoming disk jockeys at a "big-time" commercial radio station. Because most of these stations were low powered and operated by inexperienced pirates, the general lack of security would catch up with them in the form of an FCC agent or a strict ham operator in their listening area—if anyone could manage to hear them at all.

Tom Kneitel, former editor of *Popular Communications* magazine, wrote an interesting editorial in the January 1988 issue, featuring his experiences as a "low-power pirate." His station, WISP, only operated a few times in 1948 with 25 and 100 watts on 1165 kHz. Despite the lack of power and airtime, WISP received reception reports from New Jersey, Pennsylvania, Ohio, and Massachussetts for broadcasts from its New York City location (Fig. 1-4). WISP's most widely heard broadcast on New Year's Eve 1948 also

Fig. 1-4
Aside from WUMS, WISP is one of the only DX pirates that operated in the 1940s or 1950s.
Tom Kneitel

became its last—Kneitel's father walked in during the middle of it and Tom was too frightened to ever pirate again!

Numerous pirate careers have ended in similar fashion to Tom Kneitel's. A one-time-only 1984 shortwave station, WPRI, was heard with two announcers playing rock music and talking about a Chicago newspaper. Part way into the broadcast, a person who sounded as if she might have been the mother of one of them walked in on the program while the microphone was open and asked what they were doing. They said "We're just doing some recording," and continued the program after some giggling. WPRI announced that they were using an expensive Drake transceiver, but they were operating on a frequency different from what they claimed. It seems likely, therefore, that one of them pirated his father's equipment and was unfamiliar with using it. Because they never returned to the air, one can only wonder what the father did when he came home and discovered that they had been playing with his transceiver!

WBBH: The Shortwave Pirate Trigger

In the mid-1960s, few listeners realized the potential of pirate broadcasting on shortwave. Although some bootleggers were occasionally heard in the ham bands, no serious operators dared to be "real" stations and broadcast out of band or in a shortwave broadcast band.

Imagine the response from the shortwave-listening community when a new broadcaster popped up in the spring of 1966 on 4970 kHz (right in the middle of the 60-meter broadcast band, "tropical" band). This station called itself *WBBH* and programmed classical music, divided with news bulletins and a few light humor spots, beginning every evening at 7:00 P.M. Eastern time (weekend operations began at 3:00 P.M.). The station was programmed in a low-key manner, much like a present-day educational broadcaster. To further confuse listeners, WBBH claimed to be legally broadcasting from the "Courtland School of Music" in New Brunswick, New Jersey, and offered QSL cards to anyone who wrote to that address. Both the signal and the QSL card quality were described as DXers as being excellent.

With listeners generally innocent about pirate radio at the time, few realized that WBBH was actually a hoax. Legal trouble for the station began when one confused listener called an FCC field office to ask when private American broadcasters were authorized to broadcast on 60 meters (a frequency band allocated to "third world" nations only). FCC engineers had no knowledge of the "WBBH" that the listener was talking about; they thought he was pulling a prank or was misinformed.

Tuning to 4970 kHz, the FCC found that the listener was correct and that a WBBH actually did have a regular broadcasting schedule. In a short time, FCC agents began searching the New Brunswick area for the station and, to their joy, it was quickly pinpointed. Instead of finding a highly sophisticated educational broadcaster operated by the "Courtland School of Music," the FCC found a room inside the home of an 18 year old with a neat, simple, and inexpensive equipment arrangement. Instead of the expensive, professional-quality Gates BFE-50C transmitter WBBH claimed to use, the pirate had been broadcasting with a WRL Globe Scout 65A (a bottom-of-the-line ham transmitter from the late 1950s) instead (Fig. 1-5). FCC agents finished the job by closing WBBH and issuing a few warnings, but "Mr. Fisk" (the operator's air name) was never fined.

Despite the rapid demise of WBBH, many "legalistic" shortwave radio hobbyists became alarmed that anyone, let alone an 18 year old, could fool everyone into believing that his station was legal. Surely, they reasoned, a few professionals operating stations with a more-defensive approach against the FCC would be nearly impossible to catch. It is hard to tell whether WBBH directly influenced any shortwave listeners to become

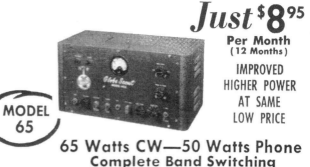

Fig. 1-5 *A 1954 ad for a World Radio Laboratories (WRL) Globe Scout transmitter, like the one used by WBBH. At only $89.95 new in 1954, a used transmitter like this could be purchased for considerably less than a broadcast-quality transmitter in 1966.*

pirates, although several stations since that time have used the original callsign. Even if Mr. Fisk had no direct impact in that manner, WBBH did become a legend in the otherwise-desolate pirate days of 1966.

Offshore Pirate Influences

One of the most important pirate influences from the late 1950s through the early 1970s was offshore pirate radio from Europe. These stations are covered further in Chapters 5 and 7, but I wanted to mention them here because they are surely at least a partial influence on many of the pirates from this time period. The European offshore pirates are particularly engaging because these stations were literally risking life and limb for the purpose of playing rock and roll music for the otherwise rock music-deprived youth of Europe. It was not until the offshore legislation began closing down the offshore stations in Europe from 1967 onward that the movement began spreading to the shore.

But in the United States, which never had a real offshore movement, stations sprung up in the basements, attics, and bedrooms of teenage radio experimenters. Over the years, I've talked to at least a dozen people who had AM or FM pirate stations between 1958 and 1968, but almost no one has any period information (recordings, newspaper clippings, magazine articles, photos, etc.) from their operations. One of the few that I've found with documentation was WKOS, Chaos Radio, which broadcast from Cleveland, Ohio, in 1965 and 1966. The teenage DJs played a diet of Beatles and other pop/rock bands from the era through their World War II surplus ARC-5 military transmitter from a bedroom (Fig. 1-6). The combination of the newspaper article and their regular broadcasts put an end to the station,

Fig. 1-6 *A QSL from the mid-1960s AM pirate, WKOS in Cleveland, Ohio.*

which was visited by the FCC in the Spring of 1966. An interview with Captain Sly is featured in the Winter 2000 issue of *Hobby Broadcasting*.

1960s Counterculture

Traditional values fell victim to the new youth counterculture of the late 1960s. The "hippies" and "yippies," the most extreme elements of the counterculture, supported many sociopolitical movements, including world peace, inner peace, free love, drugs, long hair, meditation, folk and acid-rock music, and ecological improvement. Obviously, an alternative culture of this magnitude needed a vehicle to drive its ideas to the masses. Several of the popular underground authors of the time, including Abbie Hoffman, suggested this vehicle should be radio—more specifically, pirate radio. However, few hippies had the motivation, patience, and technical knowledge necessary to operate a radio station.

One exception was the Falling Star Network with stations WKOV (Fig. 1-7) and WFSR (Fig. 1-8) on 1610 kHz, WXMN (Fig. 1-9) on 1615 kHz, and WSEX on 87.9 MHz in the Yonkers, New York area. These stations operated daily, commencing in 1970, with transmitting powers of 300 and 250 watts on AM and FM, respectively. As if daily operation with high power was not brazen enough, the original broadcasting schedule began at noon (Eastern time) and closed at 4:00 A.M. They later increased airtime to 24 hours per day, which required the services of nearly 50 part-time volunteers.

Obviously, an operation of this magnitude required a bright and motivated station manager to keep everything under control. Fortunately,

Fig. 1-7 *Mr. Natural was the mascot for WKOV, which operated from Yonkers, New York, in 1970 and 1971.*

the Falling Star Network had two. Both J. P. Ferraro and Allan Weiner collaborated on the project by owning two stations (one AM, one FM) apiece. Ferraro and Weiner (22 and 16 years old, respectively) were dedicated to the belief that Yonkers deserved to have a community broadcasting service (it had no radio stations, despite a population of 300,000) and if they were forced to break laws by being the ones to do it, so be it.

Fig. 1-8 *WFSR, the sister AM station to WKOV, was the flagship of the Falling Star Network.*

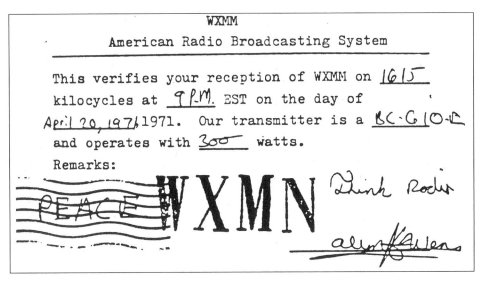

WXMM
American Radio Broadcasting System

This verifies your reception of WXMM on _1615_ kilocycles at _9 P.M._ EST on the day of _April 20, 1971_ 1971. Our transmitter is a _BC-610-C_ and operates with _300_ watts.
Remarks:

Fig. 1-9 *A verification card from WXMN. Notice the typo in the callsign at the top and Allan Weiner's signature at the bottom. The BC-610 transmitter was a large and very heavy piece of World War II vintage equipment.*

Carrying these ideals as a torch, Weiner and Ferraro set out to present a community service alternative of new folk and political music (Arlo Guthrie, Country Joe and the Fish, The Fugs, etc.) and talk about social issues. Not surprisingly, the stations operated with high power and were dedicated to a popular youth culture that had yet to be addressed by establishment radio. Even so, the FCC was anxious to dowse the young operators' interest in unlicensed broadcasting.

To demonstrate their willingness to broadcast as a licensed legal station, the Falling Star Network operated entirely in the open, with the location and announcers' identities freely revealed. It seems Weiner and Ferraro must have thought that their popularity in the New York area would force the FCC to issue the network a license for broadcasting. Certainly they weren't looking to get busted or fined by the government, a chance that was greatly increased by operating in such a blatant manner. Of course, the FCC responded. Strong warnings about the stiff penalties for pirate broadcasting were delivered to the Yonkers stations several times over their initial months on the air. But after temporarily following orders and disappearing from the radio dial, the stations would return to their regular schedules. The FCC merely continued to issue more threats and warnings to Weiner and Ferraro.

Although the circumstances are questionable, it is possible that the FCC was indeed intimidated by the popularity of the network among the media and the general public. It also could be that the agency was not used to being disobeyed; threats and warnings had been sufficient deterrents with other stations. Regardless, the FCC's reaction to the matter was much weaker and more relaxed than today's standards. Since the mid-1980s, the FCC has normally issued a Notice of Apparent Liability along with a fine of $1000 to $20,000 to each apprehended first-time offender.

As amateur radio clubs in the area found that the FCC was reluctant to close the network, they began increasing the number of complaints and vocalized discontent over the situation to the local media. With dire circumstances seeming imminent in July 1971, the operators pulled the plug on the Falling Star Network. Although it was claimed that the self-closure was done voluntarily, the FCC arrived several weeks later with U. S. marshals to confiscate equipment and apprehend the former pirates. After a court case, Ferraro and Weiner were released on probation. The two pirates later teamed up again in the Yonkers area with KPRC (described in Chapter 9), KPF-941, and Radio Newyork International (described in Chapter 6).

Other widely heard (but less notable) late 1960s and early 1970s pirates included Radio Free Harlem, King Kong Radio, WTIT, WSLH, WENJ, Radio Free Nashville, Voice of the Purple Pumpkin, Radio Clandestine, World Music Radio, Wild Turkey Radio, and WGHP.

Radio Free Harlem occasionally would broadcast rock music and stream-of-consciousness style weirdness from the 1960s through the 1980s, often in the amateur radio bands. WGHP played rock music and political commentaries with the slogan "With God's Help, Peace," in 1969. This station might have been connected with the 1980 pirate, the Voice of Togetherness, which used the slogan "Peace, With God's Help."

The "Voice of the Purple Pumpkin" name has been used by a number of different stations since the FCC bust of an operation that used this name in Baltimore, Maryland, in 1970. Also, this station sounded very similar to WSLH, which operated just above the AM broadcast band from the Baltimore, Maryland, area in the early 1970s. One magazine article from the mid-1970s, however, stated that stations using the "Voice of the Purple Pumpkin" name had been active since 1964. Although very little is known about these stations, DXer Terry Provance caught a broadcast from the Voice of Purple Pumpkin on tape in 1970. The name "Radio Clandestine" was first used by a station in 1973 and then again by station in 1979, but it is rumored that both stations were the same, with a five-year "dry spell" in between.

Activity dipped in 1975, with only three stations in operation, all on AM. WHBL transmitted from the New York City area with low power and mostly rock music. The Menomonee Warrior's Station operated from Wisconsin in the upper end of the AM broadcast band with apparently high power, because a local station received interference from the transmissions.

Most notable for the year was WHGC, a gospel music pirate that operated regularly for several months in 1975. The announcer operated entirely in the open, claiming his transmitter ran a legal 100 milliwatts (one-tenth of a watt, the approximate power of a toy walkie-talkie), which seemed highly unlikely; his signal was heard across the mid-Atlantic states. The FCC doubted the claims also, and WHGC was later traced to a church in Charlottesville, Virginia. The actual transmitter power turned out to be 50 watts.

Conclusion

The very early pirate stations were largely "wannabe" commercial broadcasters. These unlicensed stations were scattered across the country, and were quickly eliminated by the FCC in the 1930s. Later, in the 1940s and 1950s, some hoaxes hit the airwaves. By the 1960s, with the advent of hobby radio hoax stations WBBH and the Voice of the Purple Pumpkin, the medium was developed. Pirate radio had developed some purpose and style. It was time to reach some listeners.

The Dawn of the Modern Era 1976-1982

The modern era of pirate radio began in 1976 with a new crop of stations in and around the New York City area. It was not until the late 1970s that pirate radio became "free radio"—an actual movement. In addition, it was in this decade that the first newsletter devoted to free radio began, followed soon by specialty clubs for listeners to pirate radio. With increased activity, free radio listening and broadcasting suddenly became a hobby unto itself. Instead of merely being a medium for DXers to hear rare, erratically transmitted broadcasts, it became a new hobby, entailing elements of shortwave listening, the free speech movement, hobby radio writing and production, and progressive music. In this way, free radio activity became the crossroads for a handful of other hobbies and activities.

Connections to the Past

The station that helped start a new era in broadcasting, WCPR, originally began broadcasting in 1971 as WMPI through a modified Zenith clock radio. Because both operators were very youthful (elementary or junior high age), neither had connections to anyone who operated a pirate station—even though the Falling Star Radio stations (Chapter 1) were active at that point, and the AM stations were well within range of Brooklyn. The WMPI operators heard their first pirate, WGOR-FM in 1973, and called in to the telephone number that the announcer was giving out over the air.

The WMPI kids established contact with the WGOR operator, Gary, who told them to listen for WFSB (Falling Star Brooklyn) and WQLB, the Queens branch of the Falling Star Network. Before long, WMPI (later known as WPBC and WKRB) was friends with both stations. Like finding a rare fossil, WFSB and WQLB are "missing-link" FM stations that connect the 1960s Falling Star Network to the booming New York City pirate scene of the late 1970s and 1980s.

In 1973, Gary of WGOR promised that he could find a spare FM

Fig. 2-1 *The hand-drawn look: a QSL from the NYC area AM pirate, WCPR, which operated in 1975 and 1976.*

transmitter. Two years later, nothing was still available and Gary mentioned the 1943 Collins 32RA that was sitting in his basement. The WMPI kids saved up $50 to buy the transmitter and prepared to go on the air with a new name, WCPR.

WCPR and the Telco System

The most famous operation of 1976 (and probably the decade) actually began in December 1975, from Brooklyn, New York. WCPR (Fig. 2-1) became the first widely heard pirate to run a telephone call-in format, often airing live callers from across the nation. The call-in programs added a feature never before experienced on the majority of pirate stations–the listener's comments. Although a few stations before WCPR aired telephone calls, the operators were unfortunately bold (or ignorant) enough to announce their home numbers. For obvious reasons, stations that operated in this manner rarely lasted for substantial periods of time.

Because of the coverage of WCPR, the FCC was forced to scour the city for the popular broadcaster. FCC agents found that the station was located in Brooklyn, but after that, the search became nearly impossible amidst the maze of apartment buildings. With this close-in direction-finding work leading nowhere, the FCC contacted the local telephone company for some inquiries about the system that WCPR was using to take callers. The FCC discovered that the numbers used by the station were owned by the telephone company.

Fig. 2-2 *At left, the vintage Collins 32RA transmitter used by WCPR (and later, WFAT, WGUT, and WHOT). At the right is a vew from the balcony, where the random-wire transmitting antenna is running acorss the fence (at the far right) and over the courtyard below.*

WCPR's use of these numbers confused many listeners. Some assumed that a direct contact with the telephone company produced the services; others believed WCPR to be comprised of "phone phreaks" (technically inclined persons who experiment with the telephone system in unauthorized and frequently illegal ways) and that the actual lines were being tampered with. Telephone hobbyists and others familiar with the system figured it out long before, but it was not until nearly one year later that those in the DX hobby found out WCPR's secret: telco loops.

Upon requesting help from the local telephone company, the FCC found that WCPR's phone-in programs were made possible with the utilization of telco (telephone company) loops. These loops are actually a pair of ordinary numbers that are reserved by the telephone company for testing lines. WCPR called one of the two numbers (probably the lowest of the two) making the pair, such as 949-9977. Listeners were then instructed to dial the upper half, in this case, 949-9979. The two calls were then automatically connected.

Looping had been used in cities for years by phone phreaks and by prostitutes, drug dealers, and other perpetrators of underground activities. The attraction to using this system, besides the advantage of using numbers not associated with one's home address, is that calls are more difficult to trace. When the FCC visited the telephone company in 1976, little could be done to trace WCPR down. As an alternative plan, the FCC stayed at the company's office during WCPR programs, and every time a loop number was announced over the air, the agents closed the circuit.

Tired of receiving busy signals each time the loop was called, WCPR resorted to merely playing music and signing off the air a bit earlier than usual. Soon after the FCC plan was set into action, Artie Media, a guest announcer from WFSB in Brooklyn, persuaded the staff to announce their home phone number. WCPR enjoyed a tranquil signoff for that evening, but they were raided by agents from the New York City field office while broadcasting less than two weeks later.

Although WCPR was a trend-setting station, few "copycat" stations were heard during the following months. Over the rest of 1976 and 1977, activity dropped to its lowest point in several years, with no other stations being reported during that time. In late November 1977, however, WCPR returned to the air with a classic phone-in broadcast to formally bow out with a "proper" program. WCPR's farewell occurred on a sweeter note than the FCC broadcast, and it whetted the operator's tastes for more.

WFAT, The FAT one

When the 1978 New Year's Eve ball dropped in style in Times Square, New York, WFAT began regular operations from the Brooklyn area on 1630 kHz. WGOR, "Gorilla Radio," also joined WFAT on the frequency with call-in programming from the Brooklyn area (this was a new WGOR, operated by Larry Wong, not by Gary). WFAT was simply the staff of WCPR with a new name and a few extra years of broadcasting experience (Table 2-1). Both stations co-existed on a somewhat friendly basis, despite vying for the same frequency range.

Winter brought large quantities of snow and illicit air time for the East Coast in 1978 as WGOR and WFAT jumped on the air nearly every weekend—especially if the weather conditions were bad and FCC direction-finding vehicles were unable to travel. WFAT established a fetish for broadcasting during storm conditions and a fickle taste in callsigns; it was variously known as WICE, WICY, WFSR, WDBX, WPLT, WPAT, WPAP, WPOT, WPLC, WJVR, WEVJ, and WIMP during the beginning of the year. Although the names frequently changed, the format, announcers, and slogan ("free speech radio") remained the same.

Shortwave: The Voice of the Voyager

Amidst the return of anarchy to the New York City airwaves, the pirate itch spread to the northern Midwest. After an unsuccessful attempt to follow WCPR on to the top of the AM broadcast band, The Voice of the Voyager began regular Friday and Saturday night programming on 5850 kHz, just below the 48-meter broadcasting band. The Voyager began as an indirect result of the New York City activity because its operators were avid short-wave listeners and subscribed to FRENDX (the then-used title of the

WMPI	1971-1972	A low-power "clock radio" station
WPBC	1972	A low-power "clock radio" station
WKRB	1972-1975	A low-power "clock radio" station
WCPR	1975-1976	Triggered the NYC AM pirate explosion

−FCC raid against WCPR 2/7/1976−

KDBX	1976	One shot a few months after WCPR bust
WFAT	1977-1979	The best-known NYC AM pirate ever

−FCC raid against WFAT 4/15/1979−

WBEE	1977	During Hurricane Belle
WDBX	1977	Testing−saving WFAT calls for New Years Eve
WPOT	1977	Testing, also used callsign WFSR, See WDBX
WICE	1977	One-shot snowstorm station
WBAD	1978	Two-week station from John's house
WILD	1979-1980	FM precursor to WHOT
WGUT	1980-1985	New Year's Eve AM station + "snow shows"
RNI	1987	Al Weiner's offshore station

−FCC raid against RNI 7/28/1987−

WHOT	1980-1989	FM and AM. Hank's proudest "hour"

−FCC raid #1 10/25/1985−

−FCC raid #2 1/1986−

−FCC raid #3 7/6/89 1st FCC use of DEA tactics with U.S. marshalls−

RFNY	1990-present	Currently relayed by WBCQ on shortwave

Table 2-1 *A timeline, showing the different stations operated by childhood friends John and Perry (of WCPR/WFAT).*

monthly radio bulletin published by the North American Shortwave Association), which had carried reports about the New York pirates. Voyager announcers R. F. Wavelength and A.F. Gain created one of the most popular American shortwave pirate stations by combining quirky broadcasts and loop call-in programs with shortwave listening. The results were astounding; hundreds of listeners, if not thousands, tuned in, called, and reported the broadcasts to assorted DX bulletins.

The Voyager aired comedy skits along with the regular features. "Bobby the Bootlegger," a spoof on pirate broadcasting and the FCC, and the "Nighttime Melodies" music program, featuring rock music (especially that by The Beatles), were interspersed with casual talk and telephone calls. One caller discovered that the programming was aired from a Hallicrafters HT-20 transmitter, running approximately 100 watts through a dipole antenna, but the pirates refused to reveal their location, other than "from the mighty North." This mystery about the location and the identities of the operators added to the popularity of the Voyager; nearly everyone who reported the station to DX bulletins speculated on their location. There never had been a pirate so widely heard or well known among shortwave listeners, and they eagerly sought all possible information about "the new bootlegger." R. F. Wavelength said, "We had people all over the country trying to figure out just who we were. When we were on the air, we were on 'the edge' so to speak. You could think of yourself as being one step ahead of something and that added much excitement to the entire deal" (Fig. 2-3).

Fig. 2-3 *R.F. Wavelength (left) poses beside a stack of vintage equipment. A.F. Gain (right) operates the mixer under his audio stack.*

One individual who regularly reported the station was bombarded with so many questions from listeners who thought he was the pirate himself that R.F. Wavelength felt compelled to send a press release in mid-1978 to clear the DXer's name. Another mystery surrounding the Voyager concerned the arrival of many fake QSL cards to those reporting the broadcasts to various shortwave bulletins. The bogus QSLs featured several colors of ink, not just black, and included a statement attributing the station to the University of Minnesota.

Rather than hope to be granted a license in the future, the Voyager crew recognized their limits within the beaurocratic broadcasting environment and merely set out to have fun on the air for as long as they could. R. F. Wavelength once noted that 1000 verification cards had been printed up and they "hoped to give them all out before getting busted." For some reason, the operators had an almost kamakazi attitude about broadcasting; they operated with a weekly schedule on one particular frequency. The anticipated bust never occurred, but they decided to signoff permanently anyway because the two main operators were enlisting in the army and preparing to leave for college, respectively.

However, the pirates were not to casually slip away from the mess that they had created for the authorities. Two FCC agents appeared at R. F. Wavelength's door a week after the Voyager's last broadcast at the end of August 1978. R. F. Wavelength admitted operating the Voyager and the FCC agents talked a bit about the station before driving back to the field office. Although the FCC never fined the Voice of the Voyager in 1978, they threatened to revoke the amateur radio licenses held by its operators (Fig. 2-4).

Immediately afterward, station personnel busied themselves with telling the Voyager story and

2.	FREQUENCY			
2a. Authorized	2b. Measured	2c. High Low (Hertz)	3. Emission	
...ateur	5850 KHz		A3	
4. Location of Station or Name of Craft	5. Radio Service or Class of Station	6. Hour(s) of Violation (EST GMT)	7. Date(s) of Violation	8. Call Sign
Mpls.,Mn.	Amateur-Novice	See Below	See Below	WDØAYM

9. VIOLATION(S)   NON-COMPLIANCE WITH FCC RULES
   Hours (GMT)   Dates
   0355-0803   August 6, 1978
   0355-0455   August 13, 1978
   0356-0655   August 20, 1978

   1. Section 97.7(e)   The station was operated on a frequency and with an emission not authorized for the class of license held.

   2. Section 97.87(a) The station was not identified by its assigned call sign.

   3. Section 97.113   The station was used for dissemination of radio communications intended to be directly received by the public.

   4. Section 97.115   The station was used for the transmission of music.

ISSUING OFFICER Roger P. Anderson	SUPERVISOR LOCATION PL	DATE MAILED SERVED August 29, 1978

Fig. 2-4 *The August 1978 warning letter to the Voice of the Voyager from the FCC.*

readily identifying themselves by their legal names. They also disclosed the actual location (Crystal, Minnesota) and enjoyed local and minor national coverage. It was learned that the station was not really "The Voice of the Voyager," but "The Voice of the Voyageur," named after Voyageur National Park in northern Minnesota. One early listener reported the station to a DX bulletin as "The Voice of the Voyager," so the operators just decided to drop the national park identity. An interview containing more information and views of the station was published in the Summer 1998 issue of *Hobby Broadcasting*.

Late 1970s Pirate Radio

While the Voice of the Voyager and WFAT operated, a few copycat stations joined in on the fun, such as WMMR (Midwest Music Radio), Jolly Roger

Radio, Radio VOCAD, and the Voice of the Viking. Each of these stations was directed toward the shortwave listener, with articles from the FRENDX bulletin being quite popular. The Voice of the Viking planned weekly broadcasting on 7450 kHz and called DXers across the country to notify the shortwave listening community of their initial tests. To further show their commitment to the radio hobby, they occasionally used another callsign, WSWL (for "ShortWave Listener"). However, none of the stations ever matured in 1978, and only Jolly Roger Radio ever returned, becoming widely heard in 1980.

WFAT and WGOR in New York pulled together an even more impressive collection of copies, clones, and those that had been inspired to broadcast after hearing their activity. According to the station, most of these operators could thank Gary of the old WGOR for getting their stations on the air. Some of these stations included Pirate Radio New England (PRN), WEKG, WCBX, WPNJ (Paul, Nick, and Joe), WENJ (back after an absence of several years), WELO, and WLTE, all of which operated above the AM band from the New York City area. Most of these operations were low powered (25 watts or less) and difficult to hear outside of the metro area. Only Pirate Radio New England, with a professional mix of many announcers and phone calls, and WCBX, with many reminiscent looks back to the early 1970s and reruns of old radio programs, were widely heard for more than a few months.

Amidst this activity, the telephones of shortwave listeners commenced ringing to announce the return of the Voice of the Voyager. Like WCPR, the crew felt that a "proper ending" was more befitting than an FCC bust. On November 4, 1978, the crew returned to the air on a new frequency of 6220 kHz. For nearly an hour and a half, they accepted telephone calls, and did a re-enactment of the bust, performed by the "Voyager Art Players." With names and addresses of the station personnel already public knowledge, R. F. Wavelength identified the station with his own amateur radio callsign and announcers were addressed by their real names. In another daring move, when all of the loop lines were busy, R.F. Wavelength drove several miles away to a public telephone booth to continue taking calls.

Like WCPR, the crew of the Voyager just couldn't resist the airwaves and began planning a regular schedule for every Saturday night for one hour on 6220 kHz. Although WCPR split up and changed names upon returning, the Voice of the Voyager remained the same, even continuing to announce R.F.'s address and amateur radio callsign. One of the biggest events of the year for the Voyager was the Christmas/first anniversary broadcast that featured a detailed account of the station's history and the "VOV Monotones" singing such classics as "Happy Birthday" and "The Twelve

Days of Christmas."

The Voyager continued regularly scheduled programming throughout January and into early February, and then fell silent. Some listeners speculated that the FCC had finally taken action against the station, but it was later announced that the Voice of the Voyager had voluntarily ceased broadcasting. Although the FCC had just previously threatened them again for unlicensed operation, the true culprit of the shutdown was the aging HT-20 transmitter. Not built for frequent continuous (more than a few minutes at a time) transmissions, the final amplifier section burned out, rendering the equipment useless until repairs were made. This time, the FCC did not appear a week later.

Just as some stations popped up, imitating the Voyager while it existed, more hoped to fill the void left when it disappeared. The two most notable stations, Radio Liberation (a slogan occasionally used by the Voice of the Voyager) and the Voice of Venus (also "VoV"), wanted to broadcast like R.F. Wavelength. Radio Liberation copied the Voyager's style of comedy, "wild and crazy" talk, and rock music, right down to using the same theme song ("We Will Rock You/We Are the Champions" by Queen) (Fig. 2-5). Meanwhile, the Voice of Venus operators merely contented themselves by announcing a Minneapolis, Minnesota, location and claiming to be friends with the Voice of the Voyager operators.

Fig. 2-5 *This pennant was mailed out from "the swamps of Putrid, Louisiana"...which just happened to be in South Florida.*

Both stations were regularly active and widely heard across North America, especially the Voice of Venus, which habitually notified regular shortwave listeners via telephone before broadcasting. Radio Liberation announced that its transmitter site was in "the arid swamps of Putrid, Louisiana," but most of its reception reporters lived in Florida. Within a few months, the FCC took notice of Radio Liberation, DFed it to the southern tip of Florida, and closed it down. Meanwhile, the Voice of Venus continued broadcasting regularly on many frequencies in the 41- and 43-meter bands with its talk/music format through 1980.

Although pirate radio was active across the Eastern seaboard and had spread into the Midwest, the far West remained virtually unaffected. Although it was true that few Eastern pirates were readily audible in the West, the Voice of the Voyager and the Voice of Venus had strong signals that were reported across the country. The lack of activity is mysterious, considering California's reputation for alternative culture. Although the West still lags far behind the rest of the country for widely heard shortwave pirates, two stations cut an impressive profile in the 1978 scene.

The first, KDOR, operated entirely in the open on 830 kHz from Hollywood, California. The callsign represented the owner's name, Dick Dorwart, and was often announced as part of the Dorwart Broadcasting Company. KDOR operated much like the Falling Star Network in that the location of the studios and identities of the announcers were no secret. Similarly, transmissions often lasted more than six hours several times per week, and the owner attempted to have the station licensed (but to no avail). Dorwart and a few volunteers operated despite several FCC warnings and threats.

To add to this pseudo-legitimate presence on the AM dial, KDOR received promotional albums from various record companies and was listed in the local telephone book. Furthermore, Dorwart often cut promotional spots and public service announcements with celebrities. A feature story on KDOR from the Los Angeles Times quoted Dorwart as saying, "I was cutting a public service spot with Ralph Bellamy during one of the FCC investigations. I don't think they were quite ready for that." The celebrity presence on KDOR might have had an impact on the FCC's permissiveness, but the major reasons lay deeper.

Dorwart suffered from osteogenesis imperfecto, known as the "brittle bones" disease, an illness that prohibits normal growth and leaves bones susceptible to injury. While spending most of his 31 years (his age as of 1978) in a wheelchair, he began working in radio, mostly to record special announcements. But Dorwart felt he had something to say, and he began operating KDOR as an alternative to legal Los Angeles radio to the dissatisfaction of the FCC. They tried to dissuade him from broadcasting without confiscating his equipment or levying a fine.

The "kid-gloves approach" might have appeared to be a sign of good will on the FCC's part, but it was also important for public relations work. If KDOR was ignored, other legal stations would protest the matter. But if KDOR was busted, the situation could become a media event, with negative implications being placed on the FCC. So, the FCC aimed for somewhere in between; they sent threats and warnings to Dorwart whenever the broadcasts became "too regular." After three years and five closures of KDOR, the station finally quit in April 1981, with Dorwart still applying for a license—this time on UHF television.

KDOR's activity on 830 kHz might have made for an almost-legal signal on AM, but it certainly didn't set the standard for the handful of West Coast station's formats. In fact, one of the most widely heard American shortwave pirates at that time played music exclusively for the first three years of its existence. The station was first reported in 1975. It operated infrequently until 1978; then, extensive broadcasts, sometimes lasting several hours, were aired on 6420 kHz. The pirate, usually referred to as "the southern California pirate" or "the unid on 6420," played a variety of light pop music ranging from Doris Day to Barry Manilow, with some jazz thrown in.

Several times, the pirate was heard with background noises, such as turntables changing records and objects falling, which obviously meant that the station used microphones. Mysteriously, the operator never spoke, even to identify the broadcasts. This raised many questions about what the station had to hide. Was the operator just shy? Did he have nothing to say? Or was his voice easily identifiable to those who might reveal him? More importantly, why would anyone waste time and money to transmit songs commonly heard on legal radio stations?

Some of the mystery ended in early 1979 when short announcements identified the station as KVHF. Other than these identification spots every 30 minutes, no new programming was aired; the same light pop music with long breaks between each song captured the 6420-kHz frequency. Signals indicated that KVHF was located in southern California, but the station was obviously using several hundred watts because it was regularly heard across the entire continent, including the Northeast.

These broadcasts continued through 1980 with the same general format as had been noted in previous years, but live broadcasts were phased out in favor of prerecorded programming. The operator became less introverted after a few years and announced longer identification spots, some even promoting "The Tommy Johnson Show." In addition, the identifications and promos were inserted more frequently (about every 10 minutes), rather than every half hour. And the station even featured a mailbag show with answers to listener questions. If this wasn't enough, KVHF had yellow QSLs professionally printed and sent to people who reported the station to club bulletins in 1980 (Fig. 2-6). In just a year, KVHF went from being a music-only unidentified to a super-slick operation that could rival commercial stations in production quality and surpass them in overall quality.

That autumn, the station even announced a mailing address in San

QSL   OFFICIAL   QSL

KVHF RADIO
North America

'Alternative Radio'

6420 KHZ   49 meters

Fig. 2-6 *A QSL from the most active West Coast (possibly in North America) shortwave pirate ever—KVHF. Station operator Bruce estimated that he broadcast at least 30 hours per week for five years before being visited by the FCC.*

Luis Obispo, California, for listeners to use to contact the station. A month after announcing the address, KVHF was raided in Santa Ana, California. With the enormous activity generated by the station (KVHF usually operated several hours per week), it is curious that the demise did not occur sooner. In addition, KVHF was running hundreds of watts from a rare Johnson Desk Kilowatt transmitter, which explained the good cross-country signals.

Another extensive operation, RX4M, became widely heard on the West Coast in the late 1970s. With transmitting equipment running at 100

Fig. 2-7 *Despite its promises to return after being raided in 1980, RX4M was never heard from again.*

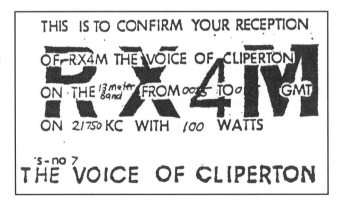

watts, the station was heard nightly as far away as New Zealand and Europe on 41 and 16 meters shortwave. The station (also known as "the Voice of Clipperton" after the famous amateur radio DXpedition) behaved much like a legal broadcaster, with a regular schedule on several different frequencies and professional-sounding announcers. These qualities made RX4M an easy catch for listeners from coast to coast (Fig. 2-7).

RX4M became one of the few pirates to imitate a legal shortwave operation. KDOR, The Falling Star Network, Radio Free Nashville, and others (especially small local stations) had imitated regular AM programming, but international shortwave programming is entirely different. Thus, RX4M regularly featured pre-television-era programming, such as *Sherlock Holmes, Jack Benny, Amos and Andy,* and *Fibber McGee and Molly.* Every week, some of the old-time radio shows would be replaced with programs with titles like "DX Forum," "Mailbag," "Let's Talk Technical," "The Good Morning Show," and the "Jerry Nelson Talk Show," recorded by their own personnel.

As can be expected, the FCC became aware of RX4M's daily operations. The agents had little trouble finding and closing the station in late October 1980; it was only a matter of waiting for RX4M to sign-on every evening, then tracking it down. RX4M was caught the day after it was first monitored by the FCC, but it had been regularly broadcasting on these frequencies for a year. Surely, the bust of RX4M could not have been a priority matter. In a press release notifying listeners of RX4M's closure, DX editor Mickey Anderson announced, "We feel that our termination was premature, as we were left with enough half-hour, unused old-time radio programs to last for the next eight years." Despite an enthusiastic "We shall return!" at the end of the press release, no similar station has operated since.

Mystery Stations

Up to this point, nearly all pirates eventually were raided by the FCC. It just didn't seem possible for a station to ever become active and achieve a large audience for an extended time. Often the FCC caught on quickly, the station slipped up and gave out vital information, or the operators became too sure of themselves to believe that they could be busted. But a few professionally styled stations began popping up in the 75-meter (3750 to 4000 kHz) amateur radio band in 1977 that defied this norm.

Hobbyists believed that each station was connected with the others (or maybe they were all run by one organization) because all were very similar in programming and technical quality. All of these pirates apparently used high power, had clean audio, played rock music and humor programs, spoofed shortwave radio broadcasting and listening, announced fake locations (usually with fake addresses), used obscenities, and operated in the middle of the 80-meter amateur band. Because few other pirates had high power and clean audio at that time, it seems likely that at least a couple of the stations worked together, if they weren't, in fact, the same station. The perplexing question is, why would anyone want to broadcast in the middle of a ham band? It is definitely the worst place on shortwave to broadcast for interference reasons and because the amateurs quickly complain to the FCC whenever someone is operating illegally in their bands.

WBLO became the first and most widely heard of the stations on 75 meters with broadcasts beginning in 1977. British-accented announcer Delty McNorton played light rock with fake commercials, comedy skits, and "WBLO, Atlanta, Georgia" singing promos. WGLI, Great Lakes Radio, claimed to broadcast "from the shores of Lake Superior" and played pop music briefly in 1979. Like WGLI, Radio Highseas International operated only in the summer of 1979. Like WBLO, the announcer had an English accent, and the program was a parody of shortwave broadcasting. Several shortwave listeners speculated in *FRENDX* on the possibility of these stations being the same. One even noted the similarities between Radio Highseas International and the 1973 pirate, Radio Clandestine, which he had read about in *Popular Electronics* magazine.

Less than six months after Radio Clandestine was mentioned in *FRENDX* by the listener, another station (possibly the original) jumped on to 75 meters with the same ID and format. It quit after several broadcasts; WBLO returned in 1980 after a few months of silence. Then in the summer of 1980, Radio Clandestine broke tradition with its 75-meter peer group and tried 6140 kHz (in the middle of the 49-meter international broadcasting band). WBLO countered that move by using 6022 kHz in the same frequency band, and was never heard again after that month.

All of the 75-meter (and former 75-meter) pirates went off the air for the last half of 1980 until Radio Clandestine returned near Thanksgiving. Upon its return, Radio Clandestine stunned shortwave listeners with transmissions in the middle of every broadcasting band from 90 to 19 meters. Listeners across the country noted announcer R.F. Burns's rock music and slick fake advertisements for such things as "industrial-strength baby cleaner" and "Marijuana Helper."

Although Radio Clandestine originally announced a "Post Office Box 100, New York, New York" address (similar to the 1973 station), it later requested mail through one of the major pirate maildrops. And, unlike many other "mysterious" stations, it began verifying reports in 1982. Other changes to Radio Clandestine included the addition of Boris Fignutsky, chief engineer, and Wanda Lust, secretary. These characters began working themselves into the programming in 1983 (Fig. 2-8).

Radio Clandestine continued operating and was still occasionally active in the mid-1990s. Considering its high power and frequent transmissions (as many as six times in one week), it is amazing that the station's operators were never raided. Furthermore, they didn't lose

RADIO CLANDESTINE

DEAR SHORTWAVE LISTENER:

Thank you very much for your recent reception report which I am pleased to verify as accurate.

We here at RADIO CLANDESTINE are always happy to hear from our listeners. I am sincerely interested in your comments and suggestions.

I am truly sorry I am unable to answer your letter personally. This is due to the tremendous amount of mail we are receiving. However, be assured that I do read ALL my mail personally.

I hereby confirm our transmission dated
JULY 2 1983 at _0543_ hours GMT on _7375_ Kcs.
For security reasons, no other information can be given.

Thank you again for your interest in our station and I hope to hear from you again soon.

PS: WE'VE BEEN AROUND FOR A LONG TIME!   73 for now.

R. F. Burns

R. F. BURNS,

Fig. 2-8 *Radio Clandestine began sending out this verification letter after picking up a maildrop in 1982.*

interest in broadcasting after a couple of years, like many pirates. At the end of the 1980s and early 1990s, some questions were raised concerning the sites of the Radio Clandestine broadcasts. Between 1988 and 1990, more programs appeared in the 41-meter "pirate band" than had in the past from the station. Additionally, these later transmissions were often weaker than usual and were sometimes of variable frequency. It is likely that these programs had been relayed from tape recordings by other stations. The status of the "real" Radio Clandestine remains unknown.

The Favorites of 1980-81: Syncom and Radio Confusion

The Voice of Syncom and Radio Confusion, two other stations that began at about the same time as the second (or return of) Radio Clandestine, also received high acclaim. Radio Confusion, the "Crazy World Radio Network," operated much like a legal shortwave broadcaster, except less frequently. The station operated for about one hour every month on frequencies around 7550 and 13950 kHz. Although operations this irregular would have drawn a small audience, Radio Confusion maximized its following by announcing the next schedule each month in *FRENDX* and other radio newsletters, such as *The Wavelength*. Fortunately for Crazy Roger and the rest of the crew at Radio Confusion, the FCC apparently never read those copies of *FRENDX*. It was a rather risky move for the station because the FCC had commenced operations against stations in the past based on information in *FRENDX*.

Radio Confusion was the first American pirate to actively operate in the higher frequency ranges. Tests enabled the station to be heard with good signals across North America and Europe. It was also the first U.S. pirate to be heard via relay throughout Europe (by such stations as Empire Radio in the UK). Radio Clandestine also began using high frequencies at about the same time, but their broadcasting was hardly "regular." RX4M and Radio North Star International moved up later that year and the following year, with equally impressive results. Unfortunately, few stations since this time have ever actively used frequencies higher than the 41-meter band—even though a worldwide audience is possible with a relatively low amount of power.

Besides using interesting frequencies and pre-announcing broadcasts, Radio Confusion's programming, high-quality QSLs, and contests augmented their popularity. The program varied little from the now-traditional pirate fare of fake commercials, mailbag shows, news, comedy skits, and rock music. The production quality of Radio Confusion was much better than that of the average pirate, yet it still rated below that of a commercial station. Thus, its popularity was derived, not so much from the production style, but from the content and service that it provided listeners (Fig. 2-9). Unlike many pirates, Radio Confusion adhered to its schedules and tried to serve the needs of its listeners. When Radio Confusion announced in a shortwave bulletin that they would be broadcasting on a certain date, they always did so.

At the beginning of 1981, Radio Confusion was plagued during a

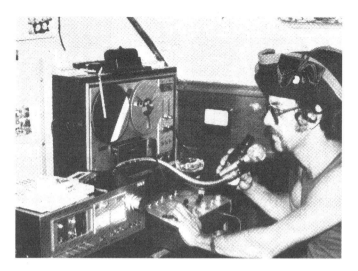

Fig. 2-9 *One hand on the mixer and the other around a beer—Crazy Charlie at the controls of Radio Confusion in 1980.* The Wavelength

broadcast by loud humming and muddy audio from their failing transmitter. After the broadcast, Crazy Roger announced that they would return in a few months with a new transmitter. True to their word once again, Radio Confusion returned to 14625 kHz and made one last broadcast that was heard as far away as South Africa, Denmark, and West Germany. The Crazy World Radio Network closed down stylishly, announcing that they felt it was time to either stop or be busted.

The other class act, the Voice of Syncom, began broadcasting to serve the shortwave listener in February 1980. During the early broadcasts, DX features and tips were the most notable program segments. Syncom

Fig. 2-10 *Syncom, a pioneer of parallel frequency broadcasting, also had a penchant for broadcasting in the standard Europirate frequency band (6200-6300 kHz).*

often announced tests from other pirates, closures of some pirates and clandestines, and details about rare shortwave stations that were then audible in North America. Later on in 1980, the programming was expanded with many comedy skits, a variety of music (including classical, jazz, and rock), telephone calls, and relays from other pirates. Some of the stations from the United States that were relayed included WJAM, Radio Telstar, Radio Free San Francisco, and the Voice of Venus, Radio Impact, Capital Radio, Radio Quadro, and European Music Radio from Europe were also relayed.

Just as Radio Clandestine and Radio Confusion made creative frequency changes to improve their stations' reception, Syncom (Fig. 2-10) used a few techniques that have rarely been matched since. The first, using parallel transmitters, is a common practice of legal shortwave stations, but had never been done by shortwave pirates before Syncom. They tried it once on their return to the air in March 1981, operating with parallel programming on five different frequencies in the 41- and 48-meter bands. Some of the frequencies lagged by others by as much as a minute, so it seems that some of the programs were belatedly relayed by other stations. Another technique, shortwave stereo, was achieved by transmitting the left channel of a program on one frequency and the right on another. If the listener owned two receivers, the programs could be heard in stereo.

The professionally printed QSL sheets and miniposters added to the popularity Syncom had attained from the DX programs and transmitting tests. Even with the large following, the station had a few enemies. Chuck Felcher's (the main operator) outspoken remarks about his critics (described in more detail in Chapter 9) created some animosity. Although the critics had originally made a false assumption about the station, Felcher's comments made them angrier still. Syncom claimed that the FCC became interested in the station during this period. Therefore, it operated less frequently until it finally ceased operating in the autumn of 1982.

The Rest of the Movement

Most of the stations covered so far in this chapter have been popular and/or experimental. They tried many creative approaches to pirate radio, and an actual "free radio movement" was built from the resulting tests. The stations mentioned up until this point had been the leaders, but many other pirates followed in the movement with varying degrees of popularity.

One of the more notable but less creative stations was WDAB, which operated for several months at the end of 1979. The announcers worked at a licensed commercial station in Florida, so WDAB had a slick FM sound. Main announcer Big Ron pulled the station off shortwave in early 1980 after he received a notice from the FCC informing him that the station's operations were illegal (Fig. 2-11). WDAB didn't realize that a notice of this sort meant

FEDERAL COMMUNICATIONS COMMISSION

FIELD OPERATIONS BUREAU

December 12, 1979

ADDRESS REPLY TO

CERTIFIED MAIL 1-12835
RETURN RECEIPT REQUESTED

Suite 601, ADP Building
1211 N. Westshore Blvd.
Tampa, FL 33607

WDAB Radio
% The Free Radio Campaign
RFD 2, Box 542
Wescosville, PA 18106

RE: Case No. TP-80-17MC/80-W-76

STATION MANAGER:

This office has received a report concerning the unlicensed operation of radio transmitting apparatus by you.

Under the Communications Act and the Commission's Rules and Regulations, radio transmitting apparatus, other than certain low power devices operated in accordance with Part 15 of the Commission's Rules and Regulations, may be operated only upon issuance by this Commission of a station license covering such apparatus.

Unlicensed operation may subject the operator to serious penalties provided for in the Communications Act, including for the first offense a maximum fine of $10,000 or one year imprisonment, or both, and for subsequent offenses a fine of $10,000 or two years imprisonment, or both. Because unlicensed operation creates a definite danger of interference to important radio communications services and may subject the operator to the penalties provided for in the Communications Act, the importance of complying strictly with the legal requirements mentioned above is emphasized. Unlicensed operation of this radio station should be discontinued immediately.

You are hereby requested to submit a written reply to this office at the above address, within 10 days, concerning the unlicensed operation and discrepancies as set forth above.

Sincerely yours,

Ralph M. Barlow
Engineer in Charge

FREE RADIO CAMPAIGN - USA
R.D. #2. BOX 542
WESCOSVILLE, PA. 18106

Fig. 2-11 *A copy of the original notice from the FCC that scared WDAB off the air in late 1979.*

that the FCC could not find the station and was forced into trying scare tactics.

Another station from 1980 received a more final notice from the

FCC. Jolly Roger Radio had operated for close to a decade in the Bloomington, Indiana, area on the FM band. Even during that time, the station had received national publicity from the threats the FCC was making to its operators. From the time Jolly Roger Radio first began broadcasting on shortwave on Halloween, the operation was a sort of radio suicide mission. The station operated for at least ten hours straight each time they checked in on 6210 kHz.

Many pirates that have operated very frequently and were subsequently closed by the FCC were incompetent in technical and programming matters. Jolly Roger Radio programming was professional, with call-ins, Irish folk music, and interesting promos. Evidently, the operators hoped that the activity and media attention would force the FCC to hand them a license. The plan didn't work as they had hoped, and the FCC forced Jolly Roger Radio from the air one night after they had been broadcasting for more than 30 straight hours.

Some of the stations that took part in the movement in 1980-81 included WONS, WARG, Radio Harmonica, Radio Joy, Radio VPR, WRAM, Radio Indiana, Moonshine Radio, Pioneer Radio, Green River Radio, Voice of the Pyramids, WOOF, WPOT, Radio Music International, Radio North Star International, Voice of Michigan, Radio Free New Jersey, WOIS, Radio Xenon, and Radio Kansas. According to the loggings in *FRENDX*, most of these were called "boring" by the average listener, but every one of these had at least one interesting characteristic.

WONS and WARG always promised to send QSL cards, but never did. Radio Harmonica chose a silly name. Radio Joy played classical music. It wasn't until after dozens of hours of broadcasting that anyone could positively identify Radio VPR (mistakenly heard as "VCR" and "VTR") (Fig. 2-

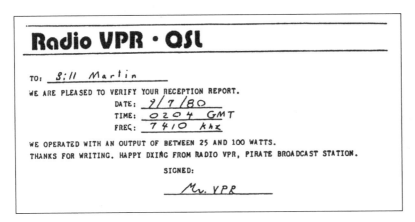

Fig. 2-12 *All questions concerning the actual name of Radio VPR were finally put to rest with their QSL card, sent to listeners several years after the station disappeared.* Bill Martin

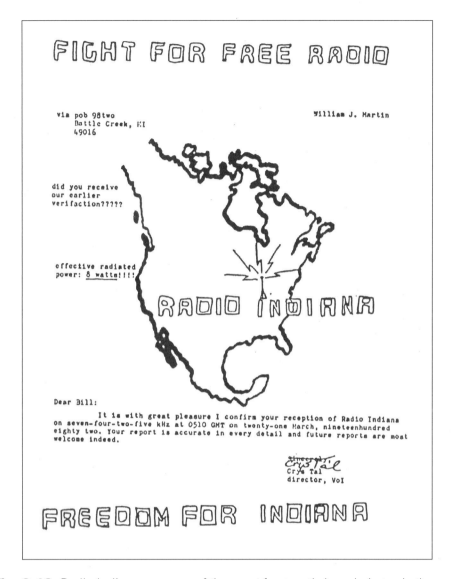

Fig. 2-13 *Radio Indiana was one of the most frequently heard pirates in the very early 1980s and it broadcast in many different frequency bands across the shortwave spectrum.*

12). WRAM played album sides, then disappeared and returned, and was finally busted in 1983. Radio Indiana always signed off with "Auf Weidershen, goodbye" (Fig. 2-13). Moonshine Radio, Pioneer Radio, and Green River Radio (The Alderaan Broadcasting Company) featured more interesting press releases than programs. The Voice of the Pyramids played heavy metal music and relayed the Voice of the Voyager. WOOF took its name from P.D.Q. Bach. WPOT was a smoother version of WARG. Radio Music International's press release announcing their decision to quit broadcasting was longer than the time they spent on the air. Radio North Star International sounded like KVHF, used 13787 kHz, and was raided in 1983. The Voice of Michigan played continuous *Star Trek* episodes on 3580 kHz. Radio Free New Jersey relayed New York Yankee baseball games (as if no one would be able to hear the game otherwise). WOIS played new wave music. Radio Xenon was heard in Europe. And Radio Kansas was one of the first pirates to use an upper sideband (USB) transmitter.

An Actual Movement in 1982

The year 1982 became an even more exciting year for pirate radio than 1980 or 1981. Those years had been the first that a nationally organized free radio effort was made. Then the first issue of *The ACE* (a monthly publication of the Association of Clandestine radio Enthusiasts) was mailed out in April. *The ACE* revolutionized free radio with an entire newsletter dedicated to the movement. For the first time ever, a large amount of up-to-date pirate information was available on a regular basis (*The Wavelength*, the most complete source of information before *The ACE*, was much smaller, was published irregularly, and often included more news from Europe than from North America).

One of the most notable stations for 1982 was actually a returning pirate. The Voice of the Voyager signed back on in January, this time on 6840 kHz. Their reception from the listening community was much different from the first time that they were on. The station was no longer a novelty and several pirates were now more professional, so the Voyager lost much of the enchantment that it once held for listeners. In fact, some people reported it as a fraud and called it "very unprofessional." After just a few months, the Voyager was closed down by the FCC again. This time, R. F. Wavelength was fined $2000 and A. F. Gain was fined $1000. They were warned that another offense would bring them a year in jail and a $10,000 fine.

The most notable new station of 1982 was KQSB, one of the most professional and well-liked hobby pirates of the decade. Phrank Phurter and Uncle Ralph stuck with the common pirate format of comedy skits, fake commercials, and rock music. But the professional production and excellent writing quickly projected KQSB above the ranks.

One important change caused by *The ACE* could be seen in the manner that KQSB operated. With the increase in information about pirates, listeners knew the best times and frequencies to check the airwaves. Often, a DXer could catch a broadcast within a few minutes of the station sign-on. KQSB normally kept its programs down to 20 minutes or less to

Fig. 2-14 *Radio Free San Francisco's QSL, featuring the sleeve art from the Dead Kennedys' single "California Uber Alles."*

avoid FCC detection. Many radio hobbyists heard the station despite the short programs, something that could not have been possible a few years earlier.

Another interesting station, Radio Free San Francisco, only operated in 1982. Known as "the voice of the heterosexual revolution," Radio Free San Francisco offered a more caustic style of humor than most pirates (Fig. 2-14). In addition to playing lots of rock music, they also made political jokes and featured editorials on shortwave radio. ZRRZ (its self-assigned call

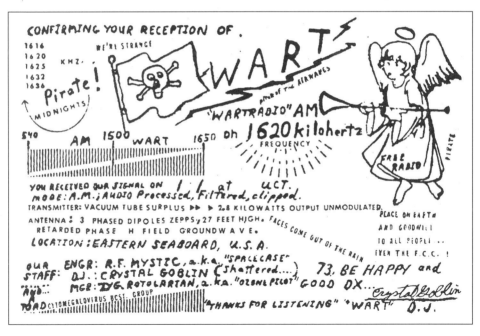

Fig. 2-15 *Going crazy with rub-on letters: the QSL card from WART, which brought AM pirate radio back to New York City in 1982.*

letters) was active at different points in the year, but it completely disappeared in the autumn.

The New York City pirate scene also reactivated with the appearance of WART and KW Radio, both widely heard just above the upper limit of the AM broadcasting band. WART often experienced technical difficulties with its reconditioned amateur transmitter, so listening to it often required patience. However, the station did feature loop call-in programs and regular tips about hearing other pirates (Fig. 2-15). KW Radio operated only at the end of 1982 with call-in shows, but with the addition of a comedic influence. Announcers Johnny Jo and Barbara Baintree often closely imitated celebrities while answering calls. Their low-key and occasionally derogatory humor was popular among many who heard it during the brief time it was in operation.

Other less-notable 1982 stations included WHUFO/WHFO/WCRS, Radio Free Radio, KCFR, Radio Toronto, ZKPR, KPHU, and the Voice of the Purple Pumpkin. These stations were, on the whole, more interesting than those from 1980-81, but still less notable than the 1982 stations included thus far. Nevertheless, each had at least one characteristic worthy of mentioning. WHUFO/WHFO/WCRS changed its name to avoid confusing listeners. Radio Free Radio played new wave music and comedy skits in USB, and many listeners complained that SSB was an unlistenable mode. KCFR almost never issued QSLs. Radio Toronto sounded like the MacKenzie Brothers. ZKPR hated elitist DXers. KPHU had a shortwave call-in show. And the Voice of the Purple Pumpkin was either a spoof or a return of the original station.

Pirate Radio
Since 1983

The number of listeners to pirate radio, FCC enforcement activity, and programming trends in pirate broadcasting have all varied significantly since 1983. During the years from 1983 through 1985, activity greatly increased on the shortwave bands and decreased on and around the AM broadcasting band. Some trends included a regular use of the 7355- to 7450-kHz (41 meter) range, while frequencies in the 6200- to 6300-kHz (48 meter) range died out. The practical location for pirates has always been 41 meters because signals are received better over long distances and amateur transmitters require no modification to operate in this range. The 48-meter band is generally avoided because it contains several legal ship-to-shore outlets; occupying these frequencies could cause harm to the ships and serious legal problems for the pirates.

Pirate festivals, which originated in July 1982, became an important segment of the free radio hobby. This is when several stations get together and broadcast back-to-back on the same frequency. On some weekends, over a dozen broadcasts from various stations would be aired. Each station would air its program and then sign off, allowing the next station to sign on. This has become one of the most popular tactics used by pirates because listeners can spend several hours at one sitting listening to different stations. At the time the second edition of this book was being written, a pirate festival (Pirate Radio Insanity) was held on December 31, 1994/January 1, 1995 around 6956 kHz. Stations used sideband (mainly upper sideband), and several stations participated.

Another trend of recent years is that pirates are becoming more knowledgeable about their hobby, and programming is now more professional than it had been in the late 1970s and early 1980s. Most pirates from the late 1970s merely played rock music and made simple announcements. Usually only one or two stations per year offered professional programming, and that was often uncreative. But some of these broadcasters from the 1970s have stayed with the hobby long enough to learn improved recording and programming techniques.

Extensive praise and criticism of program quality in *The ACE* also led to more professional and creative shows. "Veried Response," a QSL and editorial column, offers a yearly popularity contest for choosing the most and least regarded stations. The resulting competition and suggestions also enable operators to understand what pirate listeners want to hear. To help stations achieve their programming goals, information on recording techniques and where to find rare and/or alternative records, audio equipment, and transmitting equipment are also included in "Veried Response."

Pirates Since 1983

Pirate stations on the whole, continued to grow in sophistication with both programming and technical operations throughout the 1980s. The most notable broadcasters in 1983 included KPRC, KQSB, and Radio Clandestine, which all operated more professionally than many legal American stations. Radio USA, Radio Amity, and WRAM also operated extensively over the course of the year, but with rather amateurish programming. FCC raids finally eroded more of these pirates' delusions of an uninhibited free radio service by fining WRAM and Radio North Star International in the Summer of that year.

The year 1983 proved to be a crossroads for many other stations. Well-known pirates such as Pirate Radio New England, Voice of the Pyramids, WOIS, and WART gradually disappeared during the year. However, a fresh crop of pirates led by the Voice of Democracy, the Voice of Laryngitis, the Voice of Tomorrow, WDX, Radio Paradise International, and Radio Free Insanity (Fig. 3-1) arrived in the second half of the year to replace those that had fallen from the hobby. Radio Morania, Radio Angeline, and WBST (interesting but rarely heard stations) also began in 1983.

Radio Morania was a particularly interesting broadcaster. First heard on the air in 1983, the station operated frequently throughout 1984 and 1985, transmitting the same program each time. A spoof of shortwave propaganda stations, Radio Morania described chocolate mining, Morania's 6000-pound spaghetti harvest, and the Moranian Hit Parade with the #1 song in the fictitious country, "I Fell In Love With a Green Turtle Fly." Although the pompous, 1950s-style programming was humorous, the lack of a mailing address antagonized some listeners. It was not that the operators did not "try" to announce the

Fig. 3-1 *A QSL from the short-lived Radio Free Insanity, from Bloomington, Indiana.*

address. Every time the address was about to be given, various types of interference were intentionally introduced on the program to destroy the details. The mystery of Radio Morania was solved several years later when a hobbyist wrote in to a popular radio magazine that his tapes were being rebroadcast over shortwave. He had produced the tapes in the late 1960s for fun and later sold the recordings to the public in 1972. Over a decade later, some owners of these cassettes began relaying them all over the shortwave bands.

Pirate activity in 1983 dropped temporarily in September and October, directly after the bust of Radio North Star International. The mortality of the free radio movement was rediscovered; it could be stopped not just by a bust itself but also by the possibility that the FCC's surveillance activities were on the upswing. Oddly enough, the bust of WRAM in June caused no repercussions throughout the radio community. Evidently, the shortwave community thought that WRAM's careless operators (announcing a local address, making frequent broadcasts, and transmitting over extended periods of time) caused their own downfall. Other pirate stations felt this bust would have no impact on them.

Pirate Radio in the Mid-1980s

Again, pirating gained momentum into 1984, the most active year ever in North America. Much of the year's activity was caused by KQRP (Fig. 3-2), which had possibly the most active year in North American shortwave pirate history. The Midwestern stations began extensive operations early in January on or near many international shortwave broadcast bands. KQRPs varied frequency usage encouraged other operators to follow suit.

Within a few months, 6200 to 6300 kHz regained its former

Fig. 3-2 *KQRP didn't last for long before being closed by the FCC in Arkansas in 1984, but it made a plenty of broadcasts.*

popularity, and 15050 kHz became a new active pirate location. KQRP closed operations in 1984, the station made more reported broadcasts than any other North American shortwave pirate until 1991.

Other stations, however, helped build up 1984 as a monumental pirate year. Radio Clandestine offered more broadcasts that year than any other since it began broadcasting. The Voice of Laryngitis became the most popular hobby pirate ever for its legendary comedy productions, while the Voice of Communism, Radio North Coast International (Fig. 3-3), and Radio Sine Wave also offered intelligent, humorous parodies. The Crystal Ship, KPRC, Tangerine Radio, and the Voice of Tomorrow all pushed various political views. Samurai Radio, WDX, and WMTV all presented programming centered around telephone conversations. And what would the hobby be without those "simply music and announcement" stations? Zeppelin Radio Worldwide, WKUE, New Wave Radio International, KLS, and KOLD were a few of the more widely heard pirates with less-creative format styles.

Activity continued at the same high level into 1985, and some stations even seemed to think that they were immortal. Because no pirates had been caught since mid-1983, operators became reckless with their broadcasting schedules. KROK, Secret Mountain Laboratory, Union City Radio, WKUE, Radio Woodland International, the Voice of Laryngitis, KNBS, and Zeppelin Radio Worldwide all averaged at least two broadcasts per month. However, KRZY, the obvious replacement of KQRP, overshadowed them all, amassing a total of 36 broadcasts in January and February 1985. Other operators certainly must have felt that with KRZY booming out every other day (plus the activities of additional pirates), the FCC would not have the time or resources to catch everyone.

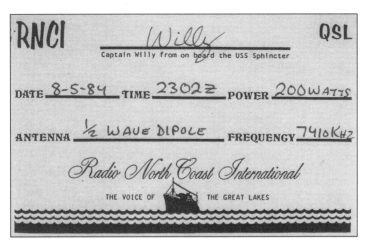

Fig. 3-3 *Captain Willy's Radio North Coast International, from the U.S.S. Sphincter, was one of the most-professional pirates of the 1980s.*

Theories on KRZY, the FCC, and the rest of the free radio scene were, for the most part, true. KRZY's overactivity drew the FCC's attention away from the others. It was busted in early March, and the free radio scene continued at nearly the same rate. This might have been because neither the FCC nor the pirate hobbyist announced KRZY's demise. Anyone who might not have realized that the station was raided was probably expecting the organization to be caught soon anyway because of the high activity.

After several months of heavy broadcasting by free radio stations, the FCC cracked down. KKMO, a new pirate with a virtually all-music format, was caught in late August. FCC agents nailed another new station, KBBR, in early September. Initial direction-finding showed KBBR in the Ozark Mountains in Arkansas, the same area where KRZY was caught earlier in the year. A close-in investigation proved that both stations were operated from the same household. Rather than try to immediately eliminate KBBR and possibly lose evidence, the agents procured a search warrant. Several days later, with the search warrant and a federal marshal, the FCC closed KBBR.

As in 1983, free radio experienced serious repercussions from these FCC attacks. According to listeners across the country, only four broadcasts from as many stations were audible in October. But just as pirates appeared to be dying out, many old favorites reappeared around Thanksgiving and Christmas, including Free Radio 1615, WDX, KROK, Radio Clandestine, the Voice of Communism, the Voice of Laryngitis, WKUE, WMTV, WPBR, Radio North Coast International, Radio Angeline (Fig. 3-4), and Zeppelin Radio Worldwide with extensive holiday programming.

These operations, however, merely became a "swan song" for

Fig. 3-4 *From time to time throughout the 1980s, Radio Angeline appeared with strange programming dedicated to his lost love, Angeline. The back of this card proclaims "Love is not what you think it is."*

the pirate activity of the past few years. Subsequent months showed a gradual slump, with fewer pirates appearing less frequently. The next downward shove came from the FCC in the form of a press release announcing the demise of KRZY/KBBR months after the original bust. Included in the letter was a list of 16 cities where the FCC claimed it had detected pirate activity. Radio magazines and newspapers across the country published the list, which effectively scared many pirates off the air.

The last and most deadly blow to pirate radio came from the pirates themselves and their listeners. A reverse KQRP/KRZY cycle kicked in after the busts. Since fewer pirates were operating, fewer listeners tuned in. Likewise, since fewer listeners were tuning in, fewer pirates operated. Although this effect continued in 1986, two stations, WHOT and CFTN/TNFM/KQRO, bought new equipment that enabled them to be heard over great distances.

WHOT had operated from the Brooklyn, New York area on FM throughout most of the 1980s. All station personnel had had experience with some of the older New York City pirates, so the programming was very professional with many telephone call-ins. TNFM/CFTN also began as an FM pirate in the early 1980s from Salt Spring Island, British Columbia, Canada. The operator later created KQRO with the acquisition of a shortwave transmitter. After several years of infrequent broadcasting as KQRO, personnel decided to combine the two stations and broadcast with FM and shortwave in parallel. The combination rivaled RX4M and KQRP/KRZY/KBBR in terms of frequent broadcasting.

Transmissions from WHOT and TNFM/CFTN encouraged listeners and pirates, mending some of the devastation caused by the FCC busts a year earlier, but still the free radio scene limped behind what it had been. Just when listeners were hoping the pirates would return, the situation worsened. The FCC helped locate TNFM/CFTN and contacted Canadian authorities. The station then received a letter form the Canadian Department of Communications (DOC) notifying them of the illegal operations. The DOC also asked the owner of the station to destroy his equipment! Although it is doubtful that he ever destroyed his own equipment, TNFM/CFTN was pulled from the air at the end of 1986.

To make matters worse, WHOT disappeared from above the AM band as 1987 was beginning. Unlike many stations that disband after being caught, lose interest, or experience equipment failure, WHOT left the AM band because of poor listener response. It could also have been that the widely heard transmissions made WHOT too much of a high-profile operation.

At this point, North American pirate radio virtually froze over. During several months in 1986, only one of two different stations were reported by listeners. These figures appear especially depleted in comparison to high points in December 1984 and April 1985, when 25 different pirates were heard each month. Most of the notable activity for 1987 resulted from the transmissions of Radio Newyork International, a pirate ship broadcasting from international waters nearly 13 miles from the coast of Long Island, New York.

Other than Radio Newyork International, the North American pirate scene was operated by a skeleton crew of older stations including the Voice of Tomorrow, Radio North Coast International, Zeppelin Radio Worldwide (Fig. 3-5), Radio Angeline, and KNBS. Although all of these pirates made little more than cameo appearances, a few new stations began in the second half of the year. The Voice of Free Long Island, WCPR, and a return of WENJ with much higher power fueled a new movement from the New York City area.

The new stations breathed some life back into free radio. Unfortunately, so many DXers had stopped listening for pirates that at first other stations didn't want to waste the time broadcasting. More stations slowly began returning to the airwaves, nevertheless. Just as activity was picking up in mid-1988, the Association of Clandestine radio Enthusiasts faced major staff problems and publication of *The ACE* was delayed over three months. The lack of information again discouraged listeners and pirates alike. But *The*

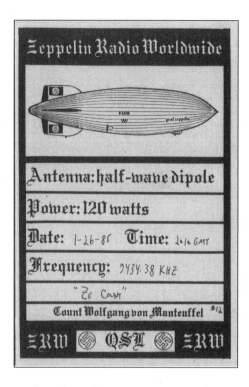

Fig. 3-5 *From 1984-1988, Ze Count of Zeppelin Radio Worldwide broadcasted an odd mix of professional ZRW blimp-related IDs and 1950s and '60s rock music.*

ACE returned with a new name *(Free Air)*, a new format, and a renewed enthusiasm for the hobby in September 1988.

A Renewal as the Decade Ends

Pirate radio in 1988 lacked active stations and listeners; the few people involved were often inexperienced and filled with an almost naive enthusiasm. Instead of the older, jaded veterans from the early 1980s, the newcomers produced somewhat sloppy programming. However, most stations were just excited that someone was listening and most hobbyists were just happy to log a pirate.

In the midst of this young scene, one of the most influential pirates of the 1980s and 1990s was born. On Halloween 1988, WKND ("Weekend Radio") fired up a homebrew 60-watt transmitter on 1621 kHz that was heard across much of the Northeast. At the time, WKND was significant because the programming was interesting, because it brought some new blood to the scene, and because people in the Midwest could finally hear some pirates (at that time, few stations were being heard west of Pennsylvania). The stations soon began to cover the eastern half of the country as Radio Clandestine, Falling

Star Radio (a reincarnation of KPRC), WBRI, and Radio USA reactivated. WENJ also acquired a shortwave transmitter. Three new stations that started early in the year were WJDI, WKZP, and Free Radio One. At the end of 1988 and the beginning of 1989, WKND was active on 1621 kHz and Radio Clandestine, Falling Star Radio, and WENJ were widely heard several times nearly every month on shortwave.

Just as the scene was picking up, one old problem reappeared and another was created. WRFT ("Radio Free Texas") began broadcasting with a schedule that was published in several radio newsletters. Of course, the schedule drew in a number of listeners, but it also attracted the ever familiar government regulatory commission that pulled the station's plug in approximately three months. This raid threatened the security of the other operators because they suddenly knew that the FCC was actively pursuing pirates again. Although the loss of WRFT left a hole in the pirate scene, the other stations barely paused.

A more damaging hole was left as the *Free Air* began arriving later and later each month. Finally, it stopped arriving altogether. As a response to the absence of *Free Air*, I began publishing an ugly little logsheet called *Pirate Pages (PiPa)* in April 1989 to keep the listeners informed about pirate activity. By June, *The ACE* had returned with president Kirk Baxter taking over as publisher. Within a few months, Don Bishop was writing excellent, in-depth articles for *The ACE* about the FCC, pirates that had been closed down, etc. Thus began what was to be the greatest era in pirate radio to date.

By the end of the year, the scene literally exploded. It was 1984/85 all over again. Even more regular stations took to the air, such as East Coast Pirate Radio, Jolly Roger International, RFM, Radio Mexico, Secret Society Radio, WGAR, the Voice of Monotony (Fig. 3-6), the Voice of Radio Free Indiana, the Voice of Stench, WHBH, WRNR, WXZR, and Zodiac Radio. Also, a number of stations that had been inactive for years returned, including Samurai Radio, KNBS, Tangerine Radio, WYMN, WBST, and Secret Mountain Laboratory. Free Radio One, a right-wing talk show operation with announcer Dr. David Richardson, planned to have outlets in every state in America within a few years. Although these plans never materialized and Free Radio One disappeared before the end of the year, the station managed to become the most active pirate from any year since 1985. The other most-active stations included Radio Clandestine, Hope Radio, and Radio USA. Hope Radio, a newcomer to the scene, blasted across North America, sometimes using as much as 1000 watts of high-fidelity AM. Although the early

Fig. 3-6 *The Voice of Monotony broadcasted irregularly outside of the standard pirate bands (near 6850 and 7515 kHz) in 1989 and 1990.*

Hope Radio programs were amateurish with long airings of comedy albums and entire articles read from *Popular Communications,* the station later helped set the pace for North American pirate radio in the early 1990s.

1989 was an interesting year in pirate radio because the stations and trends that occurred shaped the scene for the next four years. Radio Clandestine and WENJ operated regularly on 7415 kHz. Before long, pirates across North America were scouring hamfests for 7415-kHz transmitter crystals. Zodiac Radio from California (one of only a few West Coast pirates since 1980) sealed up the bid for this frequency range to be the pirate location by using 7417 kHz for its many long broadcasts in 1989. A year or so later, books, magazines, and radio shows were advertising 7415 kHz as the pirate frequency. It remained so until adjacent-channel interference forced most pirates off the frequency in 1993.

Two 1989 trends that didn't last were "QSO" (two-way contact) broadcasting and operating during weeknights. During July, Radio Clandestine was operating every weekend and during a number of weekdays in between. One Tuesday evening in the summer, Radio Clandestine broadcast for a few minutes. This broadcast was followed by several minutes of programming from Radio Garbanzo and KRUD. Before long, the stations were exchanging segments and requesting songs for each other. By the end of the evening, five stations had been on the air and the round-robin exchange was highly successful. Weeknight broadcasting ended (for the most part) because few listeners checked out the bands during the work week. QSO broadcasting evidently ended, though, because the timing and scheduling was too difficult.

Pirate Radio Enters The 1990s

1990 and 1991 ripped along at a fantastic pace. Hope Radio aired more broadcasts than any other station since KQRP in 1984. One Voice Radio, an all-medical news pirate, broadcast frequently throughout the year. Radio Clandestine, however, reduced its output and dropped out as a major force in 1989 to being barely active for the following years. Although Radio Clandestine was missed, the new stations were quick to replace it.

Samurai Radio, one of the few North American pirates to have operated in three different decades, returned in 1989 after a five-year hiatus. In addition to making standard broadcasts, Samurai Radio (Fig. 3-7) also participated in the first joint broadcasts with other pirates, including Radio USA, the Voice of Monotony, and the Voice of the Purple Pumpkin in 1989, 1990, and 1991. These broadcasts were notable because they were the first joint broadcasts between various stations, because they often used parallel transmitters (as many as three on the air at the same time), and because the programs lasted for as long as four straight hours.

Aside from the live joint broadcasts, most pirates began to prerecord to achieve better production values. As a result, a number of stations were airing well-produced broadcasts of 60 minutes or less in length, except for the 1991 and 1992 shows from WSKY and Radio USA, which often lasted between two and four hours.

Even though 1990 and 1991 were two of the best "glory years" for this present pirate radio cycle, the years were not without FCC enforcement actions. In 1989, WRFT, WJPL, WHOT, XERK, WNYS, and WNPR were all raided by the FCC. The 1990 and 1991 raids included

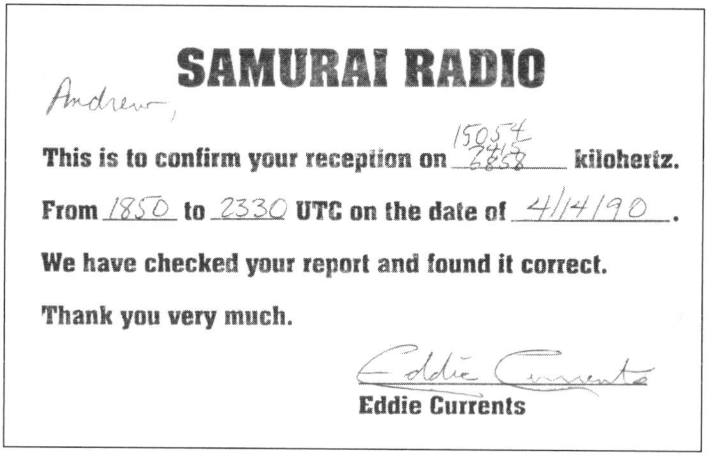

SAMURAI RADIO

Andrew,

This is to confirm your reception on _15054_ _6858_ kilohertz.

From _1850_ to _2330_ UTC on the date of _4/4/90_.

We have checked your report and found it correct.

Thank you very much.

Eddie Currents

Eddie Currents

Fig. 3-7 *A Samurai Radio QSL from its Easter weekend 1990 joint marathon broadcast on 6858, 7415, and 15054 kHz.*

WKND (twice), Zodiac Radio, Secret Society Radio, WTNU, KUSA, CHGO, the Voice of Oz, WLAR, KMUD, Fourth of July Radio, an unidentified Voice of Laryngitis relay station, the Chicago Tunnel Company, and Hope Radio.

The beginning of 1992 started off much like 1991, with all of the major broadcasters still dominating the scene. Hope Radio was still active; Radio USA and WSKY (Fig. 3-8) were still running marathon broadcasts. However, this group of active stations quickly scattered. In late February 1992, the FCC used a large publicity campaign to accuse one listener of operating Radio USA. In response, Radio USA cut back and made only a handful of broadcasts for the next eight months. WSKY mysteriously disappeared in early Spring. A ham radio operator in New York was accused of operating Hope Radio and the station dropped off the air in February. Rumors ran rampant that Omega Radio was also closed by the FCC in early spring.

Although it might seem as though the pirate scene would have withered after these losses, the opposite happened. A new crop of stations turned to pirate radio, including Voice of the Night, Radio Chaos International, WEED, WVOL, WHIZ, Anarchy 1, EBO Radio, WRMR, and WKIK.

Unlike the generally friendly and courteous operators that had broadcast in the past, a new 1992 station, the Voice of the Night, seemed to thrive on the concept of negative publicity. LAD, the 13-year-old station operator, intentionally jammed and threatened a number of other pirates. After public and private threats by other operators and by shortwave listeners and a rumored FCC bust, The Voice of the Night's operations were considerably reduced and became much less caustic.

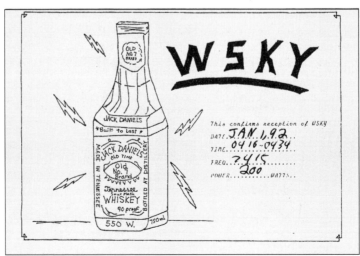

Fig. 3-8
WSKY started with hours of Doors records, but later moved on to fast-paced live rock'n'talk shows with Mike Richards, Doug Barley, and Sherry.

Radio Chaos International and WCYC also operated in close contact with the Voice of the Night (except without the negativity). Radio Chaos International was closed by the FCC and equipment was confiscated from a New Jersey ham in March. Although WCYC was neither threatened nor busted, their operations were reduced after the demise of Radio Chaos International.

For the first time ever, a group of widely heard, regularly active stations in the western United States took to the air. Jolly Roger International, WEED, and Anarchy 1 all operated regularly (especially toward the last half of the year) with high-quality programs and signals that were consistently heard from coast to coast.

When pirate radio from the East Coast took off in the late 1970s, operation just above the AM broadcast band on frequencies like 1620 kHz was far more common than operation on shortwave. However, that scene dwindled to nothing by 1990. WRMR entered the picture as a notable exception in April. The station was heard as far west as Wisconsin from its New York location!

In the late Summer and early Fall, EBO Radio and WKIK dominated with large numbers of programs. Like most new, very active stations, both concentrated more on activity than on production values. By late Fall, WKIK was operating for more than a few minutes at a time, but their transmitter began drifting at faster than 1 kHz per minute! Because of negative comments about "overusing" 7415 kHz and poor audio, EBO Radio dropped out of the 41-meter band in October in search of better, clearer frequencies where people wouldn't be annoyed about "frequency hogging."

Halloween 1992 was a huge pirate holiday, with more than 20 different stations piling onto the bands in less than 24 hours. Although a few new stations appeared, mostly veteran stations were in the spotlight (as opposed to the activities of the previous eight months).

The early Winter months of 1992 returned to a pattern similar to the one that existed in very early 1992. Stations with scripted, prerecorded programs were much more common than had occurred in the summer. CSIC and WARI continued their regular broadcasts; WSKY, Radio USA, and Action Radio (Fig. 3-9) reactivated; and Anarchy 1, WEED, and Jolly Roger International fared even better from the West with the increased broadcasting range during the Winter months. Although some jamming continued throughout the end of 1992, the amount was much less than that which had occurred during the Spring and Summer months.

By 1993, the impact of the loss of some of the highly active radio stations was beginning to weigh heavily on the scene. Although

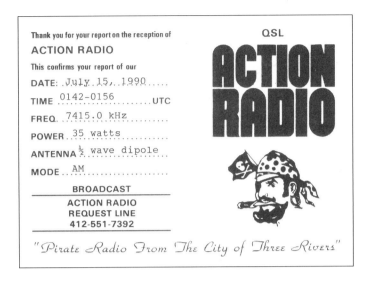

Fig. 3-9 *Action Radio originally claimed a Pittsburgh location, but later announced that it was "off the west coast of Nebraska."*

Thank you for your report on the reception of

ACTION RADIO

This confirms your report of our

DATE: July 15, 1990

TIME 0142-0156 ...... UTC

FREQ. 7415.0 kHz

POWER 35 watts

ANTENNA ½ wave dipole

MODE AM

QSL

ACTION RADIO

BROADCAST
ACTION RADIO
REQUEST LINE
412-551-7392

"Pirate Radio From The City of Three Rivers"

everyone seemed to have negative comments about the Voice of the Night, more listeners were glued to the so-called "pirate bands" during the first half of 1992 than at any other time before or since. The pirate scenes of 1993 and 1994 were on the opposite side of the curve from 1989. In 1989, a large number of new pirates with live and amateurish programming were operating. By 1993 and 1994, nearly all of the active pirates were veteran stations with higher quality programming. As a result, while the number of broadcasts per month was very steady throughout the year and the overall quality of the broadcasts improved, the number of broadcasts declined by nine percent.

During this time, a large portion of the activity shifted to the Midwest and West. A number of stations, including some of the most active 1993 and 1994 pirates (such as WEED, which claimed to be "broadcasting from the great Southwest") did not broadcast from the "pirate golden triangle"—the area between Cleveland, Boston, and Baltimore. For a number of years, pirates from this roughly triangular area dominated the pirate scene. Even though shortwave piracy was finally nationwide and shortwave listeners across North America could receive pirates at local quality, nearly all of the pirate listeners were located in the Northeast (where piracy was slumping), so the number of listeners continued to decline.

Probably the greatest cause of declining listenership was an almost complete reversal of the media coverage of the hobby that occurred in 1992. Radio Newyork International and the "Signals" radio program (both aired by legal shortwave outlet WWCR) had aided the hobby by both featuring pirate radio loggings and comments in 1992. Both of these radio outlets were absent in 1993.

In spite of the poor listenership, activity in 1993 and 1994 remained high, most pirates began decreasing the length of their broadcasts, however, to approximately 20 to 40 minutes. Thus, although the number of broadcasts remained high, the amount of time that was available to listen to pirates was actually decreasing. WEED, WLIS, Radio Airplane, the North American Pirate Radio Relay Service (NAPRS), Solid Rock Radio, and Radio Doomsday were the regularly active stations during these years.

Fortunately for pirate listeners, few stations were raided by the FCC in 1993 and 1994. Both of the raided stations had an interesting history, however. The first was a station relaying WMXN, a commercial FM station from Virginia on 6250 kHz. A number of listeners across the East Coast heard the station very weakly, but could not determine its true identity or origin. Finally, on January 12, 1993, the relay station was closed down in northeastern North Carolina. According to the press release and other rumors, the operator was a licensed amateur who wanted to hear the local FM station in his house, so he relayed it via a five-watt transmitter, which got out much further than he had expected! The other action occurred in New Jersey against WPIG. WPIG operated many times over the course of January and February 1994 while announcing his home address and telephone number. The programs consisted of announcer Ira singing a variety of acapella songs about pigs and telling stream of consciousness stories. When the station was finally raided by the FCC, rumors circulated that Ira was living in a group home, and consequently, the agents only warned him against continuing the broadcasts.

Crushing blows to pirate radio listeners in 1993 and 1994 came from a lack of publicity and adjacent channel interference from licensed broadcasters. Newly licensed WEWN began broadcasting on 7425 kHz with 500,000 watts in 1993. Unless you had an excellent receiver, the signal washover from WEWN wiped out most pirate signals on 7415 kHz. The Voice of America on 7405 kHz in the evening and on 7415 kHz in the afternoon doubled the problem. Several pirates hunted new territory. After some groundwork by Radio Airplane and WEED, 7465 kHz was adopted as the new pirate channel. But few listeners (other than the diehards) discovered this frequency. Before long, another licensed station was on 7465 kHz, so some of the pirates moved to 7445, 7470, and 7405 kHz. These frequencies also succumbed to heavy interference, so most stations finally moved to 7385 in the Summer of 1994. When Radio For Peace International moved from 7375 kHz to 7385 kHz, some pirates moved again, this time to 7375 and 7490 kHz. By early 1995, most pirates were either operating on 6955 kHz or on other frequencies throughout the 41-meter band.

The result of all this shuffling across the radio spectrum was the loss of most of the listeners. Rather than receiving 20 to 70 letters for a widely heard hour-long program in the Saturday evening prime time for pirates (such as in 1991 and 1992), most stations only pulled in about three to ten letters. Basically, only the inner core of pirate listeners were able to keep up. Because much of the excitement was gone for the listeners, some of the pirates resorted to creating extra novelty "stations" in order to continue receiving reports in the hopes that it would build more excitement for the listeners. Instead, it had some of the same effect as inflation on the economy. Hearing the same station broadcast under five or more different names just devalued the entire pirate radio experience for everyone.

At the time the second edition of this book was being written, hours of interesting and well-produced broadcasts were on the air every weekend. However, they were often difficult to find, and few new listeners were able to tune in. Additionally, poor radio signal propagation, which also negatively affected pirate listening in 1987, hampered listening.

But not everything that occurred in 1994 was negative. Some of the most momentus achievements in the past decade were in this year. For example, New Year's Pirate Radio Insanity I was the first scheduled shortwave pirate event since the St. Patrick's Day 1983 festival. WJLR, WEED, Voice of the Runaway Maharishi, UNID, The Great Southland, Radio Free Euphoria, Radio Fluffernut, Radio Doomsday, Radio Airplane, Hit Parade Radio, and CSIC all transmitted programs within a several-hour time frame.

Possibly the most important feat of 1994 was Shortwave Liberation, a 30-day pirate radio tour de force. For 30 straight days, pirate stations across the country chipped in to make at least one broadcast per day. Thirty straight days of pirate radio might sound a bit passe in 2001, but remember that this was in a period with some FCC enforcement and there was good reason to believe that all of the activity could cause a raid. At the time, some people criticized Shortwave Liberation as being a foolhardy event that would force the FCC's hand. In *The ACE*, columnist John Arthur stated that it "was about the dumbest concept I've ever heard of" and that the station operators were "accomplishing nothing but sticking their necks further into the

FCC noose." Indeed several of the major participants of Shortwave Liberation were visited or raided by the FCC within the next year. Whether or not Shortwave Liberation played into the FCC hands is doubtful, considering none of the participating stations were raided within five months of the event.

Shortwave Liberation was the brainchild of Kirk Trummel, a long-time ham radio operator who also co-founded the Free Radio Network. Trummel, then 30, was an avid experimenter who moved from project to project without ever staying in one place for long. In 1993 and 1994, Trummel was getting the most out of pirate radio, listening to plenty of stations and also broadcasting—sometimes as Nemesis from his flagship station, Radio Doomsday, but also under a dozen or so other assumed names. After Trummel's sudden death in August 2001 from liver and pancreatic cancer, friends on the Free Radio Network posted a eulogy, describing his pirating career. Although Trummel didn't single-handedly broadcast Shortwave Liberation, he did make a number of transmissions and filled in with some programs when other pirates couldn't go on the air. Shortwave Liberation ultimately solidified the scene and attracted pirates and listeners alike to 7385 kHz, which was important later in the year.

Soon after Shortwave Liberation, Radio For Peace International, a licensed shortwave station from Costa Rica, moved part of their schedule to 7385 kHz. After losing 7415, 7425, 7445, 7465, and then 7385 kHz to licensed stations, it was clear that 41 meters was too sought-after for any frequencies to remain open for any length of time. Several pirates began monitoring for clear channels and moved to 6965 and 6955 kHz just in time for the 1994 Halloween holidays. The first station to use the range was the one-shot WLBG ("We Love Bob Grove") on October 22, but it was followed by numerous pirates, such as Radio Airplane, which aired its "Radio Scareplane" holiday special five times over the week.

Pirate Radio in 1995 and 1996
6955 kHz, King of the Dial

The hard work of the frequency-hunting stations in 1994 started to pay off in 1995. With the active pirates relatively unified, the transition of listeners from 7385 kHz to 6955 kHz was smooth and fast. By 1995, almost every North American broadcast was near 6955, aside from some tests on the old pirate home of 7415 kHz and the new high-frequency testing ground of 13900 kHz. Listenership was still down throughout much of 1995, but the listeners were back by 1996.

The biggest news of 1995, now that the frequency-hopping had settled down, was the FCC raids at the beginning of the year. John Cruzan, who later developed the Free Radio Network, was raided by the FCC for unlicensed broadcasting. The FCC confiscated a shortwave receiver and audio equipment, but found no transmitters. Kirk Trummel was also visited by the FCC, who accused him of pirating but came up empty. Other stations were rumored to have been either raided or under the watchful eye of the FCC. The loss of a few stations and the silencing of some others limited broadcasting, but hardly eliminated it. Some of the most active pirates through 1995 were NAPRS, Radio Free Speech, WREC, Free Hope Experience, WLIS, WRV, KDED, Up Against the Wall Radio, Radio USA, WPN, East Coast Music Radio, and Radio Bob's Communication Network.

The NAPRS and Radio Free Speech racked up an astounding number of broadcasts, and both were widely heard—astounding for RFS, which transmitted with a mere 15 watts. Possibly the highlight of the year was when the federal government closed down on November 17, and all non-essential personnel were sent home. Bill O. Rights from Radio Free Speech had been keeping an eye on the news and busily created fresh "government shutdown" programs. The programs were heard no less than five times throughout the day. Evidently, Radio Free Speech wasn't alone. An unidentified station broadcast a two-hour loop tape of Radio USA, The Voice of Revolutionary Vinco, Radio Airplane, and Christian Rock Radio for no less than six hours through the morning and early afternoon hours. Other stations got into the act later in the evening, but Bill O. Rights was the only operator to offer special programming for the day.

1995 might have been a year of "healing," but 1996 was a year of growth. Some of the new blood infused into the scene, included WPRS, Mystery Radio, Voice of Juliet, Jerry Rigged Radio, Radio One, Radio Two, Radio Three, Radio Nine, KAOS, Up Your Radio Shortwave, WARR, KTLA, Radio Xanax, WRRN, Radio Tellus, Radio EuroGeek, WMPR, and Anteater Radio.

Some interesting trends began in 1996. One was all of the Radio <number> stations. Radio One started broadcasting with a professional mixture of 1950s rock music and DJ banter. Radio Two started as a parody of Radio One. This parody was soon followed by Radio Three, Four, and Nine. For a few years now, all of these stations have been gone, except for Radio Three. Someone later dissected and reassembled Radio Three programs with Top-40, children's music, polkas, etc., inserted in place of the rock music. These programs are parodies of the parodies and five years after the originally parodied station has disappeared, they can be a little tough to follow!

Radio EuroGeek grabbed the attention of the international DXing community with a special DX program placed on the frequency of the annual, single-afternoon broadcast of Radio St. Helena. Dozens of listeners from around the world were surprised and amused to hear the unexpected pirate broadcast.

For more information concerning the other pirates previously mentioned, *see* the station listings.

1997-1999
The Ten-Kilowatt Flamethrower

In May 1997, a huge AM signal appeared on the 6955 kHz. The announcer called himself "Dr. Tornado" and he mostly played rock and novelty music on Radio Metallica Worldwide. The programming was all live and Dr. Tornado wasn't particularly concerned about sounding professional. These types of stations aren't particularly unusual, and WARR and Radio Tellus were both somewhat similar. However, the RMWW was incredibly powerful and the AM audio was smooth and clean. It sounded like some disgruntled employee overtook one of the Radio Canada International transmitters.

It wasn't a fluke and Dr. Tornado reappeared night after night in May 1997. By the end of the month, he had broadcast on 18 different days. The longtime pirate listeners had seen stations operate in the same style as RMWW many times in the past. Some, like WPIG, managed to last a month. Others, like Radio Free North America, only lasted about one week before the FCC came knocking. In June, Dr. Tornado wised up and cut back to "only" 14 different days of broadcasting. But in July, RMWW put in literally dozens of broadcasts on 26 different days. Dr. Tornado was more powerful than a one-man Shortwave Liberation!

And the power claims of Radio Metallica Worldwide never appeared to be an exaggeration. Numerous times, listeners simultaneously reported the station in California and Maine with an SIO of 555. And when reception reports of strong signals (strength 4 of 5) rolled in from countries as widespread as Greece, Scotland, Paraguay, and New Zealand, it guaranteed that this was no commercially made amateur radio transmitter (Fig. 3-10) or transceiver.

All this activity and great signals attracted new stations to the air and drew older stations from the ether. Radio Free Speech broadcast its Halloween program 10 different times throughout the week of Halloween; Radio Eclipse transmitted its season finale and season

Fig. 3-10
Both a cozy way to stay warm in the winter and convenient for heating supper--the powerfully hot final tubes of the Radio Metallica Worldwide transmitter in action.

premiere programs each eight different times; Radio USA returned with some long programs that were heard all over the world; and Solid Rock Radio, Radio Azteca, Free Hope Experience, WREC, and NAPRS were all up to peak activity levels. Some of the new stations, such as Anteater Radio, WMPR, Mystery Radio, Radio Nonsense, and Radio Eclipse broadcast regularly throughout the week, as opposed to the long tradition of weekend-only transmissions in North America.

In the Fall and Winter, Radio Metallica Worldwide broadcast more talk-based programs, featuring in-studio political commentary and banter from Dr. Tornado, Senor El Nino, and occasional guests. The apparent increased time spent on programming corresponded with a decrease in time actually transmitting. RMWW transmissions fell back to about six to eight per month, still an impressive number— especially considering the powerful signals. Somehow the station remained unraided, after at least 100 broadcasts with approximately 10 kW. Were they especially clever, lucky, were they really broadcasting from a ship off the East Coast (as they sometimes claimed on the air), or was Dr. Tornado related to someone in a high position at the FCC?

In many ways, 1998 was just an extension of 1997. Except for the disappearance of Radio USA, the notably active stations from 1998 continued through 1998. Three of the new stations, WACK Radio, WLIQ (and associated stations WPAT and WUNH), and Voice of the Pig's Ear, helped take up the slack from Radio Metallica Worldwide.

But the Voice of the Pig's Ear would be remembered for more than just keeping pirate listeners busy in late 1998. Despite the odd name, the operator broadcast hobbyist patriot programming (but more personal and less serious than United Patriot Radio) in October. As Radio Metallica Worldwide had done the previous year, the Voice of the Pig's Ear broadcast nearly every evening throughout the month (although for shorter periods of time and with much less power). Unlike RMWW, the operator of the Voice of the Pig's Ear was raided by the FCC near Halloween, in his first month of broadcasting. In addition, three other shortwave pirates felt the wrath of the FCC over Halloween. The NAPRS, one of the five most active pirates in the 1990s, was closed in Massachussets, as were two unnamed stations in Texas and Illinois.

The sudden attack on some of the most active stations stunned the scene and activity immediately dropped. But so much momentum had built up in 1997 and 1998 that pirate activity continued and immediately started to grow back. KMUD, raided by the FCC in the Berkeley, California area in the late 1980s returned to the air in early 1999 to present West Coast DXers with some low-power broadcasts to tune in. Blind Faith Radio and Free Radio America aired a steady flow of album rock music. And Radio Metallica Worldwide returned to 1998 levels of activity. As George Zeller noted in *The ACE* at the time ". . . broadcasts that are genuinely randomly spaced and of fairly short duration (in the 15- to 45-minute range) remain virtually unenforceable by the FCC as a practical matter. . . . My point is that pirate broadcasting without FCC enforcement is still quite possible in the USA, and thus we are likely to see plenty of it once station operators re-assess the current tactical situation."

In addition to the North American pirates, the scene received a huge boost from Europe in the form of the Shortwave Relay Service (SWRS). This pirate, thought to be broadcasting from several locations around Italy, specialized in relaying international hobby radio programs on 11470 kHz. The SWRS was heard all over the world with several hundred watts of power. In addition to the hobby programs and European pirates, SWRS relayed Andino Relay Service, WUNH, Radio Blandengue, Radio USA, Rock-It Radio, WLIQ, WPOE, KIWI Radio, Scream of the Butterfly, and others. It was an example of the greatest attribute of shortwave: pirates were beamed in from thousands of miles away.

By the time the SWRS faded away in 2000, the pirate radio scenes in North America and Europe had become intertwined along the fringes of activity. Throughout 2000 and 2001, Alfa Lima International ran tests to North America and Oceania nearly every weekend for as long as 12 straight hours. ALI was consistently heard across the East Coast and often deep into the continent. Encouraged by these efforts and also from the regular European pirate DXers, such as Dave Valko and David Hodgson, Radio Borderhunter, Radio Black Arrow, Radio Foxfire, Radio East Side, Classic Rock Radio, Mike Radio, Free Radio Service Holland, Radio Dr. Tim, and others, made widely successful special tests to North America in 2001.

Perhaps the most significant test broadcast occurred in September 2001, as this book was nearly finished. Radio Borderhunter tested to North America on 15795 kHz from Belgium. As the broadcast continued, Hans continued to drop the power. At what is generally considered to be a legal amount of power for unlicensed broadcasting, 100 mW (1/10th of a watt), Borderhunter was still audible. At only 30 mW (1/30th of a watt), the station was still somewhat copyable in Pennsylvania and Tennessee. This is a remarkable achievement, not just from an engineering and DXing perspective, but it also raises some important questions. Namely, will any kind of low-power unlicensed shortwave service be possible and will pirate radio in the future be less constrained by borders and continents than any time in the past?

Outlook

As described later in Chapter 12, I've heard many who believe that the Internet will soon supplant radio as a whole and pirate radio will essentially disappear. If anything, I believe that pirate radio is being helped by the Internet and pirate radio audiences seem to be on the upswing. North American Pirate activity and audiences are better than around 1994–1995, when stations were trying to find a new pirate channel. Also, American DXers are hearing more European pirates than ever before and the South American pirate scene has emerged since 1997. I believe that shortwave pirate radio activity will be on the upswing in 2002 and 2003. As noted in the previous paragraphs, I believe that the international pirates from Europe and South America will have a larger impact on American pirate radio than at any time previous. I believe that the reverse will also be true—more DXers from other regions will also be tuning in American and Canadian pirates than ever before.

"Biographies" of Pirate Stations

Rather than narrative descriptions of stations that operated after 1982 formatted in the same manner as those in the first and second chapters, the following is a biographical listing that should be more helpful. Many of the stations that have been active since 1996 still play a major role in free radio. This listing provides an easy reference for active or reactivating stations as well as a history of stations that have operated since 1996. However, the list only includes stations that made several broadcasts. For biographical listings of stations from 1988 to 1995, see the previous two editions of *Pirate Radio Stations* and also back issues of the now-defunct *Pirate Radio Directory*.

The following list is arranged by station name, in alphabetical order. Essential details have been included, but some information and details about stations that were only heard once or twice have been excluded. Also, pirate radio is fun and you can't take it too seriously. Most of these biographies are written with that in mind. For example, when I wrote that Attencion 69 was broadcasting for reveling spies on New Year's Day, it's a joke. This is an obvious pirate hoax, not a real spy station. However, these descriptions would be no fun if I debunked every false claim throughout this listing. Creative radio is based on imagination and if someone can broadcast from the Titanic with the use a little smoke and mirrors, then more power to them!

Mailing Addresses

Most shortwave pirate stations have a mailing address for reception reports and other correspondence. North American stations, with very few exceptions, request three first-class stamps to cover their postage costs if you would like a QSL card. The following listing of stations includes an abbreviation for the postal address on the first line. A list of the abbreviations is:

AL	Box 11522, Huntsville, AL 35814
BRS	Box 109, Blue Ridge Summit, PA 17214
GA	Box 24, Lula, GA, 30554
NE	Box 641981, Omaha, NE 68134
NY	Box 1, Belfast, NY 14711
ON	Box 293, Merlin, Ontario NOP 1WO, CANADA
PGH	Box 25302, Pittsburgh PA 15242
RI	Box 28413, Providence, RI 02908

If no address is listed for a station, then it currently has no mailing address. Also, the addresses listed are the last-known addresses of these stations; some of these pirates have been inactive for some time and have undoubtedly fallen out of contact with their maildrops. As could be expected for such a medium, information can change on a moment's notice, and typically without warning.

92.5 Pirate Radio

Despite all of the FM pirate activity of the past six years, very few of those stations are ever relayed or transmitted in parallel on shortwave. These stations would benefit from the national exposure, but such famous pirates as Free Radio Berkeley, Radio Free Allston, Radio Free Gainsville, MicroKIND Radio, WPPR, San Francisco Liberation Radio, and many others, never took the shortwave dip. One station to try shortwave was 92.5 Pirate Radio in Melbourne, Florida in September. DJ Richie Dingle played '60s pop and rock, such as that by The Beatles, Shirelles, and Beach Boys. For a first-time broadcast, Richie did well, with signals being reported all over the East Coast, right up through Canada.

Alan Masyga Project   RI

Among other things, 1996 was the year of the Alan Masyga parody stations. Who? What? Why? Alan Masyga is a dedicated shortwave radio listener from Minnesota, who has written to many different pirate stations. Unfortunately, some of the operators have complained about Alan's handwriting—saying that it is unreadable. By the mid-1990s, Alan's handwriting had become legendary in the North American pirate radio scene and many pirates joked about it over the air. This status seems to have led to the creation of the Alan Masyga Project—a cross between Alan Masyga and the Alan Parsons Project. The station was active throughout 1996 (many would say overactive!) with Alan Parsons music and mumbling from a male announcer. Of course, the real Alan Masyga has nothing to do with this station, but it was a strange twist in the 1996 pirate scene! *See also* KAMP, WAMP, and the Fake Alan Masyga Project.

Altered States Radio   ON

Announcer William Hurt continued the tradition of Altered States Radio as an occasionally active pirate—with usually only about three to five broadcasts heard per year. Presumably, the station name was derived from the movie, *Altered States*, and the theme of the station revolves around the strange and supernatural in media: everything from *Altered States* to the *X Files*. It might seem a bit out of character, but plenty of hard rock and heavy metal music is aired. To add to the mystique of the station, William Hurt offers a series of "dead rock stars" QSLs. In one of the coolest pirate contests ever, the station offered a copy of *X-Files* comic book #1 for the best reception report. It was great to see a station putting out such an effort to make the hobby fun!

Anteater Radio   NY

Although Anteater Radio only made a few broadcasts in 1996, it became one of the most active stations in 1997-98. Peter Worth, station operator, started making test broadcasts and QSOs in November. By the end of December, the station had aired a number of programs consisting of commercially recorded fake ads and a variety of music, including plenty of Christmas novelty songs. The station claims to broadcast from a tractor trailer, "The 18-wheel pirate ship" and Peter Worth has commented that he hopes to broadcast from every state in the US. In a recent letter, Worth said that he has broad-

Anteater Radio

cast from South Texas to New York City and even British Colombia, so it appears that he's well on his way.

Attencion 69

Spy numbers stations and pirate stations have mystified many of those who tune across the shortwave bands. Of course, there are some pirates who try to blur the lines. Attencion 69 is one of those line-blurring stations. Over the past several years, you will occasionally hear what appears to be a spy numbers station, but instead of numbers, you will hear obscenities or the names of sexual body parts repeated. Other numbers station parodies have appeared in the past, such as the Mexican food station, but this one also calls "Attencion 69" between uncensored verbage. The 2000 broadcasts were limited to one show on January 1, evidently to direct some spys who were revelling on New Year's Eve.

Blind Faith Radio

Blind Faith Radio is one of the few regular broadcasters to come out of 1998. Doctor Napalm continued broadcasting his format of '70s and '80s album rock throughout 2000, although his activity dropped quite a bit from the previous year. In 1999, Blind Faith Radio was the fifth most active station with 31 broadcasts. Doctor Napalm also continued with his penchant for broadcasting around noon on Sundays, which he did for half of his programs in 2000. Like Radio Metallica Worldwide, the good doctor rarely plays music by the station's namesake band. Some of the bands of choice in 2000 included ZZ Top, Judas Priest, Yes, Jefferson Starship, and Golden Earring. The station QSL is a color print of the banned Blind Faith album cover. The noon broadcasts from BFR are typically only heard in the Northeast, but the later broadcasts have been reported across North America.

Boredom Radio

Sally created Boredom Radio in the late summer of 2001. The station was unusual because it was apparently run soley live by a woman with an Australian accent. The station featured '80s pop music, obviously run through the microphone of a transceiver. The most common transmission times were late at night, between 0400 and 0700 UTC. Sally also participated in several QSOs with other pirates.

Carribbean Sound System MA

In addition to the Jamaican claims of 6YVOS, another North American pirate claimed to be broadcasting from the Caribbean in 1994 and continued irregularly through 1997. Count Whip broadcast mostly ska and reggae music "from a cruise ship in the Caribbean." As far as I know, Carribbean Sound System is the only pirate in the world with a virtually all-ska music format (including plenty of music by The Toasters) and the only station that broadcasts from a cruise ship. In the summer of 1996, Count Whip announced that his ship had the misfortune of becoming stuck in an iceberg, but at least the ship appears to be built much better than the Titanic. The Caribbean Sound System was heard regularly throughout the second half of the year, between 2000 and 0300 UTC.

CELL

Heard occasionally since 1996, CELL ("Cellphone Radio") is one of the most interesting novelty pirate operations ever. Announcer Sprint Contel airs a brutally tight format of snippets of actual cellular telephone conversations, comedy skits, and brief announcements. The telephone conversations typically consist of arguments between lovers or spouses, which were often both lively and entertaining. As fun as listening to CELL is, remember: the cellular telephone is a two-way radio. Be careful what you say; Sprint Contel might be in your neighborhood, recording your voice for a future program! Used to verify reports in *Pirate Pages* and *The ACE*.

Cat in the Hat Radio See CITH

Cellblock 13 MA

Cellblock 13 was poorly heard for each of its broadcasts throughout the last quarter in 1995. But the station returned with better signals in 1996. Main announcer, Warden Cleaver, made several broadcasts throughout the course of the year. The Cellblock 13 broadcasts relate to prison in some manner. During one broadcast, a story about George Zeller's "Outer Limits" column being circulated among prisoners was read over the air. In April, Mr. Cleaver returned with a special program for late tax filers, and even some special advice from Harry Block. I don't know about you, but I wouldn't accept tax advice from a jail-based radio station.

CITH RI

Cat in the Hat Radio broadcast a number of times in June and July 1996. The programming consisted of audio from professionally produced Dr. Seuss stories—either the records or the soundtracks from the videos. Am I missing an in-joke here or did CITH broadcasting for the sheer novelty?

Crazy Celt

The Crazy Celt is more of a simple high-frequency test than an actual pirate station. He appeared on the FRN Grapevine and he e-mailed several people news of his test broadcasts above the CB band in March 2000. Several people heard the brief music broadcasts on 27740 kHz USB. Apparently satisfied with his success, the Crazy Celt disappeared and no other stations have been reported in this frequency range. As a result, there is no information about whether the Crazy Celt

was just a CB outbander playing around or if it was an experiment from an established shortwave pirate.

Crooked Man

Many pirate DXers feel that The Crooked Man is the most bizarre pirate station that has ever been heard on shortwave. Its operations have been highly sporadic since its first transmission in March 1985. The format is an insane monolog delivered with a weird slight echo effect. Disjointed descriptions of varied historical events that were allegedly witnessed by the crooked one himself are strangely woven into an almost coherent diatribe. After damage to his head during a fall from the Hindenburg, The Crooked Man determined that Barbara Bush is actually Queen Elizabeth. God, blackmail, and sanctuary themes are discussed in settings, such as mental hospitals, the Vatican, and the Philadelphia Navy Yard.

From the background sounds and the audio quality of the announcer, it appears that the Crooked Man is actually a recording of an amateur radio operator who didn't have the strongest grip on reality. 1960s rock music, such as "Hang on Sloopy" and "I Am the Walrus" were mixed in, and echoed IDs "The Crooked Man Calling" then topped off the program. This program has been appearing on shortwave for the past 15 years and it is likely to reappear in the future.

CSIC BRS ON

For the past five years, CSIC has been rarely active. Station announcer Pirate Rambo made his first broadcast in April 1990, and was frequently active up until 1993. Since the creation of the station, CSIC has averaged over 20 broadcasts per year, however, in 1995, Pirate Rambo only mustered five

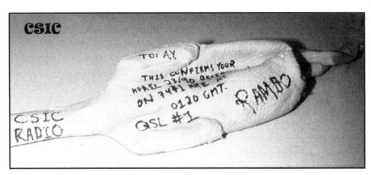

programs (one up from 1994). CSIC made the transition to 6955 kHz, although it still used the 41-meter frequencies of 7413 and 7376 kHz. CSIC always plays the novelty song, "Psycho Chicken" as an interval signal when it signs on and signs off.

Its programming is a creative mix; rock music and comedy are common, often with a Canadian focus. Also, the station was once one the biggest sources of pirate relays in North America. In previous years, CSIC announced all of its relay broadcasts, but from 1993 onward, there was a swing toward not having the relays announced. This policy makes it difficult to know just how many broadcasts were actually made by the CSIC transmitters during that year, but there were assuredly several dozen.

CSIC has sent out over 500 QSLs to its listeners, and it was one of North America's most widely heard pirate stations of the early 1990s. The station awards an actual rubber chicken QSL for every 50th report. Even if you don't win one of the rubber chickens, you will still receive one of the many different QSL sheets in the CSIC series.

CSIC reactivated in late summer 2001 and made several poorly heard broadcasts on 15055 kHz. Pirate Rambo promises plenty more.

Deliverance Radio

This is Deliverance Radio, as in the movie, *Deliverance*. If you're into bluegrass, inbreeding, and making pigs squeal, then Deliverance Radio is your station. Actually, the programs were very brief, consisting almost entirely of a few minutes of squealing, "Dueling Banjos," and brief IDs, as opposed to spending a lot of time on a creative indictment of life in rural Georgia. For a station like this, I'm not sure if I missed some sort of in-joke or dig at someone, or if everything was to be taken at face value. The station was active in 1998 and 1999, but made only one broadcast in February 2000. Like pork rinds in a health-food store, the end might be near for this novelty.

East Coast Beer Drinker
See ECBD.

ECBD   BRS
The East Coast Beer Drinker has continued to pour one more for the road for the past decade. ECBD was most active in 1990 and 1991, but has operated very irregularly since the middle of the decade. In 2000, he appeared with several shows over the Labor Day weekend. Ever since the station's debut, the format has been the same: rock music and comedy clips presented by the often-drunk East Coast Beer Drinker. As you might expect, technical problems are common, but the station usually has a strong signal—it even received a signal

report with a clear tape of the show from Finland several years ago. For the past six or seven years, ECBD has operated very irregularly, and typically between 0400 and 0800 UTC.

ECMR   NY

If the station East Coast Music Radio isn't familiar to you, then how about KMRZ, WGNK (Radio Free New England) or WARI (Alternative Radio International)? Unlike most pirate operators, which either use one station name or operate several different stations (concurrently) with different formats, Dr. Lobotomy (of KMRZ) changes the station name about every year, yet he retains his own name and the station format each time. In 1992, the WARI was the most active shortwave pirate station in North America. KMRZ, the new station name for 1993, started strong and was very active through May, but the station suddenly disappeared at that point. Dr. Lobotomy was off the air throughout 1994, but he returned in 1995 with a new station. The programming on the Dr. Lobotomy stations generally consists of "college rock" (Happy Mondays, Tears For Fears, They Might Be Giants, etc.), reggae, and some classic rock sprinkled with simple announcements. In 1995, Dr. Lobotomy broadened his horizons by airing a few Hawaiian music specials. The two big surprises for 1996 were that Dr. Lobotomy only appeared for one broadcast and the station name remained the same.

E.H. Pirate Relay Service

The E.H. Pirate Relay Service is an extention of Christian Rock Radio and Radio Free Information. Because it takes a bit of time and money to setup a new program, no new programs have been produced since early 1997. EHPRS only relays programs on the higher shortwave bands, where smaller antennas are most efficient. Over the last four years we've tried 11, 13, 15, 16, 19, and 31 meters. The best luck was on 21860 kHz and 19000 kHz, both in USB mode. When propagation ever gets better, they plan to return to 19000 kHz and 21860 kHz.

Another reason to use the higher frequencies is to skip over many of the FCC monitoring stations. Their signals skip over many of the domestic listeners, but they have gained praise from DXers who are 1000 miles away and further because they don't hear many of the 6955 kHz stations. Their slogan is, "The Alternative To 6955 In North America!" EHPRS uses the uncommon P.O. Box 422, Wellsville, NY 14895 maildrop.

Fake Radio USA

The North American pirate scene generally consists of free-radio enthusiasts who attempt to present interesting programming. Unfortunately, the "free" aspect of pirate radio occasionally attracts people who are not emotionally stable and/or who are malicious by nature. The fake Radio USA (known as *KGUN America* and *WJTA* in the early 1990s) has rarely been active since 1996. This "station" attacked several different pirate listeners with off-color comments about homosexuality, Nazism, and commercialism. In 1994, it jammed a number of other pirates, including WLIS, Radio Doomsday, Action Radio, Christian Rock Radio, and Up Against the Wall Radio. Rather than broadcast programing, the station often merely repeated insults ad naseum. New Year's Day 2000 featured another of the station's "Nazi salutes." *See* Radio USA.

Fake Radio Three   *See* Radio Three

Fight For Free Radio

A more appropriate name for this station would be "Fight Free Radio." Those "in the know" have stated that "Fight For Free Radio" is a broadcast from Radio Bob, who was noticeably miffed after being removed from his position as moderator of the Free Radio Network (www.frn.net) Grapevines. He was removed after extensive verbal altercations with board posters. Soon afterwards, Fight For Free Radio made several brief broadcasts in

October, consisting entirely of protests against the FRN and character indictments of the new moderator (and former *ACE* publisher), Pat Murphy. This is one of those stations that you can only appreciate if you like hearing "dirty laundry." Incidentally, the slogan "Fight For Free Radio" (FFFR or 3FR) has been around for decades, probably going back to the European offshore days of the 1960s; its use ultimately has nothing to do with the station of the same name.

The Fox   BRS NY ON

If pirate radio operators could be symbolized by an animal, the best choice would probably be a fox. The fox survives in the wild in part because of its cunning, and its evasiveness. This sort of mindset obviously triggered the operator of The Fox to thusly name his radio station. One program from The Fox in 1993 featured a long, preacher parody entitled "The Little Highway to God Chapel," in which Pastor Bob worked in free radio names and topic. Another series of shows from the station were part of an "all-American coast-to-coast relay," where The Fox was heard on 3885, 7417, 7425, 7440, and 15050 kHz over the course of a weekend. The approach in 1995 and 1996 was totally different again. This time, The Fox presented the Voice of Helium— an all-gas novelty program with an announcer with a helium-altered voice. Some of the segments included "Hooked on Helium," DX Indigestion, and much more.

Free Hope Experience   BRS

Broadcasting "somewhere in the backwoods of North America," the sophomore year of the Free Hope Experience was, by no means, a sophomoric effort. The station was one of the most active and most widely heard—making, on average, several broadcasts per month. Some of the broadcasts were reported by nearly every reporter to an *ACE* or *PiPa* issue.

Station announcer, Major Spook, lives up to his name by airing a peculiar mix of music, movie clips, comedy, and UFO lore. With many pirates, you just know that the station is working on comedy and to make people laugh; with the Free Hope Experience, you often wonder, "This is weird. Are these guys serious?" The regular, offbeat, well-produced programs received many positive comments from listeners around the country.

Because of transmitter difficulties, Free Hope Experience tested new equipment at times through the summer. During these test periods, Major Spook was often heard in contact with Joe Mama of

FHx reports alien sightings and abductions

MR. ANDY YODER
THIS IS TO CONFIRM YOUR REPORT OF A BROADCAST FROM THIS UFO SIGNALING STATION. CONGRATULATIONS UPON YOUR RECEPTION OF OUR PROGRAM ON 6955 KHZ FROM 2330 TO 0015 7/4-5 19 97 UTC.

QSL # 140

IF YOU HOPE AND DREAM HARD ENOUGH, THE EXPERIENCE IS ALL YOUR'S!!!

UFOs ARE REAL

Don't Try This at Home

Being a UFO Researcher is a Sure Way to Lose Friends

Radio KAOS. Free Hope Experience was primarily active during the first half of 1996 at times between 0000 and 0630 UTC, although a few broadcasts were aired earlier in the day.

Free Radio 888

Free Radio 888 operated like a regular broadcast station, then suddenly disappeared, leaving everyone wondering if they had been raided, if the equipment failed, if they were on vacation, or if they were borrowing someone's transmitter for a week or two. The station broadcast in

March 1998 during the heyday of Radio Metallica Worldwide, and like RMWW, it announced a huge power (4.5 kW). Unlike some of the tremendous power claims from pirates, this one had something to back it up: it was heard from California to Connecticut with SIO:555 signals and it had one good signal report from New Zealand. The format was album rock from KISS, Foghat, Steppenwolf, and others. It disappeared after making eight broadcasts in two weeks.

Friday Radio   RI

Some people get really happy when it's Friday—hence, the familiar phrase "Thank God It's Friday." Friday Radio appeared in October of 1995 for, what was reported in the 1996 edition of the *Pirate Radio Directory*, a one-shot broadcast. This was in error because Friday Radio appeared from time to time throughout 1996, of course, always on Fridays. Last year's format of heavy metal and jubilant comments continued; some of the bands aired included: Van Halen, Guns'n'Roses, Ozzy Osbourne, Nazareth, and AC/DC. What a way to start off a weekend!

Great Southland   ON

Stations from other continents have become popular for other stations to relay in North America. Over the past few years, stations from Russia, New Zealand, England, Sweden, Switzerland, Holland, and Germany have been relayed via transmitters in "the colonies." The Great Southland, with host John Quigley, features all-Australian rock music and the station claims to be from Australia. However, the station is not active on that continent, so it seems that

The Great Southland is probably a North American impostor (like the Voice of Scotland), or possibly the product of a former Aussie. The only broadcasts from the station in 1996 were heard at 0030 UTC in the middle of October.

Ground Zero Radio   BRS

"Broadcasting from an abandoned missle silo somewhere in the USA," Ground Zero Radio is a new station that began operations in July. However, GZR initially tested as XB-37 in 1999. Station operator Texas Pete plays heavy metal, including songs by Nazareth, Guns'n'Roses, AC/DC. TP also mixed it up with some talk about China

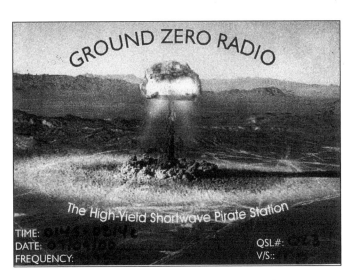

and terrorists during one broadcast, and some fake ads. The GZR QSL is a very nice full-color printed glossy photo card of a nuclear test blast in the desert.

Happy Hanukkah   ON

This station's seasonal format, first transmitted in December 1992, is a reasonably well-produced mix of pop music by Jewish artists. Most of the programs have featured music from the Broadway musical "Fiddler on the Roof" and Christmas novelty music. Sly humor is added by a male announcer, Pirate Judah, who obviously has a Jewish accent.

Some music, such as "Let It Snow" were played at variable speeds in 1993. Happy Hanukkah was mostly inactive again in 1996, but returned for six broadcasts in December. In 1995, the only programs were heard via the NAPRS relay transmitters, but no relay transmitter was announced in 1996.

He-Man Radio   BRS
Appropriately, He-Man Radio's first program was heard on April Fool's Day in 1991. Since then, announcer He-Man; his son, He-Man, Jr.; and Boy Roy were continually active on the pirate bands throughout the early 1990s. The station always broadcasts in upper sideband, which it calls "the manliest of all modes." Over the past nine years, He-Man Radio has had one of the strongest pirate signals in North America. Since its inception, the station has mailed out approximately 500 photocopied QSL sheets and neon green QSL cards.

He-Man creates a light and intentionally clumsy parody of sexism. He discusses professional football and baseball, but He Man moves on to many other topics. Aside from all this, the station plays '60s rock and bagpipe music and it uses an old rock instrumental tune by Booker T & the MGs as an interval signal.

He-Man has been less active since 1995 because the transmitter, less manly than he, broke down. He-Man's only broadcast in 2000 was on the evening of New Year's Day to wish everyone Happy Y2K and to provide an update of the Rose Bowl score. For Christmas 1999, He-Man returned with a look back at the history of HMR, but he had to leave a little early because either Santa or the FCC was approaching his Ohio hilltop. Maybe the FCC agent was disguised in a Santa costume?

Hitchiker's Guide to the Galaxy   BRS
Your name is Arthur Dent and your house was about to be bulldozed, but suddenly, you are taken into outer space, only to see that the Earth has been destroyed and the aliens saved you from it. Now what do you do? First, you star in a BBC radio and TV series, then you create a pirate radio station about your intergalactic adventures. The cult sci-fi comedy book series, *The Hitchiker's Guide to The Galaxy,* makes for an interesting pirate station, which was relayed by Radio Free Speech on occasion in 1996. The station even had collector's series of 12 different QSLs (one for each episode), but only a few shows were aired. The Hitchiker's Guide was only heard on 6955 kHz, but you'd think that the frequency should have been 4200 or 4242 kHz. Oh well, bring a towel!

Hotel California SW   RI
In the days when Take It Easy Radio was broadcasting nearly every weekend, several other parodies abounded, all with the names of different songs by The Eagles. Hotel California Shortwave was one of the least known among these stations. J.R. Henley (Don's brother from Texas?) appeared several times in December 1997, with tape-looped tests of IDs over pieces of the song, "Hotel California." *Free Radio Weekly* editor Niel Wolfish wondered aloud "What's next? Ecos del Tequila Sunrise? Voice of the Runaway Witchy Woman?" But, alas, the joke stopped here.

Howard Stern Experience   RI
Personally, I think that the self-proclaimed "king of all media" already has access to too many media outlets, as his self-description attests. Nonetheless, someone felt the need to create the Howard Stern Experience in December of 1997. The station consisted entirely of segments from the Howard Stern radio program with a few computerized ID and address announcements. This program was aired a number of times over the course of a weekend or two. It just makes you wonder why.

Indira Calling   RI
Radio, being an audio-only medium, is entirely dependent on voice characteris-

tics to convey different characters. Unlike TV, where voices can be fairly similar if people look significantly different, radio voices must vary in pitch and accents. Enter Indira Calling, an Indian-themed pirate with an announced location in Calcutta, India.

The show early in the year included Vijay Nehru playing such Hindi music as "Jim Kootchey Kong" song, "Help Me," "Ja Bala Dey, Ba Dey Ha Ney,"and a strange song about a man wanting a shirt. The primary show from the station in 2000 was the Beach Party 2000 program with the Beach Boys live from Bangalore. The Providence maildrop address is always given with the city announced as "Calcutta."

Indira Calling's seven broadcasts made it one of the most active stations in the first half of the year, but no more were heard. Because it was heard fairly often in the first half, there's no reason to give up hope. Maybe Vijay's just low on new ideas.

Jazz

Do you like to pick up generic chips, pop, and toilet paper? Do you avoid the bright and flashy colors for the faithful old black and white. The Jazz is for you. It's now more than musical genre, it's a pirate station. Of course, Jazz played jazz. The station IDs were very simple, usually consisting of a comment, like "It's Jazz, man." As you can expect, the QSL sheets are black and white—a photocopy of the Utah Jazz basketball logo. Of course, Jazz stuck to the generic frequency of 6955 kHz. In 1996, the station was less active than its rookie year of 1995; it was only heard on New Year's Day and for a broadcast in March. When active, it verified reports in *The ACE*.

Jean Chretien Station

Jean Chretien, who receives far less press than George W. Bush or Al Gore, is the Prime Minister of Canada. The Jean Chretien Station features audio clips from the Prime Minister himself, but I have a sneaking suspicion that he isn't running this pirate from the back seat of his limo. Taking it further, I would bet that his Labour Party has nothing to do with it, either. The pirate features an interval signal of "We will not let them tear them down" repeated. One program in October 2000 also included some Canadian weather reports and announcements for local Ontario events, including a puppet show.

Jerry Rigged Radio   RI

For a weekend in early November 1996, Jerry Rigged Radio aired a premier 21-minute program several times. Because the program was relatively brief, and a few odd frequencies were chosen (6950 kHz and 13900 kHz in a WREC relay), the station was not reported by a large number of listeners. Still, those who heard it, said that that the program was fun and well done. The programming mostly consisted of techno music and some comedy, such as the warning signs of the dreaded disease PRS (Pirate Radio Syndrome). It seems doubtful that they will have any problems from the FCC: It was announced that they are "broadcasting from a cornfield somewhere in the Midwest" with "antennas protected by killer attack cows."

Jimmy the Weasel *See* WRX

Jolly Green Radio
In early March 2000, Jolly Green Radio was reported with a good signal in Alberta. Unfortunately, no details were noted and nothing else is known about the station.

KAMP   BRS
KAMP broadcast regularly in the mid-1990s, but has been rarely heard since. The station aired lots of "alternative" and novelty music, including that by the Eurythmics, Nirvana, Weird Al Yankovic, and Adam Ant. The Jaunuary 2000 program was the last since the station was regularly active in 1996. In general, most listeners seemed to enjoy the programming, but they had problems understanding announcer I.M. Nutz, who has a tendency to mumble on the air. Nutz, who seems to have a love/hate relationship with pirate broadcasting, said that he was calling it quits in late 1996. So, the single program heard in January 2000 was probably just an unauthorized relay.

KAOS *See* Radio KAOS

K-2000   MA
K-2000 is in the mold of some of the other DX parody stations, such as the Voice of Revolutionary Vinco, Radio Blandx, and Radio Azteca. However, judged by any standard, K-2000 has one of the most professional sounds in pirate radio—very slick with pro-sounding announcers and singing station ID jingles. Some of the programming included an ad for FCC Clearinghouse Sweepstakes, Dick Smith of the FCC busting the station, and the Continent of Media DX program. The big feature was the Trial of the Century program *The ACE* columnist John Arthur on trial, like O.J. Simpson. The newer programs included the cast of *Seinfeld*

traveling to the SWL Winterfest (Kramer writes down the serial numbers from all of the radios) and a long infomercial about the Dave Valko (a noted shortwave DXer) Hair System. As you can see, like the other stations of this format, K-2000 features many "in" jokes about the shortwave listening community/culture. A knowledge of this information is essential to understand and enjoy the programming from K-2000. K-2000 was noted regularly throughout 1997—often via relay, according to some pirate operators. Since then, it has disappeared—even though it had been announcing that it would reveal the true meaning behind "K-2000" in 2000.

KIPM   NE
With a callsign that represents "Illuminati Prima Materia," if you bet that KIPM is different than commercial radio, you are certainly right! For years, the most creatively programmed pirates have been some of the least-active broadcasters because the emphasis is on programming,

not transmitting. KIPM is one of the few stations to defy the rule. The station broadcasts regularly and occasionally airs numerous programs back to back, marathon style.

Instead of broadcasting music or skits, KIPM consists of one man, Alan Maxwell, who tells strange stories over light, sometimes psychedelic music. All of the stories told are in the first person, so evidently Mr. Maxwell has plenty of activity occurring in his life. "He, Who Shrank" is about a man who continuously shrinks, passing through several universes. In "Electromagnetic Madness," a contact with an alien goes awry, resulting in the death of the alien and the man drills a hole in his head. Another episode featured simulated traffic between Syncom and alien Arecibo Alpha One. This show convinced some callers to the Art Bell Show that humans had finally made contact with aliens.

In all, the KIPM stories are somewhat like Garrison Keillor doing "Tales of the Unexpected." Although the station is light in production elements, it's dense in content and quality.

The e-mail address is: kipm_outerlimits@hotmail.com.

K-mart Radio *See* Blind Faith Radio

KMUD

KMUD is an old-timer with a sketchy past, thanks, in part, to a West Coast location and a very low-powered transmitter (anywhere from 10 to 25 watts). The format from the mid-1980s to 1990 had been rock music with slogans for "the muddy sounds of K-MUD" and "the mighty voice of mud." During its last broadcast in

October 1990, KMUD was raided by the FCC while relaying alternative FM station KPFA in Berkeley, California. Then KMUD, which had rarely been heard so far east as even Colorado, disappeared.

But KMUD was not gone for good. The station reactivated in 1999 and became one of the most active and interesting stations in 2000. It was announced that some of the shows were being transmitted from the desert. One of the most fascinating broadcasts, perhaps in North American pirate history was the

KMUD Mojave Desert phone booth show. In late March 2000, KMUD broadcast from the incredibly remote Mojave Desert phone booth, which was receiving dozens of calls per day from around the world because of its cult status. KMUD fielded these calls and broadcast them live on the radio. In addition, CBC (Canada) had a TV crew videotaping the booth and KMUD's broadcast. Unfortunately, the broadcast can't be duplicated because the phone booth was removed shortly after the KMUD broadcast. But even though the phone is gone, KMUD has continued to be one of the most active pirates of 2000.

KNBS   NY

KNBS is one of those niche stations that are usually mentioned when people discuss the variety of programming that is available via pirate radio. Unlike many pirates that cover broadcast topics, KNBS features the very narrow field of marijuana (pardon the pun). KNBS (the name of which refers to the active ingredient in marijuana) is operated by the California Marijuana Cooperative, which advocates the decriminalization of what has long been the biggest agricultural revenue producer in that state. Even those DXers who do not support the pro-marijuana advocacy that dominates KNBS have found the station's professionally produced shows to be quite entertaining.

The station has been rarely, but regularly, active throughout the 1990s. To the best of my knowledge, KNBS appeared in all nine editions of the *Pirate Radio Directory*. 1996 was a typical year for KNBS in the 1990s—only four programs were reported, mostly in the late Spring and Summer. Unlike 1995, when KNBS was heard numerous times (but by few listeners) on a variety of frequencies, the 1996 programs were all on 6955 kHz and some were widely reported. It's not a matter of whether KNBS will be heard in 2002, but how many times.

KOLD   MA

Not everyone in the world likes rock music or its derivative styles, but nearly all pirate broadcasters go with rock, anyway. KOLD is the only big band music pirate station of recent years. Station announcer Aldo Batista kept his broadcasts simple and primarily just aired music and brief announcements. This station is evidently a separate entity from the KOLD that aired pop oldies in the mid-1980s. Unfortunately, KOLD appears to be a simple novelty pirate, rather than someone who is a big band fan, dedicated to bringing the music to the pirate radio masses. Only a few KOLD programs were reported throughout 1996—most being heard around Halloween week.

Still, the station was a fun change while it lasted.

KRMI

KRMI, Radio Michigan International, was one of the more active stations throughout 2000, but most of the broadcasts occurred in the Winter and Spring months. Interestingly, KRMI uses a "K" prefix (Michigan is "W") and celebrated Canada Day. Perhaps they were desperate for a holiday to celebrate on the air or maybe Michigan has been secretly annexed by Canada. The standard KRMI format is long-time North American favorite: rock music, parody songs, and fake ads. Some of the music included that by AC/DC, Pearl Jam, and The Cult. The parodies included "Pick My Nose Again," "Free My Willy," and the early 1980s pirate favorite, "Shaving Cream." KRMI announced on the FRN that the transceiver had been stolen from his car in the Summer of 2001 and the station would be off the air until a replacement was found. He returned in August 2001, so either the old transceiver was relocated or a new one was purchased. E-mail is: KRMI6955@hotmail.com.

KSSR *See* Stereo Sound Radio

KSSV

Very little is known about KSSV, which was heard on 1620 kHz throughout the beginning of September 1995. The station was sometimes heard for hours at a time with rock and pop music and very little talk. The only listener who reported the station in the radio bulletins is located in southern Virginia, so it it unknown if KSSV was a low-powered local pirate or a regional station that just wasn't discovered by many people.

KTLA   RI

Many station operators begin pirating when they are older—after working in commercial radio, picking up an amateur radio license, etc. One interesting change from all of that is DJ T, the operator of KTLA, who reported that she was a young teenager. Throughout 1995 and the first half of 1996, the programming consisted of standard pop/rock oldies from the 1950s, but halfway through 1996, the format switched to recent "alternative" rock (including Green Day, Smashing Pumpkins, Pearl Jam, etc.). KTLA was regularly heard throughout the first half of the year (while the station was in the 1950s pop format), but became much more elusive in the latter half of the year (when it swtiched to the 1990s "alternative" rock format).

MARS *See* Montana Audio Relay Service

Microdot Radio   NY

Microdot Radio reappeared for a few more broadcasts in 1996, like so many of the rookie 1994 novelty pirates. Station announcer Michael Rahdot said he is the American cousin to the Maharishi (of Radio Free Euphoria and Voice of the Runaway Maharishi fame) and he followed through with pro-drug programming. Aside from these comments, Rahdot played album rock and heavy metal by Black Sabbath, Green Day, Alice in Chains,

etc. Except for one broadcast in October, all other Microdot Radio broadcasts in 1996 were aired on August 4.

MIDI Radio

MIDI Radio is a novelty station, in the same vein as WCPU from the mid-1980s and Mystery Radio from the mid-1990s. All of the music on MIDI Radio is computerized music played back from a MIDI (Musical Instrument Digital Interface) system. If you didn't know the MIDI term, you might be more familiar with the little synthesized bits of music that play on some people's Web pages. Infact, the Internet has thousands of MIDI music files, so probably all of the songs on MIDI Radio were downloaded from the Web. The only program from MIDI Radio heard in 2000 was a 10-minute show in February that featured MIDI versions of "Sailing" and "La Bamba."

Montana Audio Relay Service   ON

The Montana Audio Relay Service was a novelty broadcast that appeared several times in the middle of May 1996. The station supported the Freemen and the Unabomber on the air and urged listeners to read the Unabomber Manifesto. The music during the program was all novelty songs—some from the early days of music recordings (1920s—1930s) and others were more recent (The Chipmunks). Perhaps the most notable aspect of the station is the coverage for its broadcasts; several of the programs were heard from South Dakota and Texas to Europe (including several listeners in Germany and Scotland) with good signals. Unless some spectacular news events unfold in Montana, it is doubtful that the Montana Audio Relay Service (MARS) will ever return to the air.

Mystery Radio   MA

After the mysterious beginnings of Mystery Radio in 1995, the station was heard frequently by most North American pirate listeners between 1996 and 1998. Despite its big signals and high rate of activity, it's still tough to say that Mystery Radio was at the forefront of the pirate radio scene. It's more like the fog in England; you see it frequently, but you still can't get a grip on it. Mystery Radio's programs consist entirely of synthesized music with brief IDs by The Shadow. One broadcast that was repeated several times was the Led Zeppelin muzak special—a bizarre tribute to the now-defunct band. Mystery Radio programmed instrumental mood music that was perfect at night, with the lights off. Unfortunately, Mystery Radio disappeared back into the mist in late 1998.

NAPRS   NY

North American Pirate Relay Service became a sort of North American version of the Northern Ireland Shortwave Relay Service (NIRS) or Radio Waves International in 1992. From 1993 through October 1996, station operator Richard T. Pistek rose up to meet the aspirations of the station in the *1993 Directory*. The NAPRS became one of the most active pirates in North America in 1993 and the station relayed dozens of programs from other stations throughout the year. If the NAPRS aired any relays without their own identification, they would actually be responsible for many more broadcasts than were registered. If this is the case, the NAPRS might actually be the most active North American pirate of 1993 and 1994. According to our records, the NAPRS became the most active shortwave pirate ever from North America with 64 logged broadcasts in 1995!

It is difficult to cover the personality of the NAPRS because almost no actual NAPRS programming was ever aired. However, the station did relay: Radio Azteca, Primitive Radio, WLIS, Rock-It Radio, Northern Music Radio (Europe), Mystery Radio, Radio Titanic (Germany), Radio Communication International (Germany), Starshine Radio (Sweden), Radio Sparks (Switzerland), KIWI Radio (New Zealand), Radio Dr. Tim (Germany), Sunshine Radio International (Germany),

and others. Richard T. Pistek's work for his role in bringing European pirate radio to audiences in North America deserved quite a bit of recognition.

Instead of looking forward to an exciting 1997 from one of the most prolific broadcasting (even if it produced almost no programming) pirates ever in North America, NAPRS voluntarily closed after its much-repeated and widely heard fourth anniversary broadcast. In honor of the special program, NAPRS sent out nice QSL certificates with a gold seal. So long!

Omega Radio   BRS NY

Since its first transmission in October 1990, announcer Dick Tator of Omega Radio has provided just about the only American religious show on shortwave radio today that does not ask its listeners to send money. Omega Radio remained rather poorly heard in 2000, with only one program being reported by one listener in April. Earlier in its history, Omega Radio used the theme from "Rawhide" as its sign-off and relied on

Radio USA's transmitter for relays. In late December 1991, it began operating its own transmitter and changed its interval signal to a series of rock chords from the song, "Spirit in the Sky" by Norman Greenbaum. In the early 1990s, Omega Radio was regularly active, making about 10 to 15 broadcasts per year from his Heath HW-101 transceiver.

Tator typically provides calm religious commentary from a conservative point of view. Omega is the last letter of the Greek alphabet, and you can transpose the the name of the station to mean "end-times radio." Although many of the programs from years past were dominated with Biblical prophesy and end-times philosophies, the late 1990s programs have consisted of Christian rock music shows (including a feature on the Cornerstone music festival) and some comedy segments. In 1996, an additional address in Michigan was announced, but no address was heard in the 2000 program.

One Voice Radio   NY

Originally known as *One More Voice from America Radio*, One Voice Radio made its first official shortwave broadcast in March 1990. It subsequently became the second most active North American pirate station in that year, only trailing Hope Radio in the top slot. The station's calm announcer, Joe, was heard dozens of times with medical news (read from a variety of medical journals), and a bit of pop, blues, classical, and reggae music. At the end of 1990, Joe announced that the station was closing down, and it did stay off the air for several years.

But at the end of 1994, Joe returned with his old format of health tips. To let everyone know that these were new pro-

grams and not old shows being relayed, Joe announced that it was return program #1A. The only new show from One Voice Radio in 1995 featured talk about cancer, aging alcoholics, and kidney stones. Since 1996, One Voice Radio has made only rare appearances. The only program from One Voice Radio in 2000 was heard for about 15 minutes over New Year's Day. Among other things, Joe mentioned the miscarriage rate in working women and alternative electricity generation. The tape on this program was speeded up to a chipmunk-voiced pace, so either OVR was having some technical problems or someone else was having some fun replaying an old tape.

ORTQ   MA

ORTQ (Office de Radiodiffusion Television du Quebec) is a parody broadcaster that resulted from the conflicts between Quebec and the other provinces in Canada. Along with French-language music, the station also aired French ads for Northwest Airlines, Pepsi, and McDonalds. The first broadcst from the station was aired a number of times in the second half of June. The next show, the ORTQ Christmas program, was rebroadcast several times in the second half of December 1996. This show, with host Michel Machaud, featured Christmas music in both French and English. Despite the Quebecan origins of the station, the mailing address is located

ONE VOICE RADIO
Health Information Radio
Serving the U.S.A. and Canada

in the very non-French state of Rhode Island. How can they stand for such cultural impurity?

Outlaw Radio   RI

The Outlaw created Outlaw Radio in 1994, and was a regular for several years. Throughout the first half of 1996, Outlaw Radio was widely heard with regular programming. Many of these programs appeared to be repeats because many featured the same segments. The station programming has primarily been 1970s psychedelic and album rock, such as Pink Floyd, Amboy Dukes, Mountain, The Doors, Led Zeppelin, etc. A few of the common comedy segments included fake ads for Monistat 7 and The Sodomizer, and "the Clintons on Outlaw Radio's Comedy Central."

Partial India Radio   MA

Partial India Radio, as opposed to All India Radio, rose to the occasion of providing DX-parody-crazed listeners with another programming fix in the late 1990s. Very similar in humor and production skills to Radio EuroGeek and K-2000, the two programs from Partial India Radio were some of the highlights of the year for many pirate listeners. Harold Krshna and Sanjay broadcast a bizarre combination of news events, including coverage of the Indy 500 and Jerry Berg's QSL collection being held hostage, not to mention an ad for Reincarnation Instant Breakfast Drink. The first program was repeated several times in February and the second was aired three times in the middle of June. Most of the programs were very widely heard—one show was heard by a number of listeners in Germany!

Pirate Radio Boston   MA

Pirate Radio Boston is one of the regional pirates that has operated sporadically over the past several years. Main announcer Charlie (C. Q.) Loudenboomer and his assistant, Mr. X, air mostly older, light pop and some reggae music with talk. In addition to some of the light, chatty

comments, the announcers had some harsh words for stations that haven't QSLed, such as VOX America. The music in 1995 included reggae, The Macarena, and some Christmas novelty music. Most of Pirate Radio Boston's output for the year was aired in December, when the Christmas program was broadcast seven times, with good results across the Northeast. Old-timers in the DX hobby might remember the name "Charlie Loudenboomer" for his amusing and often controversial editorials in FRENDX (now NASWA). See Chapter 4 for a section on the life of Pirate Radio Boston by Charlie Loudenboomer.

Polka   MA

Another new generic station/format name for 1995 was Polka. Guess what? They play polka music: "The Beer Barrel Polka," "Hokey Pokey," you name it. Like the other generic name/format stations, Polka mostly faded away in 1996. Fortunately, Polka airs a few skits along with the music, including the live broadcast from Lambeau Field and the obligatory "Go Packers, go!" and a fake DX program. So, where was Polka when the Packers won the Superbowl in January 1997?

Primitive Radio   NY MA

According to the Primitive Radio infosheet: "The first broadcast of Primitive Radio was on November 16, 1991. The show continued on FM over the years with both live and recorded shows of music and literary readings. On September 26, 1994, Primitive Radio made its debut on shortwave. Primitive Radio has always tried to close the gap between modern music and literature, and to show that they have the same ideas, moods, rhythms, and verve. I have no formal learning in either literature or music; I follow my primitive instinct of what sounds, moods, words, and rhythms belong together." Holden Caulfield's shows were always interesting and well done, so it would be nice to hear them much more regularly. The Primitive

Radio programs were heard several times in April only, but one broadcast was heard by many listeners across the Eastern half of North America and in Germany. It hasn't since been heard.

Radio Aesop

Aesop's fables have been popular stories for centuries. Surely, somewhere along the line, they were broadcast via the radio. But I don't recall ever hearing Aesop's fables on licensed stations in my lifetime, so I suppose reading the fables on the air fills a viable programming niche. But Radio Aesop operated so sporadically that even the least child-like listeners didn't complain. Besides, the station offered plenty of wisdom that more shortwave pirates should have followed in 2000, such as "Be sure of your cause before you quarrel." The contact address has been: radioaesop@yahoo.com.

Radio Airplane   NY

Radio Airplane is one of the legendary pirates of the 1990s. Over the past few years, Radio Airplane developed into one of the "classic" pirates: the station was regularly active, had a large signal range, produced high-quality programming, and the station responded very well to its listeners. The signal strength from the Radio

"...if my muses be true to me, I shall raise the despised head of poetry again, and, stripping her out of those rotten and base rags wherewith the times have adulterated her form, restore her to her primitive habit, feature, and majesty, and render her worthy to be embraced and kissed of all the great and master-spirits of our world." From the Epistle of Volpone by Ben Jonson (1607).

This QSL #28 confirms that Andrew Yoder of Pennsylvania heard PRIMITIVE RADIO on 9/10/95, 1330-1356 UTC 6955USB.

Airplane was generally quite good across all North America.

The station significantly reduced its output in 1995, and it virtually disappeared by 1996. One wdely heard program was aired in February and one very poorly heard program was broadcast on Halloween. In the past, Radio Airplane was heard from a Piper J3 Cub broadcasting from "the clear skies over North America." Classic rock music and commercially and station-produced comedy was the standard fare on Radio Airplane. The programming on the 1996 show was centered around censorship. The broadcast ended with the cryptic message from the

FCC: "Radio Airplane is bad."

In addition to a large assortment of professional-looking QSL letters, cards, and certificates, Radio Airplane had a number of nice stickers available. Also, Captain Eddy has responded excellently to listeners who send taped reports—often returning the cassettes with studio copies of the broadcast. Unfortunately, it appears that this classic pirate is gone for good.

Radio Azteca   NY

After hitting the airwaves in 1992, Radio Azteca truly came into its own in 1994. The station's excellently written DX jokes and parodies were widely heard across the continent, as a number programs were aired (including some relays by the NAPRS and Radio Doomsday). Main announcer Bram Stoker (who apparently DXes on the side) assembles short high-quality programs of fake radio news and tips. As expected, Bram has continued in 2001 with more top-10 lists, including "Best Disney DX Movies" (#1 was "Jeff White and the Seven Dwarfs"), "Bumper Stickers We'd Like to See" (#1 was "Honk if you love Glenn Hauser"), and "Signs You're a Burned-Out DXer." "Ask Dr. Radio" and "Mail Scrotum" (Mailbag) also returned for another year of quips, jabs, and yuks.

The broadcasts have always used the trumpet fanfare from *The Rocky & Bullwinkle Show* as an interval signal, and some of the other musical intervals from the cartoon are also used. Considering the quality of the programs and the radio news content, Radio Azteca is a favorite of DXers in North America.

For those of you keeping score at home, the Radio Azteca shows aired in 2000 included #35, 36, 37, and "Music and ID Special" #3. As you can see by the program numbers, the music and ID specials are a relatively recent phenomena. Regular Radio Azteca programs are nearly 100 percent talk based, but the music and ID specials are mostly music and, well, IDs. From what I can tell, Bram only plays music that is: at least 30 years old (preferably older), sexually explicit, and

filled with puns.

Some listeners have noted that the Radio Azteca staff talks mostly about radio and sex—and sometimes seems to have difficulty distinguishing between the two. A new slogan for 2000 was "Radio Azteca: the station that lifts and separates." As one reporter commented several years ago on the top-10 reasons to keep your daughter away from DXers, "The #1 reason should be Bram Stoker!"

Radio Beaver   ON

Ever since its first transmission in September 1990, Radio Beaver has been a very pleasant addition to the North American pirate bands. Announcer Bucky Beaver's high-pitched voice is familiar in a well-produced mix of entertaining programming. This overtly Canadian station's activity dropped off considerably from years past. Radio Beaver is normally easily identified by its theme music from the "Leave it to Beaver" television program, and by its slogan, "If you're really Canadian, show us your beaver!" After falling silent in 1994 and 1995, Canadian politics spurred Bucky to return. In the broadcast aired in January 1996, the station pushed for Canada to split from Quebec. But when the station returned in August, Bucky's mood had declined and he pleaded for the Edmonton military to drop a nuclear weapon on Quebec. So much violence from a dam-building rodent! Bucky hasn't been a very busy beaver since 1996.

Radio Bingo

Like a weird cousin of the numbers station parodies, Radio Bingo is exactly what its name connotes: a radio version of a bingo game. Unfortunately, it's not a real bingo game because although the numbers are called, no bingo cards have yet been sent out and no one has reported winning any prizes. ACE publisher, John Arthur, is consistently announced as being the winner, so before we jump to any conclusions, perhaps we should check with John to see if he has been winning all of those prizes (he might have a 2002 special edition Radio Bingo VW Beetle in the driveway). Radio Bingo returned again in 2001 with plenty more "contests" throughout the year. The "programs" vary and, at times, the announcer will either call out to use the blue, green, or orange bingo card. Sometimes bits of programming from other pirates, such as Radio Metallica Worldwide, are dripped into the mixture.

Radio Biscuit

Radio Biscuit was only reported with one test broadcast in the afternoon hours in June. The one reporter was located in coastal Virginia, so it's probably safe to assume that the transmitter was in the mid-Atlantic states or possibly in the Carolinas. Aside from the test messages, the only programming was the song "California" repeated a number of times. It was not mentioned whether the signal was hot, buttery, flakey, or crisp.

Radio Bob's Communications Network *See* RBCN.

Radio Cobain

Is there a point to this station? I've heard this a few times, but I've never heard the Radio Cobain ID. Instead, I've heard a lot of bits and pieces of pirate radio audio—especially "What a f——— idiot he was," which I believe is a reference to Kurt Cobain from a Radio Three program. I don't know if this station has actually ever identified itself or if the listeners just pulled out a dissected slab of audio and chose it as the closest thing to an ID. Regardless of whether or not Radio Cobain

is the actual name, it's a disjointed mess of repeating audio clips. It's somewhat in the vein of Radio Tornado Worldwide, but much less cohesive.

Radio Doomsday   RI

Radio Doomsday has had an extremely tumultuous career as a pirate throughout the mid-90s. Doomsday has consistently been a source of excellent productions, loaded with creativity. But unlike most popular pirates, Nemesis (the station operator) enjoyed stirring up the short-wave listening scene as much as anyone. One of the most controversial activities in

1994 was the QSL policy of the station. If listeners didn't include comments about the programming, Nemesis wouldn't verify the reports and he mailed back a two-page letter outlining how to write a reception report to a pirate station. Many long-time pirate listeners were insulted by these letters, and they vented their feelings. To pull the focus of the listeners off of the QSLs and onto the programming, Radio Doomsday stopped verifying. This action further upset the listeners and Radio Doomsday called it quits; at the end of one program in September, Nemesis committed suicide.

But pirates are hard to kill. While Captain Eddy from the Radio Airplane was busy conjuring up spirits on his Halloween show, his found the spirit of Nemesis and brought him back to life, so long, as he would verify all old reports. And Doomsday continued with lesser amounts of activity into 1995. Like Radio Airplane, who brought Nemesis back to life, Radio Doomsday nearly disappeared in 1996. The only real program from the station was aired in February.

In spite of the controversy surrounding Nemesis and Radio Doomsday, the station was a top-performer, with excellent programming, strong signals across much of North America, and creative QSLs and pennants. In August 2001, when DXer Kirk Trummel passed away, friends announced that he was actually Nemesis of Radio Doomsday.

Radio EuroGeek   RI

Very similar in humor and production skills to Partial India Radio and K-2000, the excellent Europirate parody, Radio EuroGeek, made shortwave history with its first program. The station aired its program on the frequency of 11092.5 kHz USB—just before Radio St. Helena was about to air its famous annual broadcast on the same frequency! The program featured plenty of Europirate-style music (such as that by Abba and Nena), along with Euro news, weather, and sports; a parody of Media Network, where Jonathan Marks is kidnapped and Harpo Marx takes over, and even an ad for the TV show *Third Reich from the Sun*. The broadcast was adored by the dozen or so pirate listeners who heard it, and even by the many unsuspecting shortwave listeners, who were waiting for St. Helena.

Radio Exotica 2000

In October, one station became known amongst listeners as "Radio Exotica 2000." Those who heard the station had a tough time pulling a clear ID out of the mud, so it's tentative. The programming was certainly exotic. Aside from plenty of sound effects, the only piece noted was a long feature on how to call crows in for hunting. Radio Exotica 2000 appears to

have plenty of potential if the operators choose to continue broadcasting.

Radio Free Euphoria   NY

Pot-heads of the world unite. Captain Ganja has broadcast sporadically with well-produced programs of drug-related music and comedy skits (both homegrown and commercially recorded). Radio Free Euphoria was heard on occasional weekends (typically with two or three broadcasts per weekend) throughout 1996. Like such older now-inactive stations as KNBS and WARR, Radio Free Euphoria is a pro-marijuana station, however, it is the most humor-based of the handful of marijuana-formatted pirates.

One of the other personalities on the station, the Maharishi Hashishi, has also gained a main role on RFE—in fact, he too has expanded beyond the bounds of RFE and he is also heard on The Voice of the Runaway Maharishi. This was expanded even further with the one-shot station, The Voice of the Runaway Mahiroshima, the Maharishi gone mad!

Radio FCC

Although apparently not related to WMPR because of its USB transmitter, Radio FCC reminded listeners of the station because of its format of music and brief synthesized female IDs. Otherwise, the station aired a variety of pop and rock music. The station has been occasionally active since May 2001 on 6950 and 13913 kHz USB.

Radio Free Speech   BRS NY

The format of Radio Free Speech revolves entirely around the subject of free speech. The focus of the programming is on editorials and mailbag programs with DJ Bill O. Rights, an extremely pro-sounding announcer with a "perfect" voice. Typically novelty songs are also played with plenty of fake ads, such as: Lee's Press-On Teeth, Iraqi Chiropractic Centers, Park & Learn Community College, Looney Tunes Frozen Dinners, Died-of-Natural-Causes Beef, Sugar Frosted Dead

Weasel Puffs, John McEnroe Tennis Camp, and much more. Despite the "downloaded from the Internet" sound of many of these ads, Bill has stated that most were created by him. And after you hear RFS, you don't doubt it.

The signal from Radio Free Speech is peculiar—the station's transmitter has beautiful AM audio, from only 15-watt Grenade transmitter. So, the best reception is in the Northeast, where the signal often booms in. Outside of this

---

RADIO FREE SPEECH

FAREWELL QSL

You heard the Radio Free Speech "Farewell" broadcast and now are the proud owner of an "FCC Loves Radio Free Speech" official photograph.

You heard our broadcast on: _3-8-97_
On 6955khz in beautiful AM
From the times of: _2023 — 2041 UTC_

We were using our 10-watt Grenade transmitter, fed into a cut to frequency dipole, at an elevation of 5,967 feet, overlooking Missoula, Montana in the Rocky Mountains of the USA.

Radio Free Speech - P.O. Box 1 - Belfast, NY 14711
P.O. Box 109 - Blue Ridge Summit, PA 17214

Bill O. Rights

---

region, RFS is rarely heard with better than fair signals, although Bill has had some momentous occasions with the little transmitter, such as making the trip to Europe and California. Radio Free Speech has popularized the Grenade, which has become the topic of many conversations and reverse engineering projects.

Radio Free Speech has a number of different QSL sheets, a regular newsletter, full-sized vinyl bumper stickers, and even fluourescent station rulers, making it one of the best verifying U.S. pirates ever. However, in 2000, Bill announced that maildrops aren't safe and he announced a no-QSL policy in his July broadcasts.

An unfortunate and strange situation developed in September, when Radio Bob (*see* RBCN) began somehow accusing Radio Free Speech of censorship

on the frn.net Internet site. Several times someone (evidently RBCN?) aired Radio Free Speech in SSB to jam other pirates. Remember, if it's not in AM, it's not Radio Free Speech. *See* RBCN and Fight For Free Radio.

Radio Fusion Radio   RI

To date, there have been very few overtly black pirates from North America on shortwave. Black Liberation Radio gathered great attention for its FM operations in Springfield, IL, but they were rarely relayed on shortwave. Radio Fusion Radio added rap music (typically not heard on shortwave) to the 1996 pirate scene. In addition to music from Public Enemy, Run DMC, NWA, Beastie Boys, and Dr. Dre, the station played skits by the Jerky Boys. The programming in 1996 was very similar to that of 1995; in fact, all of the voice announcements appear to be the same, only the music changes. However, one innovation of the Summer of 1996 will likely be remembered for years: the all-Macarena show. This program was about 25 minutes of the Macarena repeated over and over again. All-Macarena drew responses ranging from anger and disgust to curious amusement. Radio Fusion Radio was frequently active throughout 1996 with a very good signal.

Radio Garbanzo   NY

During the years since its first pirate transmission in 1987, Radio Garbanzo has become known as having some of the best-produced programs in the North American pirate radio scene. Radio Garbanzo is known for elaborate skits and for playing music "that FM used to play, but threw away like an old prostitute while seeking bigger profits." Just a few of the plots in Radio Garbanzo programs have involved: station manager Buck McMoney tuning the station into a radio home shopping network, Radio Garbanzo overtaking a satellite uplink, spiking Hugh G. Gough with LSD, Buck McMoney going to jail, and finally the gang broadcasting mobile from an Ethel's Buns panel truck.

Unfortunately, Radio Garbanzo was nearly silent from about 1991 to 1998. Fearless returned with a new, music-based program that was heard by many listeners with good signals in December 1999. This program was repeated on New Year's Day. Then the station was silent until May, when it rattled off seven of its eight broadcasts for the year. Three of these were the well-produced Radio Garbanzo semi-annual Memorial Day show in which Fearless pays a respectful tribute to the shortwave pirates that have been raided by the FCC over about the past 10 years. Several of the other broadcasts aired were for the Mothers Day 2000 show, which included a timely ad for the U.S. Forest Service Fires-R-Us Los Alamos BBQ.

As you might expect, Radio Garbanzo is clever, well-produced, and crass. Fortunately for shortwave listeners, it appears that Radio Garbanzo will be running well into the '00s.

Radio Gerbil

Years ago, pirates would merely imitate other people in parodies. These days, some enthusiastic digital "tape splicers" are willing to expend plenty of effort to literally use people's words against them.

Someone labored away over some sampled Radio Azteca programs to create Radio Gerbil, a station consisting entirely of cut-and-pasted Radio Azteca audio. The broadcast started with the normal Radio Azteca interval signal, except backwards. The program also included a number of Radio Azteca "bloopers," so Bram must have a secret double agent hiding in his studio. Bram Stoker identified himself as "Bram Toe Jam" and "Bram Buttcrack, among other names. Some of the alternate station IDs included Radio Toe Jam, Radio Cheez Whiz, and Radio

OPEN SEASON DECLARED ON AMERICA'S "DEAR" FREE RADIO OPERATORS!

BAG YOURS TODAY BEFORE BIG GUNS FROM THE FCC RENDER THE SPECIES EXTINCT...

OFFICIAL RADIO GARBANZO VERIFICATION:

DATE (UTC): 7-19-89
TIME (UTC): 0430-0442
FREQUENCY: 7415 KHz

QSL #: TWENTY-NINE

TO: ANDY YODER
         Beaver Falls, PA

Fearless Fred
Program Dir.

THANK YOU FOR TAKING THE TROUBLE AND EXPENSE TO REPORT!

Montezuma's Revenge. Most of the broadcasts of this 17-minute program were aired in late 1999, but one widely heard transmission was aired on the late afternoon of New Year's Day.

Radio Is Not Radio   RI
See Radio USA, Radio Is Not Radio, Fake Radio Is Not Radio, and Fake Radio USA.

Radio KAOS   NY

Along with Up Your Radio Shortwave and a few others, Radio KAOS (also just known as KAOS) was one of the top new stations of 1996. The station was consistently active throughout the year (starting in early February) and was heard with strong signals from coast to coast. Announcers Joe Mama and Roger Wilco produce fun programs of commercially recorded ads and skits, interlaced with hard rock music, and talk and comments from Joe and Roger. Typical bands aired on Radio KAOS include: Pink Floyd, Aerosmith, Led Zeppelin, Alice Cooper, Lynyrd Skynyrd, ZZ Top, etc. Some of the ads included: Spatula City, Mother Perfume, Vibra Shorts, Al Capone's Glove Company, Zipper Airlines, etc. And finally, some of the comedy skits were from Monty Python, Cheech & Chong, Jerky Boys, Stephen Wright, Dennis Miller, etc. In its rookie year of broadcasting, Radio KAOS provided a fantastic fun and regular service.

KAOS broadcast regularly until April 1997, when it announced that it was making its last broadcast. A few months later, Joe Mama returned under the name Radio Nonsense, which lasted until Halloween 1998, with similar programming. After a few years of silence, Joe Mama returned with a live show in September 2001.

Radio Metallica Worldwide
BRS

The most widely heard, most powerful, and most active North American pirate station of the late 1990s was Radio Metallica Worldwide. RMWW started broadcasting in the Spring of 1997, and it often broadcast nearly every day at various periods throughout the Summer. The station operated at a lighter pace in 1998 and 1999.

Radio Metallica has always claimed an output power of 10 kW. Unlike some inflated power claims from pirates, this rating seems to be fairly accurate. For example, one afternoon (EDT) broadcast on 6955 kHz AM was heard with very strong signals simultaneously in New York and California. During the nighttime broadcasts, reports of clear broadcasts have arrived from New Zealand, Germany, Hungary, South Africa, Australia, Greece, and other countries. The AM signal strength and audio quality helped RMWW to receive literally hundreds of reception reports in 1997 alone.

Programming from Radio Metallica Worldwide is always live or live to tape. So, you won't hear the audio snippets, fake ads, or other productions that are popular with many other pirates. Shows from the station in much of 1997 almost entirely consisted of rock music (including their theme, "Secret Agent Man") with brief announcements from Dr. Tornado and Senor El Nino (and, sometimes, other announcers). Later, the station began airing more and more talk. These programs varied significantly in both content and style. Some seemed like drunk morning DJs gone wild and others sounded like political call-in shows. Aside from the content of the shows, the quality and clarity of thought also varies significantly, depending, in part on the quantities of alcohol inbibed.

Radio Metallica Worldwide was essentially silent in 2000 and 2001. Only three programs were heard, all in USB. In the past, Dr. Tornado had tested several times in USB with an amateur radio transceiver, but every real program was in AM. It's unknown if the 2000 broadcasts were from RMWW or just replays from a fan? Is the 10-kW transmitter under the weather or has RMWW disappeared for good?

Radio Neptune Universal Service   BRS

Radio Neptune has been in existance for several years, but nearly all of the broadcasts to date have been relays through shortwave stations in Europe. So, despite the American location, it was better known to European listeners in the late 1990s. In 2000, Radio Neptune's Universal Service was heard with two relatively brief test programs in September and October.

In 2001, RNUS returned approximately monthly or bi-monthly with programs that were very late at night (often between 0500 and 0700 UTC). Joe Mack's programs often revolve around album rock music, although he has included some other features, too.

Radio Nine   RI

Radio One certainly set a trend in 1996. The station began broadcasting in the spring and before long, a bunch of Radio <number> imitators were on the air. One of those stations was Radio Nine, which broadcasted a numbers/music theme. Some of the songs aired on Radio Nine included "Revolution 9," "Driver 8," "4 Seasons In 1 Day," "2 Hearts Beat As 1," "Less Than 0," and "The Magnificent 7." Additionally, all QSLs are numbered with 9s: QSL number 1 would be listed as "#9," QSL number 4 would be "#9999," and QSL number 11 would be "#99999999999." All in all, a cute parody.

Radio Nonsense   NY

After Radio KAOS (a.k.a. "KAOS"), Joe Mama returned in September 1997 with Radio Nonsense. The programming was essentially the same as with Radio KAOS, but the station name change was evidently because this was a solo project without Roger Wilco. Radio Nonsense broadcasted as often as 10 times per month, including numerous weeknight shows. The music was fairly common 1970s album rock and tons of fake ads and comedy from Monty Python and others. After 13 months of regular broadcasting, Radio Nonsense disappeared after airing a Halloween 1998 program. See also KAOS.

Radio Obscura

The life and times of Radio Obscura are simply obscure. The station was reported by two different listeners in August 2000—one heard a broadcast on 6950 kHz USB, the other heard 13910 kHz USB. The program featured a game-show

parody and several fake ads, but few details were mentioned in the reports. The station announced that QSLs would be offered to those who sent reports to their e-mail address, but none was noted. However, one of the reporters in Free Radio Weekly received a QSL shortly thereafter. The QSL letter was signed "Anonymous One," who noted that no other broadcasts would follow from Radio Obscura. Because of the use of 13910 kHz, chances are good that Radio Obscura was either a relay by or a test broadcast from KIPM, RBCN, or one of the other stations that used this frequency regularly in the Summer.

Radio Tellus   RI

All of the poor listeners in the West and South, who were rarely treated to good signals from pirate stations, hit paydirt in 1996. In addition to the great signals from Radio KAOS, WARR, and some of the veterans (Outlaw Radio, Mystery Radio, etc.), Tellus Radio began broadcasting regularly in September with lots of heavy metal music: Queenryche, Dio, Metallica, Rush, Guns'n'Roses, AC/DC, etc. The station also played a few old-time radio dramas, such as The Shadow. Heavy metal and old-time radio aren't a typical mix, so it was an interesting listen. Tellus Radio was heard from coast to coast with good signals, and it was one of the most active stations during the last quarter of 1996.

Radio Three

The most infamous of the Radio <number> stations that proliferated in 1996, Radio Three has since been continuously active. Sal Amoniac, station announcer, has a very gruff voice (and uses extra voice processing) and plays "alternative" rock. Aside from those details, the only real noteworthy aspect of the station is that Sal constantly baits DXers with QSLs by asking for impossible information to be sent to The ACE, then laughing about how no one will ever receive a QSL. Evidently, the master tapes of Radio Three fell into the wrong hands. Before long, the Radio

Three announcements were all the same, but the music had been replaced by that which was evidently deemed repulsive: disco, light pops from the 1970s, and songs from the *Barney* TV show! Before long, Sal Amoniac was crying "foul" and "Totally Bogus Radio Three" QSLs were landing in the hands of DXers. Both stations have continued into 2000, but unfortunately the story of the stations is more fun and interesting than the actual broadcasts. *See* Fake Radio Three.

Radio Tornado Worldwide

Following the numerous high-powered broadcasts of Radio Metallica Worldwide in 1998 and 1999, the parody station Radio Tornado began broadcasting. A sign of the modern times, Radio Tornado mostly consisted of swirling, computer-edited montage of off-air radio clips from Radio Metallica Worldwide and other stations. Although it might have been a bit humorous for insiders, to new listeners, it was just a confusing mess of voice bits.

Radio Toronto

Radio Toronto, which does not appear to be connected with the station of the same name from the early 1980s, made its first broadcast in November 1999. On its first pro-gram, it played plenty of music by Culture Club, mixed with some fake ads for Indian Pale Ale beer and 2000 Flushes toilet bowl cleaner. It announced that it was "coming to you from the 14th floor of Pitman Hall on Toronto's Ryerson University Campus." Another Radio Toronto appeared in summer 2001 using the Merlin, Ontario maildrop, but it is not known if this is the same station that was active in 1999.

Radio Two   RI

Radio Two was the first parody of Radio One. Like the other parodies, Radio Two was performed in good taste—no attacks on Bobaloo, just some fun and even some self-depricating humor: "We may not be Radio One, but we're still going to have some fun." Like Radio One, the station mostly used 6950 kHz AM for its broadcasts, to fool listeners. Only one program, the Niel Wolfish (a long-time well-known Canadian DXer) dedication show, with plenty of Canadian pop oldies, was aired—seven times at the end of June and beginning of July. Host Yabba Dabba Doo was widely heard, and the program was an apparant success—even if Radio Two never returns.

Radio Unknown

Radio Unknown was only heard briefly with an ID and "Come Dancing," by The Kinks in February 2000. The signal was reported as being good in New England. Any other details of the station are unknown.

Radio USA   NY

Starting in late 1982, Radio USA is one of the oldest and most widely heard pirate radio stations ever to operate from North

America. In 1983, it was the third most active station, and it was occasionally active throughout the '80s until becoming regular again in 1989. After setting a North American record for most reported pirate broadcasts in 1991, Radio USA dropped back behind some of the pack from 1992 to 1996. In the second half of 1997, Radio USA was again one of the most active stations on the air, with numerous live programs often lasting up to two hours. But, again, the station became much less active in '98 and '99. In 2000, activity was again slow, only one new high-frequency test on 9525 kHz and several old programs were heard in the second half of the year.

R. F. Watts, Joe King, Mr. Blue Sky, Ginger Aile, and Moglie have been the main station announcers in the late 1990s, but they are often joined by others, including Hubie, Sam, The Grinch, Nacho, and long-time station engineer Gary Indiana rarely makes an appearance.

Typically, the station claims to broadcast "from a leaky bathtub somewhere off the coast of North America," although sometimes a northern Indiana location is announced. The station's musical format has always been obscure punk rock tunes, but these songs are really a secondary focus of the station. Well-produced comedy sketches, parody ads, and commentaries are common.

Radio USA has always encouraged correspondence from its listeners with a nice collector's series of QSLs and has distributed more than 1400 of these since it was first heard 19 years ago. *See* Fake Radio USA.

Radio Wolf International PGH

The solitary program from Radio Wolf International in 1990 became one of the best-known shows ever in North America, thanks to it being videotaped while it was being produced. This tape was later edited and sold through some sources, such as an ad in *Popular Communications*. Radio Wolf International was a "superstation" of Radio Animal of WKND (soon after he had been raided twice by the FCC), Sparky of WKZP, and Uncle Salty of the Voice of Monotony. The station's broadcast via a generator and a Johnson Viking transmit-

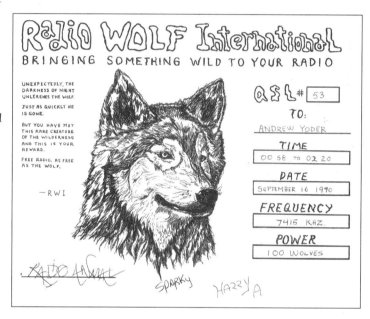

ter were also caught on tape and the show netted more than 50 reception reports from Idaho to New England. The Radio Wolf International show aired in October of 2000 was a repeat of the original program, possibly aired as a 10th anniversary repeat. If you missed this show, maybe it will be repeated again in October of 2010.

Radio Xanax RI

With a name like Radio Xanax, how could it be anything but "The Relaxation Station?" Radio Xanax was apparently a one-program novelty station; it was heard four times at the end of July with a wide variety of music and some sound effects. Of course, some of the music fit in with the overall theme of the station, including "I Wanna Be Sedated" by the Ramones and "Don't Worry, Be Happy" by Bobby McFerrin. Even if Radio Xanax never returns to the air, at least the station operator will be happy.

Radio Zebulon   GA

Apparently Radio Zebulon was another of the one-program novelty stations that appeared in 1996. These stations used to be one-shots, but recently pirates have made more-sophisticated programs and they rebroadcast them several times to get a large audience. The program featured male announcers Pirate Mordecai and Button Gwinnet with a detail-packed program: a fake mailbag program; a three-way telephone call between CELL, Major Spook, and Joe Mama; Bob Grove speaking on the evils of marijuana; Kulpsville YMCA adventures, and more. The program was aired seven times at the end of June and in early July, with good signals reported by many listeners.

RAVEN Radio

RAVEN Radio only tested for several minutes in early October 2000. George Zeller reported that they repeated the phrase "If you want uplifted, here comes the song." But the song never came, and the station went off without announcing a mail drop address.

RBCN   GA

Radio Bob's Communications Network has a long, fascinating, and now confusing history. The station dates back to 1991 when it started broadcasting with an interesting Southern-style comedy-based format that is much different than anything heard in the Northeast or, I presume, anywhere. Unfortunately, its operations were very irregular and it was rarely heard. But all of this changed in 1994, when Bob peppered the shortwave

with broadcasts on eight frequencies in four different meter bands.

1996 marked RBCN's most active and widely heard period. Some of the interesting programs in 1996 included: the Valentine's Day special, the Terry Flynn CB Memorial show, the 20th Anniversary of Mr. Bill show, and the Coca-Cola Olympics Special Broadcast. These programs were well received by listeners and RBCN was considered to be one of the better pirates on the air. But by the end of 1996, Radio Bob spread the word that one radio enthusiast had been prying into his identity, and RBCN was subsequently closed down. Changing his mind, RBCN again returned and made 12 broadcasts in 1999.

In 1999, Radio Bob hit the Internet and regularly participated on the frn.net web bulletin boards. His regularity in posting helped him to be named moderator of the shortwave board. But, Bob's board moderation also included insulting those who asked general questions, which led to a number of online flame wars. And when he was finally bumped from his position, he retaliated against those he felt were responsible for him being "censored" with a "poison mic" station/program called (ironically) Fight For Free Radio in September 2000. Other unsubstantiated rumors have been afloat that Bob has been reporting people that he believes are pirates to the FCC. This situation is an unpleasant reminder that some people are into hobby broadcasting for the ego and not to please the listener.

RFM   NY

RFM traces its history back to an initial broadcast in June 1989. Since then, station announcer H. V. Short has steadily produced a low-key mixture of rock, jazz, and new-age music and originally produced jokes and humorous announcements. Although station QSLs indicate that the call letters stand for "Radio Free Massachusetts," RFM is the only identification heard on actual broadcasts. Engineer Twilliger, communications expert Freddie

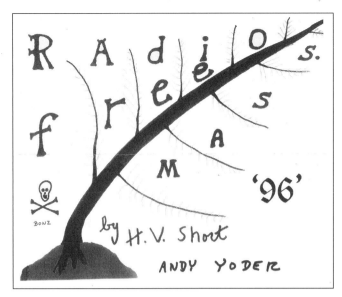

Radio—a mixture of old-time radio programs and rock music with IDs in 2001. Of course, one of the more-popular OTR shows on the station is *The Shadow*. Despite the similar name, this is not the same Shadow who operated Mystery Radio.

Sin City Station

Las Vegas is known for a lot of things, but pirate radio is not on that list. Trying to put Las Vegas on the pirate radio map, the Sin City Station was reported by one listener in Alberta in April. Programming was apparently not very high on the "things to do" list for the operator that day. It consisted entirely of recordings of commercial radio from Las Vegas, with the IDs cut out and replaced with "the Sin City Station." Too bad, this format could have lots of possibilities: for example, a spoof casino show from a nonexistent night club in Las Vegas, featuring Tom Jones and Burt Bacharach music. Of course, as on Radio Bingo, John Arthur would probably win all of the prizes.

Solid Rock Radio   NY

After placing third on the most active pirate list in its rookie year of 1993 and second in 1994, Solid Rock Radio basically disappeared in 1995. It finally returned in December 1996, after nearly two years absence. SRR moved much of its programming to relays via the licensed commercial shortwave station WRMI in 1996.

Announcers James BeBop Brown and Dr. Love played a wide variety of popular music: rap, country, soul, rock, etc. In past years, SRR replayed tapes from Black Liberation Radio, the famous political FM pirate from Springfield, IL, which had never before been heard on shortwave. The programs in 1996 featured many of

Krueger, and security manager Michael Myers round out the station staff, but H. V. Short is usually the only on-air voice. Attitude-adjustment person and all-around heavy Harry Callahan is an occasional exception to this rule, and a female named Nadine popped up six years ago. RFM used to regularly be heard on 41 meters, but it had made a practice of operating on off frequencies throughout the mid-1990s. Unfortunately, the station only made a handful of 1996 shortwave broadcasts, and many of those were just above the AM band (which mostly limited its range to New England). The station has since disappeared. RFM had an excellent set of full-color logo photo QSLs that were available for correct reception reports.

Scream of the Mosquito

The Scream of the Mosquito was obviously some sort of parody of the Scream of the Butterfly, which has been relayed from time to time by pirate stations and by WRMI and WBCQ. Little is known about the Scream of the Mosquito, aside from an ID that was heard after some music on 6955 kHz in early October.

Shadow Radio

The Shadow began broadcasting Shadow

Date: 3-23-96

Time: 21:30 — 22:00

Frequency: 9.955 KMZ VIA RMI

Reporter: Andrew Yoder

We're Glad You Caught Our Signal And Not Us!!

the same elements of the older productions, but it also included quite a bit of prerecorded Christian discussions. Although SRR had been heard most anytime in past years, all of the recent programs were heard in the morning or very early afternoon hours. As a result, the typical exteremly weak signal was confined to the Northeast.

Currently, the station is broadcasting via streaming audio at their webpage: http://www.solidrockradio.net. This is simulcast on 104.9 MHz 24 hours per day. In addition to the regular programming, SRR also relays Crazy Wave Radio, Laser Hot Hits;, Radio 3, WDRR, AM600, Tempered Steel, and more. James also mentioned that SRR should be returning to shortwave not long after this book is published.

Stereo Sound Radio

Between the Christmas and New Year's holidays, when everyone else is shopping, watching football, or still eating, Colonel Billy Bob of Stereo Sound Radio was preparing to make his 1996 debut broadcasts. His efforts were rewarded because the station was heard from Alberta to Rhode Island to Texas with very good signals. The broadcasts featured a wide variety of soemwhat-recent pop music, including that by David Bowie, Aerosmith, Robert Palmer, and Alice in

Chains. In addition, Colonel Billy Bob asked for all reports to be posted on the Grapevine section of the Free Radio Network.

Sycko Radio

There always seems to be at least one pirate every year that broadcasts like a psycho, taking plenty of risks. In 2000, the station had a name to match the activity. Sycko Radio had a flurry of broadcasts in March, when it first began pirating on shortwave, and it continued at a high rate throughout the year. Many of the broadcasts lasted from 30 to 60 minutes, but a few, such as one over the Independence Day weekend and one on a Saturday morning in October, lasted for at least two hours.

Sycko Radio continued through 2001. Unlike the first-year shows, those in 2001 featured plenty of homegrown segments, mostly related to the shortwave pirate scene. A few bits included the Adventures of Shortwave Man, preaching from Rev. Willy B. Pirate, and grandpa being hypnotized to listen to Sycko Radio.

The exact station spelling is not yet known. Presumably, it would be Psycho Radio, but *Free Radio Weekly* lists it as "Sycko Radio" and the station's e-mail address is: psycoradiohd@yahoo.com.

6YCAT

In mid-summer 1996, activity was rabid. Often a dozen or more stations were on in the same weekend. In 1996, even the short-term novelty stations made numerous broadcasts. 6YCAT was an extension of this—a Jamaican cat-themed station, based, in part, on the Jamaican 6YVOS pirate. All music was cat related, such as "Year of the Cat" and "Stray Cat Strut." 6YCAT appears to be long dead, but who knows, it should have eight more lives left.

6YVOS   NY

Announcer Pigpen Marley broadcast a somewhat peculiar mix of Grateful Dead and reggae music several times over the Summer of 1994. It returned for a few theme broadcasts in 1995—a special salute for the life of Grateful Dead guitarist, Jerry Garcia, and a Christmas broadcast. In 1996, the station remained dormant—except for a brief test broadcast in the Summer and a very widely heard Christmas reggae program. With "The Voice of Smoke" slogan, the type of music played, and the announcers names, it would be easy to assume that this is a pro-pot station. Nope. The station announced that they supported the smoking of barbecued ribs, not pot! 6YVOS, as you might guess, is an amateur callsign for Jamaica. However, don't be counting this as a log for Jamaica on shortwave any time soon. It like, the Caribbean Sound System (which announced a location on a cruise liner in the Caribbean), is one of the many novelty pirates that has been on the air in the 1990s.

Take It Easy Radio   NY

Also known as TIE Radio, announcer Desperado played music by, as you can probably guess, The Eagles..TIE played plenty of Eagles music, but also featured other Southern rock (Marshall Tucker, Steve Earle, etc.) and some old radio recordings.

Desperado was always clearly interested in enjoying himself on the air first. Sometimes he would get drunk and sing along with the songs...to the consternation of some listeners. When it was running at peak activity, it inspired the parody stations Take It to the Limit Radio and Hotel California Shortwave, which operated in November and December 1997.

Take It Easy Radio's first broadcast was in July 1997 and it made numerous live broadcasts each month for the next year. The station transmitted at nearly any time of day, but especially very late at night, between 0400 and 1000 UTC. Since 1999, it has remained essentially silent. TIE returned in September 2001 with an old program of JFK and Richard Nixon campaign speeches, so perhaps Desperado hasn't entirely given up.

Take It to the Limit Radio RI

TIL Radio was the most active and elaborate of the Take It Easy Radio parodies from 1998. "Don Pardo" hosted the program of country parody songs, such as "Trailer Park King" and "I'm Not in it for the Love, Just for the Beer," and ads for Yankee Motel yankee traps and Jerry Clower alarm. This single program was aired no less than 12 times in November and December 1997.

See Take It Easy Radio and Hotel California Shortwave

Tuna Radio

The lone Tuna Radio broadcast for 2000 (or any other year) occurred in early June, when a man signed on 6955 kHz and gave information that a longer Tuna Radio program would be on that night. It never happened.

UATWR   RI

Main announcer Owsley started the station with relays via X-Ray Yankee Zulu, but soon purchased a transmitter and began broadcasting on its own in 1994. Most of the broadcasts to date have consisted primarily of older hard rock and heavy metal, with a few listener letters and commercial comedy skits added. The 1996 activity from Up Against the Wall Radio slacked off from 1994 and 1995, but the station also relayed a number of pirate broadcasts without identifying itself, so the numbers are a bit deceiving.

Some of the highlights of the year for UATWR included the Christmas special and the Led Zeppelin Remix broadcast. The latter program was aired several times and was available for listening on the Free Radio Network's audio web page. All of the Led Zeppelin songs were drastically remixed by Owsley on the computer, and the end result showed amazing production skills!.

UATWR gradually petered out, with Owsley last being heard in July 1998

UYRSW   NY

Certainly a contender for top rookie pirate station of 1996 was Up Your Radio Shortwave, which did something that no one else has been able to do: provide a Democratic counter to Rush Limbaugh. Of course, the station didn't do this by purchasing time from hundreds of high-powered commercial transmitters or by airing conceited political commentaries. Station operator, Woody B. Serious, airs some political commentaries and plenty of political comedy music, but he is known as the master of the computerized audio editor. Woody pieces political speeches together with amazing clarity. The overall effect is hilarious. UYRSW was extremely active from August until the end of the year and was very widely heard. The station was silent through early 1997 because of amplifier problems and because of the time-consuming nature of audio editing. However, the station disappeared in 1998 and I've never heard anything like it since. It takes a rare pirate to make Ollie North say "I always wanted to be a hooter girl."

variety of rock, jazz, Afro-pop, punk, and/ or even some classical orchestra music, although mailbag features sometimes dominated the program. The music was a good choice because Leonard's home-modified amateur radio transmitter has superb audio. The Voice of Anarchy claims a transmitter location in Chicago, Illinois, reinforced by a "Chicago's own" slogan.

UP YOUR RADIO!
PIRATE SHORTWAVE

"WHERE'S MY ICE BUCKET?"
This is to certify that _____ heard "UP YOUR RADIO SHORTWAVE" on: _____
Time: _____ Frequency: _____ Mode: _____
73 es tnx de Woody B. Serious _____
QSL # _____

Voice of Anarchy   BRS

The Voice of Anarchy, which should not be confused with Radio Anarchy or Anarchy 1, began its pirate braodcasting career with a poorly heard transmission in March 1990. The Voice of Anarchy was a regular in 1991, slacked off in 1992, and disappeared by 1994. It returned in 1996 and 1998 for a few brief broadcasts. In its earlier days, the station transmitted a repeating loop ID tape by a female announcer, but Leonard Longwire was always the main host on this one.

Most progams consisted of a

The Voice of Anarchy has been by far the least political of the three "anarchy stations."

Voice of Bono   NY

For five years, the Voice of Bono was occasionally active with relays of brief programs via other stations. The standard format of the Voice of Bono is a variety of rock and alternative rock music and a bit of talk. The production and announcing during these programs are always very well done. Only once, when the station aired a live New Year's Day broadcast in 1993, has the Voice of Bono broken format or been on the air via their own transmitter. This show was plagued with audio problems, and at one point over 20 minutes of dead carrier was transmitted. Ever since, the Voice of Bono has stuck to having its short programs relayed by other stations. The station is named after Gary's dog, Bono, who also helps "sign" the QSLs. Often, old programs from the Voice of Bono are relayed; these shows still have a now-defunct Baltimore address.

Voice of Bozo

Sometimes you wonder where these pirate radio programs come from. Evidently, the operator of the Voice of Bozo spent way too much time listening to The Voice of Green Acres. He claimed to have stolen the Voice of Green Acres program, which only consists of IDs and singing along with the "Green Acres" theme song, and added a few extra bits. Some of the IDs were as "The Voice of Bozo Acres." This was a strange, extended parody of a parody that probably won't make anyone's list of favorites, but still makes this hobby interesting...or at least makes relatives look at you strangely when you try to describe this hobby.

Voice of Captain Ron Shortwave

The Voice of Captain Ron Shortwave is one of the few regulars from the class of 2000. Like many shortwave pirates, Ron stuck with a hard rock format with some talk. It gradually evolved into rock, comedy, and talk format in 2001, when Ron was often joined by co-host Major Prick.

Ron is better connected to the pirate scene than many other pirates. He often responds to questions and reports on the Free Radio Network Grapevine and was interviewed on the Allan Handelman Show (syndicated on 32 large FM rock stations in the South) in summer 2000.

VoCRSW was rarely active after the first half of 2001, but considering the past year, it seems likely that the station will continue thoughout 2002.

Station e-mail: captainronswr@yahoo.com

Voice of Christmas   RI

When the kiddies are opening up their gifts and the in-laws are getting sick on last year's fruit cake, the pirates are hitting the airwaves. Of course, it's time for Christmas theme stations. A number have operated over the past few years, including the Voice of the North, WYOY, etc. The Voice of Christmas operated a number of times in the week before and including Christmas, near 6955 kHz USB. The programming was worth turning on during the Turkey dinner; lots of Christmas music from The Chipmunks, Chuck Berry, Alan Sherman, The Beach Boys, etc. Between the songs were a few brief announcements by a man with a crackling falsetto voice, much like Bucky Beaver from Radio Beaver.

Voice of Green Acres

The Voice of Green Acres...always live to air. This one consists almost entirely of the operator singing/chanting/droning along with the lyrics to the Green Acres TV theme song over and over and over and over. This pirate station is somewhat entertaining (for the first few minutes). The only QSL in existence was written on a hotel laundry tag from The Holiday Inn in Kulpsville, PA. The two broadcasts in 2000 occurred in January and February, with the announcer telling listeners to write the station in care of WMPR (one of the most notorious nonverifiers of the past decade).

Voice of Indigestion

The Voice of Indigestion has aired several programs that have varied considerably in quality, but not in content. The music is food-oriented and much of the talk is punctuated by belches. On the New Year's Eve broadcast, much of the talk was unintelligible, buried in echo. Both of the Voice of Indigestion programs were aired in December, on 6955 kHz USB, at 1900 UTC. I guess that VoI's arch enemy would be the Voice of Tums.

Voice of the Inky Pen
The Voice of the Inky Pen appeared several times in early 2000 with a parody of Jimmy the Weasel and some other pirates. Announcer Old Inky said that he's taking quinine for his bad leg, but the side effects might be causing him to lose his eyesight. While speaking, Old Inky uttered many Jimmy-the-Weaselisms, including, "That's right, thaaaat's riiiight," and sang some Jimmy songs. The quinine evidently hasn't helped Old Inky much recently because the Voice of the Inky Pen hasn't been heard since March.

The Voice of the Lake Superior Circle Route Radio Network   BRS
The Voice of the Lake Superior Circle Route Radio Network must be one of the longest names from a pirate station that wasn't necessarily trying to have the longest radio station name. The VOLSCRN, which looks more like an acronym for a Middle Eastern clandestine station, is actually a loose federation of several pirate stations, including WLIQ, WPAT, and WUNH. It must be a large circle for WUNH (University of New Hampshire) to fit in with Lake Superior! The VOLSCRN made its only broadcast with this ID in late July, although WLIQ was heard both in North America and via European relay. All of the LSCRN stations offer well-done, rare roots rock (rockabilly, surf instrumental, etc.) music programs. See WLIQ and WPAT.

Voice of Laryngitis   NY
To say the least, the Voice of Laryngitis is legendary in the field of pirate radio...and deservedly so. No pirate station in North America (possibly in the world) has ever developed such complex and humorous

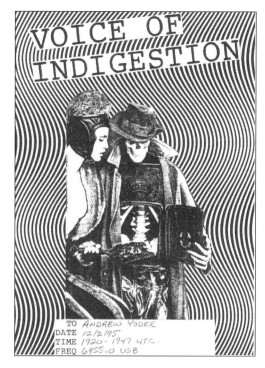

TO ANDREW YODER
DATE 12/2/95
TIME 1920 - 1947 UTC.
FREQ 6955.0 USB

original skits. In fact, most of the skits are far more complex than anything heard on shortwave or commercial radio. Because of the style of programming and the work involved, only about 20 different 45-minute programs from the Voice of Laryngitis have been created since their inception in November 1983. As a result, the station has rarely operated since programs stopped being produced on a regular basis in 1985. In light of everything, it is easier to compare the Voice of Laryngitis to The Firesign Theatre than to other long-time active pirates, such as Radio Clandestine, Radio USA, KNBS, CSIC, etc.

The Voice of Laryngitis is operated by a rather large extended family, all of whom have the surname of "Huxley." The two main announcers on the Voice of Laryngitis are the gruff-voiced Genghis and his nephew, Cowboy Stan. They are joined by Rev. Billy Bob Huxley, Bull Bruiser, Fudgie, and many others.

Although the programs vary in content, many of them are sponsored by

Friendly Freddy's Budget Burial, "Where death is cheap." Freddy's ads for kitchen cremation kits and other no-frills funerals are backed by a calliope-like organ played backward at slow speed. Another regular feature is the Rev. Billy Bob Huxley, who appears in various skits and ads for such things as a championship mud-wrestling grudge match "Battle of the Monster Ministers" with Oral Roberts. Mr. Huxley's Neighborhood (a survivalist parody of "Mr. Rogers' Neighborhood" was produced in 1983–before every comedy group did parodies of Mr. Rogers); full news coverage of the Persian Rug War; "Let's Ring The President" with President Reag– no, Barney; the Atilla Huxley Assassination show; and much more.

      The Voice of Laryngitis is well-known for their collector's series of photo QSLs–some of the nicest cards to grace a QSL album. The station also recently started producing *The Voice of Laryngitis Monitor*, a funny two-sided newsletter with news and information about the station.

      Although the Voice of Laryngitis

is apparently dead, you can still listen to the station's old bits on the Internet on www.live365.com.

Voice of Pancho Villa   BRS

Not many dead Mexican bandits are known to have a penchant for pirate radio broadcasting, but Pancho Villa is a notable exception. Now past his tenth year of pirate broadcasting, the Voice of Pancho Villa is consistently the favorite pirate station in Kulpsville, Pennsylvania. Strangely enough, Kulpsville is the only town where the Voice of Pancho Villa has been heard and the only days that the station has known to be active is while the SWL Winterfest is occurring. One year, lovely pink and purple Pancho Villa mugs were given out to all who participated at the Winterfest. Is there a connection? Probably not. Either way, the Voice of Pancho Villa is an interesting DX station. Pancho and his sidekick Cisco frequently make fun of *World of Radio* and *Popular Communications* in its yearly five-minute long (or shorter) broadcast.

      The 2000 annual pirate show consisted of the parody Worldwide Crackpot Radio (WWCR), with announcers discussing New World Order by NASWA, NARC, CIDX, and ODXA. Glenn Hauser is the first beast! The comedy bit was based on Y2K survivors, including Bob Dole, Bill and Hillary Clinton, and how to repopulate the world. One listener reported that "With that group, who would want to start over?"

Voice of Prozac   PGH

Little is known about the Voice of Prozac. In some ways, it would seem to be a sort of sister station to Radio Xanax. Both stations are named after popular anti-depressants and both feature female announcers. The similarities end there. The Voice of Prozac was regularly active through the first half of 1999 and was rarely on afterward.

      In the days when it was on several times per month, the Voice of Prozac primarily aired alternative rock. Unfortunately, it rarely was heard with a strong signal and often only a few people in the East or Midwest reported hearing each broadcast.

Thought to be gone for good by late 1999 or 2000, Prozac returned in the summer of 2001 with a test broadcast. Station e-mail addresses are: vop6955@hotmail.com and voiceofprozac@medicrawler.com

Voice of the Runaway Maharishi

See Radio Free Euphoria

Voice of the Pig's Ear

The Voice of the Pig's Ear was a hobby version of a right-wing political station. Unlike United Patriot Radio, the VoPE wasn't dogmatically pushing politics.

The slogan was "right-wing radical lunatic fringe radio" and featured plenty of discussions about the New World Order, complaints against then-President Clinton, and editorials about the Brady Bill, Waco, and Janet Reno. However, the operator also regularly QSOed with other pirates, aired a number of old-time radio programs, and sometimes dedicated songs to other pirates.

At the wrong place at the wrong time, the Voice of the Pig's Ear broadcasted nearly every day in October 1998 and was closed by the FCC in their coordinated Halloween 1998 raids.

Voice of the Rock   RI

Few pirates take advantage of advance notice for pirate broadcasts anymore. Radio Confusion used to announce all of its broadcasts with great results. The Voice of the Rock, a new, experimental station, took the chance on advance broadcasting in August 1995. The programming was nothing spectacular, just a loop tape of top-40 pop songs from the 1970s, with frequent simple IDs. According to the station, the transmitter was only outputting 10 watts of AM, yet the three-hour long marathon was heard with varying signal strength

across the entire Eastern half of North America. The Voice of the Rock claimed to be operating "from a deserted island, somewhere off the coast of North America."

Voice of Shortwave Radio
BRS

Some stations have been known to receive a few negative comments from listeners and other pirates for downloading fake commercials from the Internet and replaying them. The Voice of Shortwave Radio goes the extra mile and replays entire segments from other stations. Early broadcasts featured a mix of commercial and pirate radio sections. But some, such as a broadcast from December 1999, consisted entirely of a Howard Stern radio show, with a few Voice of Shortwave IDs cut in. The lone broadcast from the VOSWR reported in 2000 was on the odd (for North America) frequency of 6240 kHz in September. This show featured a skit about a pirate radio station that was broadcasting regularly, but no one was listening. This one doesn't sound like any other pirate show that I've heard, so perhaps it was an original VOSWR production?

La Voz del Zapatista

La Voz del Zapatista was heard several times in the first half of 2000 with good-quality, serious Zapatista (Mexico) programming. At first, it was thought to be a clandestine station, but it was discovered that only one man was programming the station and it was only being relayed on pirate frequencies. So, it was then thought to be a pirate. Later, the announcer responded to e-mail that he didn't have anything to do with the broadcast and that someone must have downloaded it from the Internet. It went from clandestine, to pirate, to a "legit" program (that someone else aired) in a matter of days. Considering that the second show was aired in the middle of a KIPM program block, it was evidently relayed by Allen Maxwell. However, the programmer behind La Voz del Zapatista was interested in his programs being aired on shortwave, so perhaps it will become a pirate "again."

VOX America

Vox America reappeared in 1995 with a handful of programs toward the end of April. As active as the station was for these two weeks, it appeared as if it would be a real force in the 1995 pirate radio scene—strong, clear AM signals with creative programs. Some of the material included Andy Looney on the Oklahoma City bombing, Jus' Disgustin' with "Cajun Cooking," outakes from Bo Jackson's UNICEF PSA, and politcal editorials about Janet Reno. Jus' Disgustin's "Cajun Cooking" program was also aired on Hope Radio years ago, so this segment was either taken from Hope, or if MJ returned under a new station name.

WACK

WACK ("Wack Radio") is one of the few stations that has worked live telephone calls into its programming since the mid-1980s. Unfortunately, WACK never places its callers live on the air, but the callers names frequently mentioned by the announcers and sometimes the answering machine messages are played back. Aside from the element of live contact, the station plays alternative rock and rap music mixed with commercially recorded fake ads for such things as Flintstones Vitamins with Viagra.

Like many other North American shortwave pirates, WACK significantly reduced its broadcasting output in 2000. They were reported for 23 broadcasts in 1999, but the total dropped to only five in 2000, and it disappeared in 2001. WACK is also one of the stations that responds to some comments on www.frn.net/vines. Although WACK didn't send out QSLs, they did send many bumperstickers out to listeners. When active, the telephone number was 888-959-8177.

WARR

"ARRRRRR, ye mateys, the radio seas are full of pirates, the scurvey lads." WARR's Captain Nobeard added to the unpredictable nature of pirate radio. WARR was "the war against the war on drugs," which is a somewhat normal pirate radio format, but unlike some of the other pirates, the Captain often sounded quite toasted. With a deep, grainy "pirate-type" voice, Captain Nobeard frequently IDed as "double-U—A—ARRRR—ARRRRRRRRRR."

The WARR transmitter was heard across North America with good signals and onto Europe, however, the audio was extremely narrow and splattery. Sometimes, it was necessary to tune across the signal a few times before determining whether the mode was in USB or LSB, and the exact frequency. Throughout its existance, WARR broadcasted plans to set up a low-power pirate TV station, support for *High Times* magazine and NORML, and a mixture of hard rock and the songs "Let's Go Smoke Some Pot" and "Up in Smoke."

WARR started in late June 1996 and was active nearly every day until August. From then until June 1998, the station operated with several broadcasts per month. During its announced final broadcast, WARR was unceremoniously wiped off the air by another pirate

WBIG   NY

As in the past few years, WBIG was big in name only. Big Mike's only broadcast was in January when he had a special on industrial music. This special was reported with a weak signal in New England. In the past, Big Mike has played a wide variety of hard rock music, including Metallica, Frank Zappa, and Green Day. Unlike many of the stations that play music on shortwave, WBIG has great-sounding AM modulation that makes listening fun. The WBIG QSL sheet features an enormous tube transmitter; with equipment that tall, it's no wonder Big Mike doesn't broadcast often—the glowing tubes would be visible for miles!

W.D.C.D.

W.D.C.D. (Wanton Display of Controlled Destruction) made its initial broadcast in late March 2000. Its claim to fame is a relay of the Voice of Pancho Villa that was heard in the Northeast on 6955 kHz. However, little was reported about the W.D.C.D. program. Bits of audio and strange electronic music were very highly produced into a bizarre mess. One of the longer pieces aired was a 1960s gynecological program played back at varying speeds over the electronic soundtrack to the movie *Bladerunner.* The programming is perfect for hearing very late at night. Unfortunately, the station rarely IDs and has only given out a voice-mail telephone number for reports, which have not yet been responded to. In 2001, W.D.C.D. was primarily relayed via WBCQ on Monday evenings at 6:00 P.M. Eastern Time on 7415 kHz.

WDRR   NY

WDRR's ("Desperate Rock and Roll") initial broadcast featured rare and "desperate" 1950s rock music, campy 1950s ads, and rantings from announcers J. G. Tiger and Jimmy Zero. Over the past few years, WDRR has occasionally surfaced for a brief, typically poorly heard broadcast. True to form, WDRR appeared for only one broadcast in April 2000...and again was only reported in the *Free Radio Weekly* by one listener. Could it just be another case of bad luck? WDRR can also be reached at desperaternr@hotmail.com.

WEAK

WEAK made its inaugural broadcast over the 2000 Labor Day weekend. One report mentioned "Mr. Longwire" and the location was announced as being Chicago. So, there's a real possibility that WEAK is either the new station for Leonard Longwire from the Voice of Anarchy or that someone wants everyone to believe that is the case. The Voice of Anarchy broadcast somewhat regularly between

1990 and 1994, but has only operated sporadically since that time (often via the Radio Metallica Worldwide transmitter). During its entire career, the Voice of Anarchy focused on playing music: everything from '60s rock to Afro-pop to classical and punk. During its first show, WEAK primarily played '60s rock music, including one song by the Barbarians. Ironically, this first broadcast was reported with a very strong signal and great audio.

WGNR or WGMR

WGNR was reported twice in February 2000 with heavy metal music, including Guns'n'Roses. Of course, the obvious ID connection would be "W-Guns-N-Roses." But, no, it can't be that easy. Another listener thought that he heard it as WGMR, "General Metalronica Radio" with techno music. Regardless, this was a little novelty station with IDs recorded to sound like those from WMPR. At the end, it was announced that the show was dedicated to Niel Wolfish. Thankfully, everyone was in agreement on this point.

WHYP   RI

And the 2000 Crown of Activity goes to: WHYP. WHYP broadcast more often than any other North American pirate in 2000, with Sycko Radio and WMFQ (respectively) not too far behind. It's always nice when the most active station also has great signals and programming to match (such as Hope Radio in 1990, Radio USA in 1991, and NAPRS in 1994 and 1995). WHYP had the complete package in 2000, but it didn't seem that way when James Brownyard first started broadcasting in 1998. Then, the station was purely a strange tribute to the original commercial station known as WHYP, a one-man station from near Erie, Pennsylvania. Evidently, the original James Brownyard was a true pioneer in broadcasting—the kind of guy who would occasionally be out mowing the lawn while the radio station would be transmitting dead air after the last song had ended. For a while, WHYP broadcasts consisted of original airchecks of James announcing the weather, mixed with other weird bits (often snippets of other pirates).

WHYP–The James Brownyard Memorial Station

Pictured above: KIPM's Alan Maxwell and Radio Three's Sal Ammoniac taking a break from the action at the James Brownyard Memorial Eat 'N Park/Iron City Beer/Rangroll/Troyer Farms/Quaker Steak & Lube Pro-Am Celebrity Croquet Tournament held last summer at the North East Croquet Grounds.

Thanks to everybody who reported us recently for our Hallowe'en Program, our replay of "WHO WANTS TO BE A PIRATE RADIO OPERATOR", our Christmas program or our Chanukkah program featuring cousin Abe Brownyard.

Be sure to tune in again! Yay yay yay! Let's go get 'im Tiger! Why not? One more chubs!

James Brownyard: May 11, 1927-November 23, 1986

In 2000, WHYP developed into a real station, playing a wide variety of music and creating plenty of skits. The skits were often based around a new extended Brownyard family, which was seemed to continue on and on, like the Huxley family from the Voice of Laryngitis. Some of the features in 2000 included a Halloween trip into the Brownyard mansion to dig into the scary record collection (one tune played was "The Zeller Mash"), a special Olympic program, and Regis Brownyard hosting "Who Wants To Be A Pirate Radio Operator" with contestants Sal Amoniac and Vijay Nehru. Some of the other parody songs include "Will You QSL When I'm 54?," "Hotel JTA," "Living La Vida Lula," "Sideband (No Carrier At All)," "JTA Went Down to Georgia," and "King of ACE." The e-mail addresses are: whyp1530@yahoo.com or whyp1530@starmail.com.

Witch City Radio   MA

Witch City Radio once again aired a few special Halloween broadcasts from an announced location in the town of witches, Salem, Massachusetts. In that sense, the station is a sort of second-generation WBST, but their crosstown rival didn't appear again this year. The climax of the program was a reading of Poe's "Fall of the House of Usher" by Vincent Price. The operator also tossed in a bunch of Halloween theme music to keep things appropriately spooky. Station operator Tommy Pickles was heard on 6955 kHz USB and 7412 kHz LSB at times between 1600 and 0300 UTC.

WJFK

WJFK is one of the strangest theme pirates that have existed thus far. As you might guess from the callsign and date of broadcasts, WJFK always appears on the anniversary of President John F. Kennedy's assaination in Dallas, Texas. This station features a repeated taped loop of the song "Abraham, Martin, and John" and a male announcer provides phonetic identifications. Listeners who reported hearing WJFK in *The ACE* and/or *Pirate Pages* received a peculiar QSL.

Probably the weirdest aspect of WJFK is its longevity: it was first heard on November 22, 1991, and broadcasted on every November 22 (and never another day) through 1998—a remarkable string of eight years. After three straight years of silence, it appears that WJFK is now gone for good.

WJTS

WJTS and HJTS (John 3:16 Radio) is a peculiar story in pirate radio. The station started to be reported nightly in mid-Autumn 1996 on 1630 kHz. The format is primarily contemporary Christian pop and rock music. The station announced an address and telephone number in Florida—by all appearances, it was a new expanded-band MW broadcaster. However, the station contact denied all aspects of piracy and said that the pre-midnight broadcasts were emanating from a ship in Central America (HJTS) and that the post-midnight broadcasts were coming from a small island off the coast of Florida. However, even though WJTS wasn't licensed and operated nightly throughout the Fall and Winter, it was not closed by the FCC. When it was active, it could be reached at: 640 Seabrook Parkway, Arlington, FL 32211.

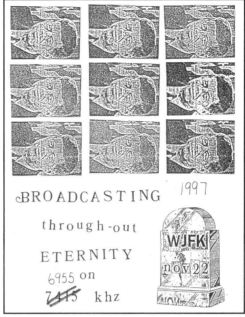

WKND   BRS

If you have been interested, involved with, or informed about shortwave pirate radio in the past 13 years, you have heard of the Radio Animal, main operator of WKND. The station helped jump-start the nearly dead pirate radio scene in 1988 and 1989, and the Radio Animal was busted two times by the FCC in 1990. In 1991, "Doghouse Productions" pirate radio news

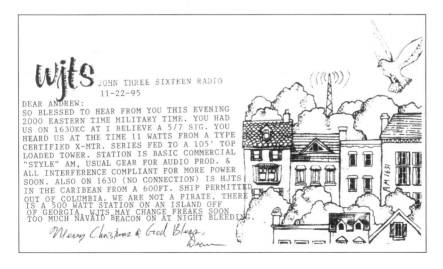

segment was heard all over the world via Hope Radio. Since the mid-1990s, the Radio Animal's name has become better known for his Grenade transmitters, an excellent, tiny rig that he engineered, but rarely sells. Sure, he has continued to broadcast, but only irregularly, and his transmitters overshadow all WKND radio activity because of their high-quality and innovative designs.

The programming features a professional mix of pop, rock, metal, disco, and rap music interlaced with much talk about pirate radio and dogs. The Radio Animal was working with Radio Albatross International, a commercial program that was relayed over Radio Copan International, until his broadcasting partner suddenly passed away in 1995. WKND has continued with regular programs on 6950 and 6955 kHz. Since 1998, most WKND broadcasts have been with Ric of Ricochet Radio, which seems to be more of a subset of WKND than a separate station.

WKUE   NY

WKUE, with hosts Mr. Koffee and Laughing Bill, were fairly regular staples on the pirate bands from 1984 to 1989. The station always stuck to a diehard format of early- to mid-'60s rock, with just a little DJ patter. The main station slogan was "74 WKUE," from back in the old days when most pirates were transmitting between 7400 and 7450 kHz. The 2000 transmission was in SSB and the old (1984-1988) Hilo, Hawaii maildrop address was announced. In the 1980s, WKUE strictly used only AM modulation, so it appears that this was an old program aired by some other station operator. At the end, a WBSA relay ID was heard, but it's unknown if this was the '80s pirate KBFA or if it was from a new station. Perhaps this activity will prompt the Mr. Koffee to return with some new programs.

WLIQ   BRS

WLIQ has been a hit-or-miss station since going on the air in 1998. It usually appears fairly late, between about 0300-0700 UTC, and appears very sporadically. DJ Jimmy Hix presents plenty of rockabilly, blues, and other forms of older "roots rock." The lone WLIQ program appeared in August on 6955 kHz USB. Like Radio Neptune, WLIQ received more mail from its European relays than via their own transmissions in North America. The official WLIQ slogan is "Your Low-IQ Station," although "We Like You" is always the first thing that comes to my mind. In late 2000, Jimmy Hix was announcing a special contest, where WLIQ program CDs would be given away to the next 25 reports sent to the station—certainly one of the best giveaways in North American pirate history! See also Voice of the Lake Superior Circle Route Radio Network and WPAT. The station e-mail is: wliq6955@excite.com.

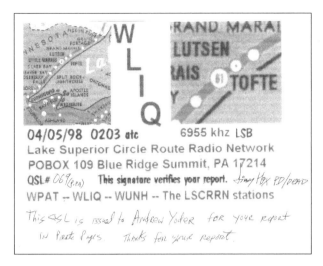

04/05/98 0203 utc   6955 khz LSB
Lake Superior Circle Route Radio Network
POBOX 109 Blue Ridge Summit, PA 17214
QSL# 069(pen)   This signature verifies your report. *Jimmy Hex PD/Dead*
WPAT -- WLIQ -- WUNH -- The LSCRRN stations

*This QSL is issued to Andrew Yoder for your report
in Pirate Pages. Thanks for your report.*

WLIS   BRS

WLIS, which first transmitted in 1990, remains one of the most unusual consistently active stations in the shortwave broadcasting bands. Jack Boggin, Ron, and Charles Poltz developed the call letters from the WLIS station slogan "We Love Interval Signals." The programs feature some rock and novelty songs, but the major portion of every show consists of actual interval signals from a variety of licensed international broadcasting stations. Local music from each station's country sometimes supplements the interval signals. The 1993 and 1994 twist on interval signals was to air many, very short tribute broadcasts to different pirate DXers.

WLIS was the second most active pirate of 1993, they tied for fourth place in 1995 with Free Hope Experience, and Jack and Charles struck it big with the #1 spot in 1998. But the WLIS output slowed in 1999 and nearly came to a halt in 2000, registering only three broadcasts. However, two of these were for the WLIS tenth anniversary program, which was heard by many DXers in March. Despite the seemingly narrow format, WLIS has managed to regularly produce and air numerous entertaining broadcasts for 10 years. Congratulations!

Nearly all of the 70 different (!)

WLIS QSLs have been humorous send-ups of various people who are involved in pirate radio listening. WLIS is one of the few stations that regularly verifies reports via many of the major DX newsletters.

WLOA, WOLA, or WOOE

This is another mystery station that was heard with a brief ID after a song. Its only broadcast was on 6950 at the end of September 2000. If the ID was, in fact, WOLA, they could have played "Wola" by The Winks.

WLOW

Announced as broadcasting live from New York City, WLOW was heard one morning in October 2000, with a variety of rock and rap music. The announcer aired a fake interview with FCC chairman William Kennard and offered some comments critical of the FCC. The announcer also mentioned that the programming was being simulcast on FM. It's possible that this was an FM pirate that borrowed someone's ham transmitter for a morning, but considering that he offered QSLs via ACE or FRN, WLOW was obviously familiar with the shortwave pirate scene.

WLWLIS   RI

We Love WLIS (WLWLIS) is one of several fan stations that made appearances in the late 1990s. Most of these were one-shots that didn't make this edition, such as "We Love Bob Grove" (WLBG), but WLWLIS was reported with a number of broadcasts in 1997 and 1998. The program featured comments from various listeners about WLIS, fake interval signals, and a few fake ads, such as one for the Scooby Doobie Brothers.

WMFQ   RI

Have you ever waited months for a QSL

card? Have you gotten really upset after a few months without a QSL? Do you become absolutely livid after years without a QSL from a station that promised it to you? WMFQ ("Where's My F––– QSL?") sums up those sentiments. Programming on WMFQ is typically rock music with a song or two from another genre thrown in for good measure. The IDs consist of a large group shouting. The station's first broadcast was in the Summer of 1998. Judging by the station's narrow novelty format, you might think that it would rarely be active. Instead, WMFQ has been one of the most regular stations to broadcast over the past three years. Maybe the goal is to broadcast until WMPR sends out QSLs?

WMPR

From 1996 through 1999, it seemed that most every weekend was time for another WMPR dance party. WMPR was one of the few total mystery stations on the air. It never QSLed; only two or three different simple, spliced-together voice IDs were ever aired; no station studio tapes have ever filtered through the pirate scene, and although its transmitter was distinctive (6955.3 AM with good audio, but slight distortion), it evidently did not relay any other stations or have any other connections to pirates. Except for a few rare instances, WMPR only broadcast the same techno dance music from the late 1980s and early 1990s.

WMOE

WMOE, as in Moe from *The Three Stooges*, started broadcasting on Labor Day 1999, and continued with an occasional program. Moe Howard's station featured some *Three Stooges* snippets, but the programming really wasn't centered around these characters. Instead, *Three Stooges* audio clips were added at times to give the station some continuity. Most of the programs featured '80s pop by The Bangles, The Clash, Tears for Fears, Tom Petty, Bryan Adams. The New Year's show also included several fake ads and tips for would-be pirates. Moe's favorite broadcast was the "no ID" special in which the station broadcasted music mixed with Stooges audio clips. There were no IDs, so the listeners had to guess.

•WMOE was active in the Winter and early Spring months, but disappeared after April 2001. According to Moe, Larry, and Curly pulled the tubes out of the transmitter in the wintertime, trying to stay warm, but forgot where they all went and Larry spilled a beer in the transmitter. Not to worry; they plan to buy some straws and resume broadcasting.

WPAT   BRS

This station is part of the Great Lakes Circle Route Radio Network, which includes WUNH and the more widely heard WLIQ. WPAT was heard occasionally with an oldies rock format (Monkees, Steppenwolf, CCR, Buddy Holly, etc.) *See* WLIQ and WUNH. Station e-mail is: j_spencer@mailexcite.com.

WPN   AL

WPN, the World Parody Network, com-

menced broadcasting on Halloween 1996. Regularly, its programs of rock music and sexually explicit comedy were heard, primarily in the East. Its numerous broadcasts that Fall placed it at #6 on the list of most active pirates for 1996. Since that time, WPN activity has been much more sporadic. In 2000, WPN's shows were only reported for brief lengths of time, ranging from 8 to 16 minutes. Given the brief programs and the technical problems on one program, few details were reported. However, at least one of the programs was the Valentine's Day show (in April) and another was dedicated to masturbation. However, WPN is probably better-known for its promotional items than for its programming or signal regularity. The station has given away dozens (maybe hundreds) of its professionally printed multicolor stickers and pinback buttons. For the past several years, the station has offered an excellent, large WPN 24-hour clock in a special treasure-hunt giveaway at the SWL Winterfest, and the winners have never yet been disappointed.

WREC   BRS NY

WREC ("Radio Free East Coast") was one of the new high-quality regular stations to hit the airwaves in 1992. Although WREC aired its first show in late September 1992, transmitter problems forced the station off the air for the rest of the year. But P. J. Sparx nursed his Johnson Viking II transmitter back to health and the station became a regular by early 1993.

From 1993 through 1995, WREC has unselfishly relayed a number of stations, including the Voice of Bono, the Voice of the Dead, Radio Marabu (Ger-

many), Laser Hot Hits (England), and others. Standard programming from WREC is "alternative rock," metal, and novelty songs, with announcements from P. J. and some cartoon audio clips. In years past, WREC aired a wider variety of music, including: rock, oldies, and country. Among other things, WREC was thoughtful and, on occasion, mailed advance notices to some of the faithful listeners.

The advance notices for the 1994 Halloween program were even written up on special Halloween party invitations. WREC reached third place for most active North American pirates in 1995. In 1999, with very little warning, P. J. voluntarily quit pirating and sold his transmitters.

WRV   NY

WRV ("Radio Virus") began transmitting regular programs in mid-December 1992. The self-described "station nobody wants to

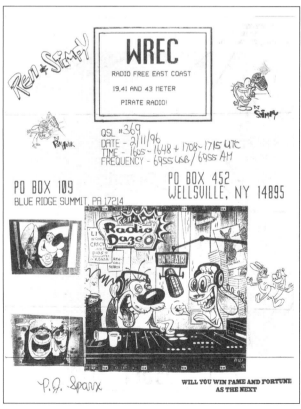

catch" was quite active over the mid-1990s with a variety of mostly punk and "alternative" rock music. After Kurt Cobain's death in 1994, WRV left the air, but it returned midway through 1995 and was much more active. In fact, it hit #5 on the most active list and it was much easier to catch than in any previous year. It was frequently heard on 6955 and 6958 kHz SSB and AM, respectively, between 1400 and 0700 UTC. In addition to the musical selections, station op, Pete the Pirate, often discussed and editorialized about the AIDS situation. In 1996, Pete also featured a tribute to the drum of Blind Melon, who had recently died.

WRX

WRX is one of the few stations better known by its DJ than its station name. Mention "WRX" to many pirate listeners and you'll get a blank look, but just say "Jimmy the Weasel" and the reaction is immediate, and typically strong. Jimmy the Weasel has been coming on the air since at least 1997 with QSOs and brief comments. But, soon, Jimmy decided that it would be best to entertain shortwave listeners with his lovely, off-key voice. Before long, Jimmy's comments about "your mama" were transformed into song and a legend was born.

In 2000, Jimmy added the WRX callsign (Weasel Radio) and some actual programming. Still, the focal point seems to be Jimmy's singing in the brief programs. One of Jimmy's new releases for 2000 was "Pee on the FCC," which has not yet reached the top-40. Many have speculated about whether Jimmy is a brilliant Andy Kaufmann-esque comedian or a guy with some problems and a transceiver.

WSRR *See* Solid Rock Radio

WVDA

The Voice of Dead Air appeared several times in 2001, but was plagued with technical difficulties. On the first program in May 2001, announcer Willy Wang attempted to relay a KIPM, but had problems with wth the CD player. In the late summer, WVDA returned, but only for a minute before they had to shut the transmitter down. The e-mail address is:: wvdapirate@yahoo.com.

Xanax Radio

Xanax Radio is not the same station as Radio Xanax. Now that we're all confused, take another Xanax and don't worry about it. Xanax Radio was just a brief test late on a July night and the "real" Radio Xanax wasn't active in 2000. The announcer said he was from Detroit and would QSL "via Pueblo, Colorado."

XEROX

XEROX ("Radio Duplicado") is an obvious parody of Mexican/Latin American stations; in fact, I am fairly certain that this station name/callsign was used in one of the editions of Don Moore's BLANDX radio bulletin parody. This radio version of XEROX last showed up for a broadcast on April Fool's Day 1996, on 6955 kHz at 2200 UTC. In 1993, XEROX featured various Latin American music; the 1994 show featured brief recordings from a Jamaican FM station; and the

1995 show featured FM reggae and an interview with *George Zeller*. Announcer Bart Sambo has hosted all programs, and you might recognize the face on the QSL from the QSL column of a popular DX newsletter.

Z-100

Here's a station with a name that represents the sound of the station. Z-100 sounds like a typical rock station, but without the ads. The station plays only the

top-selling singles by bands like Boston, The Eagles, BTO, and Steppenwolf, interspersed with pro-sounding jingles. Z-100 was extremely active in November and December 2000, but slacked off activity to a safer amount in 2001.

Z-100 often broadcasts during weeknight evenings, as well as weekends, after 0000 UTC on 6955 kHz. Unfortunately, this has been a poor choice because of splatter-over interference from La Vol Del Campesino from Huarmaca, Peru. Often, listening to Z-100 meant finding the heterodyne squeal near 6955 and trying to tune to a close frequency where you could hear more Z-100 audio and less squeal—often a difficult task! Despite using only 35 watts on a frequency with heavy interference, Z-100 has been widely heard across North America. Unfortunately, Z-100 has no mailing address and only e-mails generic Microsoft Word file "e-QSLs." The e-mail address is: bigz100fm@yahoo.com.

Zappa Radio

Last and one of the least, Zappa Radio tested briefly, late at night, at the same time that Xanax Radio was also testing in 2000. The announcer stated that he hoped to see his name in print six months later in *Popular Communications*. No other details are known.

Inside
Pirate Radio

The past two editions of this book have featured the "Inside Pirate Radio" chapter. The first edition featured WENJ and the second WCPR, two Northern New Jersey AM stations that had disappeared by 1990. The second edition featured WKND, an AM and shortwave pirate that is still occasionally active today, and CSIC, a Canadian shortwave pirate that hadn't operated in several years. However, I received a note from CSIC operator Pirate Rambo in June 2001 that the station would soon be working on the antennas and transmitting equipment and return to the air. It was the first time that I had heard from him in a few years.

After 20 years of listening to and communicating with radio pirates, I think that this is probably the toughest chapter of the book to write. I've met, talked on the phone with, communicated in chat rooms and bulletin boards, and e-mailed literally hundreds of different pirates from around the world. Many of these communications are brief, covering such important questions as "What frequency do you think would be good for making Europe at 2200 UTC?," "How was my signal when you tuned in last night?," or "Do you think I've been broadcasting too often?" With so many "nuts and bolts" questions, is it any wonder that we never get around to philosophizing over the finer points of pirate radio.

I'm sure that some pirates are less free around me also because they don't really know me. It's not like I went to school with 50 different pirates and everyone just knows me. It's kind of like being the great uncle who's invited to a wedding. Everyone's gracious, but it's not like you really know the bride and groom, no matter how friendly they are at the reception.

Finally, some things, I've been told in confidence, and I'm not always sure exactly how secretive I'm supposed to be. I always try to err on the side of secrecy, so some of the pirate stories that I've heard either won't be committed to paper or have been forgotten entirely.

A good example of a worst-case scenario for me occurred after the writing of the first edition. In preparation for the "behind the scenes" chapter, I talked on the phone for many hours with the operators of WENJ and WCPR. I even met Jack Beane at a shopping center in New Jersey and talked with him for about an hour. In 1989 and 1990, WENJ was broadcasting much too often, spend-

Fig. 4-1 *A view from behind the mic of WENJ, the biggest target for the FCC in the late 1980s.*

ing hours at a time at the top of the AM band and also broadcasting several times per weekend on shortwave. Many people told Jack to take a break, but he didn't listen. Of course, the entire state of New Jersey is accessible from both the New York and Philadelphia FCC field offices, so WENJ made a big target (Fig. 4-1).

I never heard from Jack, but according to news releases and an article by a reporter for *The ACE,* WENJ was closed by the FCC in January 1990. Jack Beane returned for a final broadcast in February 1990, where the theme was "don't trust Andrew Yoder because he'll have the FCC raid your station." Of course, I never told anyone anything confidential about WENJ, and judging by the FCC press release information, they knew a whole lot more about the location of the station than I did.

Fortunately, the WENJ raid was a fairly isolated incident, but it's still enough to make me a little gun shy, even today. So, when pirates write to me, I often ask that they don't tell me too much about their operations.

A Case of Intentionally Mistaken Identity

Surprisingly, it's often not as much fun to know too much about a pirate. This sounds kind of strange, but the elaborate image created with on-air personas and programming tends to tarnish when you realize the truth behind a station. For example, even half asleep, I wouldn't truly believe that Radio Clandestine was broadcasting from a large radio ship in the Atlantic Ocean, as they describe in their programming. However, it's much more fun to imagine that such is the case and to think about R. F., Boris, Wanda and the rest of the crew suddenly appearing out of the mist, firing up a huge glowing transmitter, broadcasting 50 minutes of '70s rock, and disappearing back into the fog.

Likewise, imagine listening to a KIPM broadcast. KIPM shows are rife with megalomania, altered perceptions of reality, and other forms of the unexpected. I would prefer staying ignorant of the KIPM operations and imagine that the broadcasts are emanating from a long-abandoned trailer in the Arizona desert. Would my perceptions of the programming change if I knew that it was being broadcast by a well-to-do accountant from suburban Syracuse? Probably.

Sometimes, though, the truth is more fascinating than the fictious identities that have been created on the programs. Recently, the operator of a station (who asked to remain anonymous) told me the somewhat harrowing details of his broadcast from the previous weekend. He drove away from his house to a site on public lands to string up the dipole. He sling-shotted the fishing line into the tree and pulled the antenna in place. Then he noticed that the antenna crossed over places where marijuana plants were growing. Obviously, if they had been planted, he could be at some risk if the "farmers" noticed him shooting wires into the trees and stomping around the area. The operator cast caution to the wind and fired up the transmitter.

Before long, the relative calm was broken by a pickup truck of drunken "yay-hoos" (as described by the pirate), who drove to a ridge beyond the transmitting site and were creating a stir in the area by shooting guns. One family quickly exited the vicinity. As the pirate listened to the broadcast on his monitor receiver, the pickup drove out from the ridge and parked on the only road into the transmitting site, blocking the path.

Fully expecting that the guys in the truck were looking to cause some serious trouble, the pirate pulled out a small knife that he had just been using to cut his antenna-support line before raising the antenna. Much to the pirate's surprise, the three guys from the pickup asked if he had any firewood at his site. He said that he didn't. While they were talking, one of the three noticed the antenna, shortwave receiver, and the small transmitter. He said "You wouldn't be one of those radio pirates that I read about in *Popular Communications*, would you?" Stunned, but still under control, the pirate responded, "And you wouldn't be taking wood from public lands, would you?" The guy responded with, "I won't tell if you won't." Then the pirate gave him a tour of the setup and discovered that the guy was high-powered CBer (illegal) and was interested in *Popular Communications* for that reason.

Of course, the pirates who travel out to make broadcasts have much better stories than those who stay at home and occasionally open the blinds to peer out of the windows. Of course, because of the locations where most pirates broadcast, the traveling pirates always have stories about people walking their dogs and joggers. Slightly less-common stories involve park rangers, police, and other generally curious people. Probably the best dog-walking story that I've heard occurred in England with Radio Aquarius in the early 1970s. The tube transmitter was unattended and being watched from a distance at a park. A man walked through with a Great Dane on a leash. He unleashed the dog, it walked into the bushes, located the transmitter, and relieved himself on it.

One peculiar non-people-related story is when Anteater Radio was broadcasting a few years ago. Anteater Radio was known as the "18-wheeled pirate station" because it always operated from a running tractor trailer. Because a truck is not an ideal location for a full-size shortwave antenna, the station usually transmitted through a whip antenna. During one broadcast, operator Peter Built noticed a strange sound. He looked out and the antenna had become so hot from the output power of the transceiver that it was melting down and dripping molten plastic insulation while he was on the air!

Certainly one of the all-time best in-the-field stories occurred near London in the late 1990s. Radio Free London, which has operated since 1968, owes its long life to creative transmitter placement. Unlike most of the U. S. mobile and stations, Radio Free London uses tube transmitters, which typically require much higher voltages and AC power. Thus, they look for sources of outdoor line power, rather than batteries. For some time, the favorite source of power for RFL broadcasts was the outdoor lighting that shines on billboards. Of course, the advertising companies don't just wire up outlets on the sides of their billboards, so Radio Free London personnel hardwired their own "extension cords" into the wiring.

In April 1997, RFL changed sites after the previous one was discovered by playing children. The new site, which was thought to be much safer, was in the woods near the M25 expressway in Kent. The power was tapped from a "Keep Left" sign and the electrical lines led back into the woods, where the transmitter, tape deck, and antenna were hidden. Unfortunately for the station, the equipment was not hidden well enough; a passing motorist noticed the antenna and, thinking that a terrorist's bomb was hidden nearby, called a warning in to the police. The police, in turn, brought in a bomb squad and closed down several exits of the highway for two hours. This was not the sort of attention that RFL was working toward—several major news sources around London, such as the *London Evening Standard* (Fig. 4-2), reported on the station "closing the highway."

Luckily for RFL, the police considered the incident lightly, rather than as a terrorist threat, and didn't fully investigate the station (especially in light of all of the evidence left behind). RFL continued broadcasting without any further troubles from the authorities until early 2001, when a staff feud disrupted operations. Although the official word is that the station is dead, it seems likely that they will return before long.

Pirate radio station plunders power from M25 Keep Left sign

A PIRATE radio station set up operation beside the M25 and tapped into the power supply of a motorway traffic island Keep Left sign. The station, Radio Free London, had hidden its aerial and transmitter in woods alongside the M25, but was discovered when a passing motorist reported the aerial as a possible terrorist device. The equipment was found after the motorway was closed by police in a security operation.

by JUSTIN DAVENPORT
Crime Correspondent

The incident is highlighted in a new report which reveals that pirate radio stations are booming pilots." Many of the stations operate from the roofs of council housing blocks. This

Fig. 4-2 *Sometimes the authorities are relieved when a pirate is broadcasting: a snip from the Radio Free London bomb scare of 1997.*

NAPRS and More: A History

Of course, some isolated stories about pirate radio are amusing, but it's also enlightening to see the full scope of a pirate's career. This pirate story is similar in style to many that I've read over the years in the sense that the concrete details are there, but it contains little information about the who, what, when, why, and wheres of it all. I've always suspected that it was a combination of the technical nature of most pirates combined with the sense of not wanting to give away much detail. Regardless, the following story provides a framework by which a number of different stations have broadcast:

This is my story about involvement in shortwave pirating and the events are as accurate as my memory is!

Pirate Radio Boston

I began my pirate career as a studio pirate in the Spring of 1992. At that time, CSIC, The Voice of The Great White North, was active almost every weekend. His powerful AM signal was well heard in the Northeastern USA. Station operator, Rambo agreed to relay other stations, so I decided to start one called *Pirate Radio Boston* using the name Charlie Loudenboomer. Our first broadcast via CSIC aired in May 1992 and was well received.

Being a hands-on type of guy, I knew studio pirating just didn't cut it for me. I had to get some sort of transmitter, but from where? I remembered a local classified booklet called the *Want Advertiser* and looked through there and found an advertisement for a Heathkit Apache. Not sure what kind of rig it was, I asked those in the pirate hobby about it and they said it would fill the bill.

I then bought this boat anchor and lugged it up to my room one rainy July day in 1992. Having never

Fig. 4-3 *Charlie and Mr. X in action or just two of the station's fans?*

tuned a rig before I kept trying to get it to work properly by reading the manual. After replacing a blown 6146 tube, I hooked it to my Alpha-Delta sloper antenna and made my first test broadcast, which was heard in Pennsylvania and elsewhere. The Alpha-Delta sloper didn't really work that well in that I couldn't really efficiently match the SWR, so I went out and bought a 41-meter dipole, which I used from home for the duration of my pirate career.

The Apache worked OK broadcasting Pirate Radio Boston and others until January of 1993 when it developed a short, arced in the high-voltage section, and actually caught fire during a transmission! I was prepared to dump it out the second story window if I couldn't put out the blaze, but I managed to. Pirate Radio Boston continued broadcasting a couple of times a year until around 1997. Charlie was joined by Mr. Excellence during the later annual Christmas broadcasts.

The Birth of NAPRS (North American Pirate Relay Service)

In September of 1992, a new station called *Radio Azteca* appeared. I wrote to the station saying how much I enjoyed the broadcast and was rewarded by receiving a studio tape of it. A little while later, I went on the air replaying the tape, announcing that this was a broadcast of the NAPRS. It was met with a warm reception by pirate listeners. NAPRS was one of the most active pirates in 1994 and 1995, relaying all Radio Azteca shows, other domestic pirates, and many Europirates, such as Heavy Dude Radio

Fig. 4-4 *One of the rare QSL shirts from NAPRS.*

from Sweden and Germany's Radio Titanic International. During this time, I became good friends with another active pirate operator, P.J. Sparx of WREC and we even broadcast the same program at the same time on different frequencies. Due to popular demand, NAPRS t-shirts (Fig. 4-4) were printed featuring the original QSL design on the front and the slogan "We Love Dick Pistek" on the back. A later design featured stations relayed by NAPRS on the back. The NAPRS continued until 1996 when I phased it out.

How Many Stations Do You Operate?

After NAPRS was off and running in 1993, I became bored again and started many more stations such as Hit Parade Radio, Down East Radio, North Jersey Coast Radio, etc. In fact, the operator of Radio Doomsday and I had a competition to see which of us could have the most stations! It was fun to see which listeners could hear the same voice and put two and two together. Say what you will; all my stations did QSL!

When I started out in 1992, 7415 kHz was the popular frequency. As that became crowded, we tried the 7465 area with mixed results. Then 7385 was used for a while until Radio For Peace International took over. Radio Doomsday was one of the first to try 6955 and it has been in use now for over five years as the frequency of choice.

The Voice of The Rock

In 1994, the talented Radio Animal of WKND was manufacturing these great little 10-watt AM transmitters that operated off battery power and were dubbed *Grenades* (a big whallop in a little package). I was lucky enough to get one and decided to go mobile from the family cottage off the New England coast. The VOR was active for two summers in the mid-'90s, but it wasn't well heard to poor propagation conditions. One listener did hear it in Germany though! The Grenade transmitter (Fig.

Fig. 4-5 *Two early versions of The Radio Animal's Grenade transmitters.*

4-5) always worked well and was sold to a Canadian pirate in 1998. I wish I still had it.

Seasonal Pirates

Several seasonal pirates that were mine included XEROX, Radio Duplicado, which poked fun at NASWA (a shortwave listening club—ed) QSL column editor Sam Barto who printed a QSL from XEROX from a reporter that claimed it was an actual Mexican station. He didn't get the connection between XEROX and Duplicado and I didn't let him forget it. There was the Halloween station, Witch City Radio, which aired every Halloween starting in 1992 until 1997. Halloween still is my favorite pirate holiday.

The Equipment

In addition to the Apache and The Radio Animal Grenade, I went through about eight transmitters, including those made by Hallicrafters, Johnson Viking and Henry Radio. I'm not much of a technician, so when they broke, I gave them away or sold them for scrap value.

In Retrospect

I was active as a shortwave pirate for about six years and enjoyed every minute of it. All good things must come to an end and one must remember that it is an illegal activity frowned upon by the FCC. The people I met through the hobby are still my friends today. Although I no longer transmit, I still listen in on pirate activity every chance I get.

Pirate Equipment

Now it's time to switch gears a little bit. Being such a technically dependent hobby, to truly understand pirate radio, you must have at least a grasp of what equipment is involved. It's not simply a matter of a pirate saying something witty, strange, or radical that makes the station. It is necessary to have at least a running knowledge of shortwave and some equipment to be capable of even getting on the air. This hurdle seems a bit too high and unwavering for the many (thousands?) of people who have said that they had an interest in starting a radio station, but didn't have the technical knowledge to back it up. Many of these people found the hurdle to be insurmountable.

The following sections briefly cover some of the equipment necessary or commonly used in pirate radio stations. It's likely that the rest of the chapter won't suit anyone. Those with a background in electronics or amateur radio will likely find it oversimplified and superficial. Those who aren't technically inclined could easily get lost in tubes, transmission lines, and impedances.

Books have been written about each topic; in some cases, such as antennas, dozens of books have been written about particular types of antennas. So, if you already know the technicalities of hobbyist radio, just skip these sections. If not, this material is intended to be used as a springboard for learning. For more information on the topic, see *Pirate Radio Operations* by Andrew Yoder and Earl T. Grey. See also some of the books referenced within each section.

The Microphone

The microphone is the device that converts audio vibrations into electronic signals. Some low-budget pirates have been known to use the microphone on a

Fig. 4-6 *How much difference will you hear between the $20 microphone (left) and the expensive ribbon mic (right)?*

tranceiver for both announcements (talking into the mic) and playing music (holding the speaker to the mic). This method sounds terrible and allows external sounds to be heard while the music is playing, so most pirates only use the microphone for announcements. And the pirate doesn't need a thousand-dollar plus Neumann studio microphone; a $20 Radio Shack mic will sound just fine on shortwave (Fig. 4-6). The audio from the microphone goes to an audio mixer.

Audio Mixer

The audio mixer is a large box that can combine a number of different audio sources. For example, many low-cost mixers can

Fig. 4-7 *The well-designed studio of Wreckin Radio International from England. Notice the flush-mounted mixer (foreground, right).*

combine four different audio sources, each of which can be varied in loudness via faders (slide potentiometers). So, the mixer is essentially a control box, but the more expensive models also include a tone control on each channel and reverb (echo) controls (either on each channel or on the output as a whole). These less-expensive mixers are sold for such applications as mixing audio in churches or schools, or for light-duty disk jockeying (Fig. 4-7).

Professional broadcast mixers are called either *consoles* or *broadcast consoles*. These units are often three or four feet across, with large, smooth "gliding" knobs and cueing switches. With the cue switches, you can place a channel in Cue mode and listen to the next audio segment without it actually going out over the air. Broadcast consoles can mix anywhere from four to eight channels through the different inputs. Although few pirate stations use consoles, they are preferable for live broadcasting because of the smoother fades, larger controls (handy for fast-paced live productions), and the cueing ability.

A few pirates have used band mixers or recording studio mixers. Both of these types typically have at least eight channels (often 16 or more) and have separate rotary equalizers and other effects on the left and right side of each channel.

Fig. 4-8 *The beautiful studio of Radio One, with a large broadcast console at the center right, an RCA ribbon microphone at the top right, and several cart machines at the top center.*

Other Audio Components

Just about any audio media is useful in a radio station. Many pirates use CD players, cassette decks, MiniDisc decks, computers, open reels (reel-to-reel), cart machines, turntables, etc. I think that this equipment is self-explanatary, except for computers and the semi-obsolete open-reel decks and cart machines.

Computers are now frequently used in radio production at all levels. FM pirates often use computers with auto playback software to make a sort of computer jukebox. Stations also use computers to download audio segments (sound effects, movie or TV audio clips, and comedy bits).

Cart (short for *cartridge*) *machines* are tape decks that play endless-loop tape cartridges (Fig. 4-8) that look a lot like 8-tracks. They use a system of control tones (not audible during playback) that automatically stop the tape. So, if a promo was recorded with a stop tone at the end), you could press the button and it would play until the end and then stop. When it was played again, it would start from the beginning the next time the start button would be pressed. If no

Fig. 4-9 *An early 1970s-era Sony reel-to-reel player/ recorder.*

tone was inserted, the tape would play as an endless loop.

Open-reel decks are tape-recording and playback machines that use large reels of tape, rather than have small reels built into a plastic case (a cassette). Despite the appearance of obsoleteness, a good open-reel tape deck (Fig. 4-9) will sound much better than a high-end cassette deck.

Transmitters and Transceivers

The keystone of a radio station of any type is the transmitter. The transmitter is the component in the radio chain that takes the audio signal and outputs a radio signal. To cover a few definitions that are sometimes confused by beginners, a transmitter can only transmit, but a trasceiver can both transmit and receive radio signals. Also, an amplifier can't actually transmit a radio signal, but it can increase the power of already-present radio signals. A transmitter that is made specifically to produce a signal that is to be boosted by an amplifier is called an *exciter*. As a result, the term *exciter* is sometimes used to describe any type of transmitter.

In the 1980s, it seemed that most North American shortwave pirates used old AM-mode amateur radio transmitters from the 1950s and 1960s. In the 1990s, primarily SSB transceivers were used. Now in the first decade of a new milennium, North American stations are using a combination of homebrew transmitters, vintage AM transmitters, and both analog and digital transceivers. In the 1970s and 1980s, nearly all European shortwave pirates used homebuilt transmitters, but today many surplus military transmitters are in operation—especially in Holland.

Homebrew transmitters are common among those with a

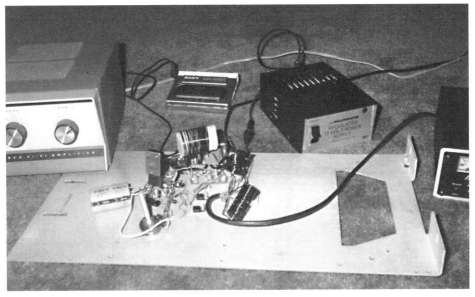

Fig. 4-10 *Rough, but ready: a prototype homemade transmitter of Radio Anarchy in operation.*

technical interest in pirate radio. For example, half of the hobby of licensed amateur radio is simply building kits or home-designed equipment and using it on the air. And such seems to be the case with many of the Dutch pirates, who often seem to have a greater interest in the equipment, quality of the signal, and where they are being heard than in the programming that they air. Because so many amateur radio transmitter designs have been developed over the years, almost any type of homebrew transmitter is out there. For example, although *electron tubes are almost never used in any commercial electronics equipment*, many of the pirate radio transmitter designs use tubes (known as *valves* in England).

In fact, some stations deliberately built historic styles of equipment—especially since the recent popularity of vintage electronics. For example, the Chicago Tunnel Company from the early 1990s built a 1930s-design transmitter on a board (bread-board style). The operator said that the project was an experiment in vintage electronics construction and operation.

Other stations enjoy building equipment for the higher fidelity and power than can be achieved with amateur equipment. In the days when Hope Radio was running more than 1000 watts, operator MJ said that he enjoyed the creative engineering opportunities of pirate radio...and also the way 1000 watts of RF would

Fig. 4-11 *Inside a Johnson Viking Challenger transmitter from the late 1950s. Notice the two final tubes near the center and the huge transformer at the right side.*

raise the hair on the back of his neck!

Somewhat along the same lines as homebrew transmitters are amateur radio transmitters from the 1950s and 1960s. These days, we rarely expect a computer or television to survive more than three years without some sort of electronics problem or failure. So, it's no surprise that a 50-year-old transmitter might have some technical problems. Thus, the old beasts are not for the faint-hearted (think of Pirate Radio Boston's 40-year-old Heath Apache going down in flames) or for those with weak backs (that Apache tips the scale at approximately 95 pounds!). But today's radio hobbyists treat the vintage radio equipment with respect that approaches that of antique cars; transmitting with a Johnson Valiant II is something akin to taking a spin around the block in a '57 Chevy. The good aspect of the nostalgia kick is that amateur radio newsletters and web pages cover all aspects of classic radio restoration, so the information is available. On the other hand, the prices of such transmitters have gone up from $10 (or maybe even "If you can carry it, you can keep it") in the 1980s to well into the hundreds of dollars today. Another drawback for pirates is that almost every 1950s transmitter operates from AC power, which combined with the hernia-inducing weight of the equipment, ensures that the transmitters will be used in fixed locations with AC power.

The next generation of amateur radio transmitter is the

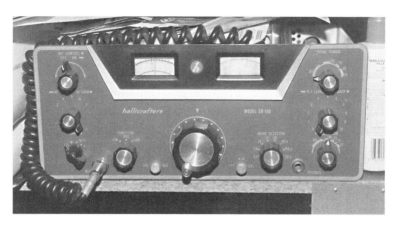

Fig. 4-12 *With a cabinet full of paperwork, this Hallicrafters SR-150 transceiver from the early 1960s doesn't quite look ready to transmit.*

1960s and 1970s SSB transceiver (Fig. 4-12). As mentioned earlier in this section, a transceiver is a transmitter and a receiver built into one package. The advantage is that some sections can be used for both transmitting and receiving, which saves both money and space. The introduction of the transceiver in the late 1950s had a revolutionary impact on amateur radio—ultimately it changed the way amateurs operated (people then communicated regularly from moving vehicles) and its economic wake propelled some small companies to the forefront and literally sunk many of the major ones that couldn't keep up with the technology.

Nearly all of the transceivers used SSB modulation, rather than the AM of the older transmitters (AM and SSB are covered in the next section). Most of these transceivers are about 20 pounds and are about the size of two VCRs stacked on top of each other. Most of these units also have a separate power supply (one for AC and another for DC) that weighs between 20 and 40 pounds. Despite the somewhat restrictive size and weight, these rigs are still capable of going out in a car for mobile operations or for broadcasting from a stationary outdoor location.

From the early 1970s onward, power supplies have typically been built into the transceiver cabinets. Over the past 20 years, transceivers with a 100-watt output (moderate power) have been built into small cabinets that only weigh about 10 to 20 pounds. All of these later rigs are digitally controlled and can broadcast on most any frequency with the snip of a wire. These later transceivers can broadcast in the AM mode as well as SSB, which adds a little flexibility.

AM vs. SSB

Over the past 10 years, the points most argued in pirate radio have not been related to the FCC, amateur radio, U. S. politics, or the legalization of low-power radio. It's been AM vs. single sideband.

All AM broadcast radio stations and nearly all licensed shortwave broadcast stations transmit in the AM mode. An AM-mode signal consists of three parts: an upper audio sideband, a lower audio sideband, and a carrier wave. The two sidebands are the entire audio portion of the signal. The carrier wave is a wide, "empty" signal. The best way to think of a carrier is when you listen to an AM station and there's a gap of dead air. You can hear the station even though you can't hear them playing any programming. If you have a communications receiver with meters, you can see the levels go way up on a strong "dead" carrier—even though no audio is being played.

In single sideband, the carrier and one of the sidebands are eliminated. The obvious advantage here is efficiency. It's estimated that between the easier copyability of the SSB signal and the power savings between the dropped carrier and extra sideband signal that an SSB signal is six times more efficient than an AM signal. So, if a station had a 100-watt SSB transmitter, they would need a 600-watt AM transmitter to cover approximately the same territory. If you know much about amateur radio or electronics, 100 watts of sideband comes standard on transceivers. However, 600 watts of AM is a serious undertaking, requiring either an expensive linear amplifier or some major work with beefy components in a homebrew transmitter.

At this point, it seems like everything is in favor of SSB broadcasting, but there are some serious drawbacks. First, not every shortwave listener uses a radio that's equipped to receive SSB. If your radio doesn't have either a SSB (USB/LSB) selector switch or a BFO, any SSB broadcsts will not be intelligible. Regardless, music in the SSB mode often sounds awful, like seagulls screeching over a garbage dump. Generally, the older American-made transceivers (Drake, Heath, Swan, etc.) sound better in SSB because their signals are a bit wider than later rigs (Fig. 4-13). Another SSB drawback is that it can be more difficult to tune in an SSB signal than an AM one. I've had some signals that I've had trouble tuning in in either LSB or USB. Sometimes, it doesn't really seem like the station is broadcasting in either LSB or USB mode, but some unknown sideband.

Overall, the difference is generally audio quality vs. signal

Fig. 4-13 *Caught in the dilemma of AM vs. SSB? Not a problem. The Drake TR-4 can do either AM or SSB with better-than-average audio quality.*

range. I'm not quite sure why it aggravates some people that other stations are using either AM or SSB, but it has been a source of contention between pirates. Possibly some of the problem is related to AM stations that don't hear SSB stations before they sign on the air. Another problem is with pirate listeners who also broadcast and choose AM only because of their frustrations with hearing SSB stations.

Antennas

Every radio station needs a transmitter to produce the radio signal, but every transmitter must have an antenna to broadcast that signal. Without an antenna, not only will the signal not be heard by anyone, but the transmitter will fail.

Here are some of the most important concepts for antennas:

* The length of the antenna is inversely proportional to the frequency on which the transmitter is operating.
* The antenna must be matched as the same length necessary for the frequency being transmitted on.
* The higher above ground the antenna is, the more efficiently the signal will be radiated.

If you want more information on the basics of antennas and construction, see *Build Your Own Shortwave Antennas* by Andrew Yoder. For more technicalities, see the Joe Carr antenna books or the annual ARRL handbooks, which are excellent references. For this book, I'll just keep the antenna discussion at an absolutely basic level.

By far the most popular pirate radio antennas are variations of the half-wave dipole. This is a simple two-element wire antenna, with one "leg" connected to the ground side of the connector and the other connected to the "hot" center. Because nearly all of the pirates in North America and Europe use the range from 6000 kHz to 7500 kHz, assume that any of the dipole antennas in the following paragraphs will be somewhere around 65 feet in length.

The standard dipole form is T-shaped, with the coaxial cable running up to the center of the T, and the two elements forming the top. Probably the least-popular versions because of the installation difficulty is the vertical dipole, which is like the dipole "T" lying on its side (few people have anything at least 65 feet in height on which to mount the end of antenna). Another variation is the sloper, which is a standard dipole in which the one side of the "T" is mounted high (to a tree, tower, roof edge, etc.) and the other end of the "T" slants down to a few feet off the ground. Slopers are usually mounted at anywhere from about 30 to 45 degrees. Slopers are especially directional (meaning that most of the signal goes in one general direction) toward the low end of the antenna

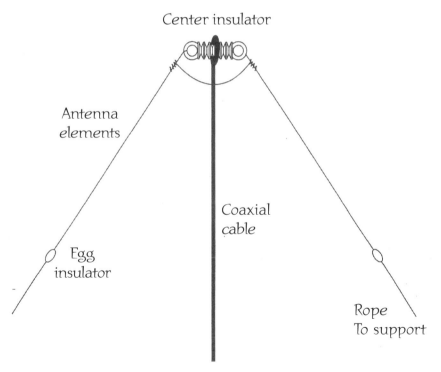

Fig. 4-14 *The basic components of an inverted V antenna.*

when a metal support is used on the high end. This directionality is useful when the station is attempting to reach a particular region (such as during a test to Europe) or when trying to keep the signal away from a particular area (such as a coastal station that doesn't want to waste its signal over the ocean).

One of the most common antenna forms is the inverted V (Fig. 4-14). The reason for its popularity is that it behaves fairly

Fig. 4-15 *Does Roger Clemens pirate in the offseason? Probably not, but his arm would be welcomed when slinging ballast over trees.*

similarly to a standard dipole, but it only requires one support and less space than a typical dipole. The number of supports is an important issue in pirate radio. If someone is going to drive out somewhere to make a broadcast, it is much easier and less conspicuous to slingshot or throw a rock over a tree one time, rather than three (Fig. 4-15). Once the "pull line" is over the tree, then it's a simple matter to pull the antenna in place.

Antennas have caused an amazing amount of frustration for radio hobbyists of all types over the years. Wires become looped and knotted around branches, the coaxial cable becomes twisted around the elements, and wires snap or become caught in the branches.

I know one hapless pirate who was setting up an antenna in

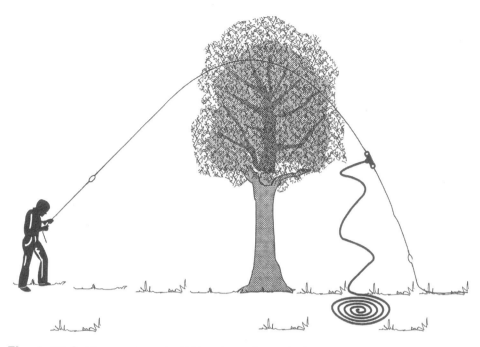

Fig. 4-16 *Pulling an inverted V in place through the tree.*

the rockless Midwest. He was in a remote location and had a few rocks along to throw the antenna up in the air. One wire looped around the limb and knotted. The line snapped, leaving the rock dangling in the tree. The knots around the next rock loosened and the rock flew out into a field and was lost. In desperation, he found a leather change bag with about $5 to $10 of coins inside. It, too, became lodged in the tree and the line snapped while attempting to pull it down. The location was finally vacated with a variety of objects dangling from the limbs, like a poorly decorated Christmas tree.

I've heard other tales of searching on hands and knees in the dark for antennas, falling out of trees, getting poison ivy in the woods, nearly falling off a snow-covered roof, and being hit by flying ballast.

Grounds

Technically, a ground is something that can be used as a reference voltage and equal to that of the earth (generally, around 0 volts). In audio equipment, grounds are used to "drain the hum out." In transmitting equipment, grounds are used (particularly with the Hertzian family of antennas) to make the antennas operate more efficiently. For these antennas, it's best to lay down ground radials (a star-shaped network of wires that are typically 1/2 or 1/4 wavelength of the frequency on which the transmitter is operating) that either on the ground or a few inches below the surface. Needless to say, antennas that require ground radials are not popular with pirates.

I know one pirate who operated from his house with a Hertzian antenna for the top of the AM band. Sometimes, he would hose down the area around his ground radials in an attempt to make his ground radials operate more efficiently. Radio Aquarius (mentioned earlier in this chapter) was preparing for one broadcast and the station engineer was running a ground wire from the transmitter to a grounding rod when someone turned on the transmitter. The radio frequency energy shot through the wire with enough of a jolt to burn his hands!

Antennas with complicated ground radial systems are typi-

Fig. 4-17 *The MFJ artificial ground.*

cally avoided in pirate radio. When they are used, they are often considered to be more trouble than they're worth. To eliminate hum, some people connect the ground terminals to a large coil of wire to make a sort of "dummy ground." Over the past few years, some of the apartment-dwelling radio hobbyists have started using the MFJ tunable ground–an electronic artificial ground that can be tuned to match the transmitter (Fig. 4-17).

Conclusion

Getting a behind-the-scenes look at a pirate radio station is kind of a catch-22. Depending on the station and its operations, it might be more enjoyable not knowing anything aside from what is broadcast on the air. Many times, it can be more fun just to let your imagination soar.

Fig. 4-18 *A pirate station packed up and ready to pirate. A Heath DX-35 transmitter is in the front center and a deep-cycle marine battery is to its right.*

On the other hand, some pirates have had some amazing adventures far from the high seas. Personally, I think that all of these trials are a fascinating aspect of listening to pirate radio and one reason why I can't be too critical of most of the programs that I hear on the radio (Fig. 4-18). The program that you have just heard could have been put on the air by someone who just had his car stuck in the mud, fell out of a tree, or spent the past 90 minutes walking through slush while the sky was dumping freezing rain.

European Offshore Radio 1958-1986

In Europe, rock music = pirate radio. It's that simple. Throughout the 1960s and some of the 1970s, if you lived in Europe and wanted to hear rock music, you either listened to the evening broadcasts from the licensed Radio Luxemburg or you tuned in a pirate station. The European continent was locked with countries that would not relinquish the radio bands without a monumental fight. Perhaps it was because of the scars from the role of World War II propaganda that these countries were so serious about controlling the airwaves or maybe they felt that they deserved to own it all. Regardless, the history of radio in Europe is far different from the grassroots "one amateur broadcasting from his hamshack" beginnings in the United States.

The power of rock music is not to be underestimated. If you ask people from Europe about when they first heard different songs in the 1960s, chances are good that it was on Luxemburg or a pirate. So, imagine if millions of people across Europe all had the experience of listening to The Beatles, The Rolling Stones, The Kinks, The Who, Dave Clark 5, and many other bands for the first time on Radio Caroline, Radio London, Radio City, and others. Imagine if rock music could not be cleanly separated from pirate radio, like Siamese twins that shared organs in order to live.

The result is that pirate radio is very popular in Europe and many of the old announcers (often called *presenters*) from the old offshore stations are celebrities. If R. F. Wavelength from the Voice of the Voyager had shown up in Minneapolis for a press conference about pirate radio in 1978, probably a friend or two would have made the trip. But in Europe, sometimes literally hundreds of people would flock to see the pirate radio announcers arrive on the mainland from the ship. There were times when announcers requested listeners to drive to the coast and shine their car's headlights across the water. In at least one case, hundreds of people showed up and lit the coast. So, Americans might have had Woodstock, but the English and Dutch had pirate radio.

Radio Mercur:
Setting the Trend in Europe

Europe set the trend of pirate radio as offshore stations began sailing into international waters in the late 1950s. Unlike North America, a continent filled with commercial broadcasters, Europe was largely filled with government-controlled stations that offered little programming for younger listeners. Whenever music programs were aired, they usually consisted of classical or other music considered too stodgy by average teenagers. These stations usually carried no advertising. Given this programming and advertising void, businessmen, musicians, and radio enthusiasts all sought the possibilities of commercial broadcasting. They all knew a land-based commercial pirate station would be quickly closed down if it operated on a regular basis. A revolutionary new idea was desperately needed to break the governments' monopoly of the airwaves.

That new idea came to life in the summer of 1958 when Radio Mercur (Fig. 5-1) began regular broadcasts on the FM band from a ship off of the Denmark coast. A Danish businessman foresaw the commercial possibilities of a broadcasting station. Undaunted by government regulations against broadcasting, he found a loophole and had a small fishing vessel, the *Cheeta Mercur*, fitted with a transmitter and antenna.

Fig. 5-1 *Radio Mercur broadcasting from its original ship, the* Cheeta Mercur, *which was later purchased and used by Radio Syd.* Martin van der Ven

This ship was anchored in International Waters, and broadcasting began. Within several days, the *Cheeta Mercur's* mast broke and it had to be sailed back into port for repairs. After repairs, the ship returned to its original site and began regular programming.

In 1961, Radio Mercur expanded to two broadcasting ships, with the newest weighing 450 tons and equipped with an 8000-watt FM transmitter. Although the Danish authorities could not touch a ship broadcasting in international waters, they did pass a law in 1962 making it illegal for a resident of Denmark to work on, assist, or advertise on an offshore radio station. This move instantly eliminated all Danish free radio. But Radio Mercur lived on in the hearts of others inspired by the activities and in Danish government radio. After legislation ended the station, a new government radio network was created following Mercur's format, and many of the old announcers were hired to assist the network. Just as the pirate activities were coming to a close in Denmark, a golden era of offshore broadcasting was beginning in England.

The most famous English or worldwide offshore station began in the Spring of 1964 as *Radio Atlanta,* This station operated from a 53-year-old German schooner formerly used by the Swedish pirate, Radio Nord. The ship was renamed *Mi Amigo* and continued broadcasting for two months until it merged with another offshore pirate, Radio Caroline. The combined operation, known as *Radio Caroline,* became the premier rock station in England. It was one of the first broadcasters to play new sounds in rock music by The Beatles, The Kinks, The Who, and The Rolling Stones. And, unlike the BBC, it played such music continuously instead of restricting it to a single "pop music hour" each day.

The new British rock explosion helped Radio Caroline to reap pirate fame and plunder, but it could not save the station from the government. After nearly a dozen offshore pirates followed Radio Caroline into international waters off the coast of the United Kingdom, government officials were unnerved. The British Broadcasting Corporation had obviously lost its monopoly throughout the country, much to the dismay of its leaders. This prompted them to start claims that the offshore stations caused interference to the legitimate broadcasters and emergency outlets along with other propaganda.

After seeing that rhetoric alone had not kept listeners from tuning in the pirates, they took the matter one step further. In 1967, the British government introduced the Marine Broadcasting Offenses Act. The act was similar to those passed throughout Scandinavia in 1962, in that any citizen of the United Kingdom who aided an offshore broadcaster could face stiff fines or imprisonment. Although British offshore broadcasting was virtually destroyed after 1967, Radio Caroline continued with foreign announcers and advertisers.

Strangely enough, Caroline lived through the end of the 1960s and trudged across the entire 1970s as well. Unlike its colleagues from international waters, Radio Caroline never quit at the sight of bad weather, low revenues, or government intervention. Even the BBCs diversification into pop music on Radio One, Radio Two, and Radio Three on AM and FM had no *effect* on Caroline's status. But 16 years of broadcasting from the *MiAmigo* (Fig. 5-2) finally ended in March 1980 when the ship sank during bad weather in the Thames Estuary.

Fig. 5-2 *The* Mi Amigo's *mediumwave radio mast was a bit conspicuous.* Martin van der Ven

Just when the authorities thought that Radio Caroline would finally be gone forever, the owners announced that the station would return with a larger ship, the *Imagine,* and 50,000 watt transmitters. This was allegedly made possible by a large backing of American dollars and advertising help from radio legend Wolfman Jack. Although it took longer than originally planned, Radio Caroline returned in late Summer 1983, with pop music interspersed with commercials that were mostly from American advertisers. For several years in the late 1980s, Caroline had a shortwave transmitter on 6215 kHz. This transmitter fell silent when Radio Caroline went off the air in the early 1990s because of the high cost of operating the station. Although Radio Caroline can no longer be heard from its own transmitter on the AM band, it has been audible via a number of different satellites in Europe. Most recently, it was being heard via the Intelsat 702 satellite.

Other Offshore Pirates
of the 1960s and 1970s

Remarkably, many of the first British offshore pirates did not broadcast
from ships...or any seagoing vessels for that matter. The stations took over
abandoned offshore World War II machine gun nests, called *forts*. Ironically,
the very structures that were installed to prevent a Nazi invasion in the
1940s allowed British entrepreneurs to invade the country's airwaves in the
1960s. Each of the forts consisted of a group of squarish metal buildings,
each anchored on a large pole and sunk into the water. Each group of forts
consisted of seven separate platforms: the control tower, surrounded by five
gun towers, with a searchlight tower off to the side. The forts had the
capacity to house 100 men and store all of their necessary supplies. According to the Radio Forts web site, the manpower on the forts shot down 22
Nazi planes and approximately 25 flying bombs (V-1 and V-2).

The forts were built tougher than a 1975 Buick. More than 10 years
after no longer being used, they were still there and the British government wondered what to do with them. No longer in danger of an invasion
from the sea, they found that the upkeep on the towers were both expensive and dangerous. Finally, in 1956, the British armed forces stripped the
forts of any expensive equipment and abandoned them. They stood soley as
a silent monument to England's struggles in World War II until 1964 when
Radio Atlanta and Radio Caroline started British entrepreneurs thinking, "If
there would just be some way to broadcast from international waters
without having to go to the expense of buying a broadcasting ship."With
six of the seven abandoned forts located in international waters, the answer
was quickly found.

Icon of early shock rock, Screaming Lord Sutch sailed out to the
Shivering Sands tower block (Fig. 5-3) and had the first of fort pirates
installed and on the air by May 1964. The station, naturally called *Radio
Sutch*, was intended to be an outlet for new rock bands that had received
little airplay. Radio Sutch was poorly organized and often ran into problems
while broadcasting. Rather than adhering to a tight format, the DJs often
played album sides without announcements. And according to the book
When Pirates Ruled the Waves by Paul Harris, it was common for listeners
to hear the DJs broadcast such messages as, "If any boat is coming this way,
we're running short of bread."

Radio Sutch was a great publicity stunt for Screaming Lord Sutch,
but the 1000+ pounds/month cost of the operation and the extra responsi-
bilities were too much to be saddled with. Sutch sold the station to his
manager, Reg Calvert, in Autumn 1964 for a little more than the original
set-up fee. Calvert then changed the station name to Radio City. All ran well
until June 1965 when a boarding party, armed with acetylene torches,

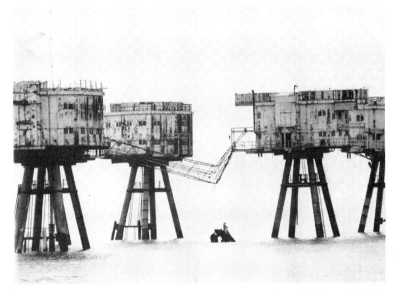

Fig. 5-3 *A post-pirate photo of the Shivering Sands Tower block, formerly used by BBMS and Radio City. Notice the collapsing catwalk between towers.* Martin van der Ven

invaded the fort and held the station's crew hostage for a week. During that time, Calvert was murdered. The British armed forces felt that they couldn't intervene with the hostage crisis because Radio City was "squatting." The events made headlines in the UK and even in some international newspapers and magazines. Even one of the boarding party members stated that it was more exciting than *The Man from U.N.C.L.E.* Indeed, one later episode of *Secret Agent Man* featured a spy who had infiltrated a fort-based pirate radio station. Who says that reality-based "ripped from the headlines" television started recently?

The next fort-based pirate station to follow Radio Sutch out to the platforms had an equally disturbing history. Radio Invicta ("the good music station") began broadcasting from the Red Sands Tower in June 1964. In December, the station's supply ship sunk and the body of the owner washed ashore the next day. Despite rescue crew efforts, the bodies of a DJ and an engineer, who were also aboard, were not located. Months later, the badly decomposed body of a man washed ashore in Spain. One of the clues of his identity was a tape recording found with the body. The tape was dried out and played back. It was a studio tape of Radio Invicta.

But, that was not the end for Radio Invicta. The station was then sold and became KING. KING soon ran into financial difficulties and was sold to

Ted Allbeury, who changed the station name to Radio 390 in early 1965. Allbeury also changed the format to light music, targeted at older listeners. Although the format change was successful, the station was not a winner in the courts. The Red Sands Tower was ruled to be inside of British territorial waters, thus subjecting it to the laws of the United Kingdom. After a court battle in 1966, Radio 390 closed in September 1966, returned in December, and closed for good in July 1967, after receiving another several fines.

The other fort-based pirates included Radio Tower, which tested, but failed commercially and never made it into the regular broadcasting phase. After it concluded broadcasting from Sunk Head Fort, the government demolished it. Contrary to the life of Radio Tower, the owner's life on the fort of Radio Essex/Britain's Better Music Station (BBMS) has lasted far longer than the broadcasts. Radio Essex broadcast from October 1965 until September 1966, when the station was slapped with a summons and fined. It continued under the BBMS name until December, when it ran out of money. Its fort, Knock John Tower, was ruled to be within British Territorial waters, so the owner, Roy Bates, moved the station to Roughs Tower, but never returned to the airwaves. In 1967, he renamed Roughs Tower as *The Principality of Sealand,* which has since issued passports, minted money, and printed stamps. Now in existence for more than 30 years, the sovereign nation of Sealand adds more to the curiosity and intrigue of the radio forts.

Back to the Sea

Of course, the success of Radio Caroline quickly lured other pirates out to the sea. Some of the other offshore pirates included Atlanta Radio, Radio London, Radio Scotland, Radio England, Radio 270 (Fig. 5-4), and Britain Radio (United Kingdom); Radio Nord and Radio Syd (Sweden); Radio Hauraki (New Zealand); Radio Veronica, Radio Northsea International, and Capital Radio (Holland); and Radio Antwerpen (Belgium).

Nearly every one of these broadcasting ships shares similar turbulent histories. For example, almost all of the broadcasting vessels either went adrift, and was towed into a harbor and impounded, or ran aground. Most of these outfits lost an antenna in the high winds. Because of the unsettled waters and huge expense of losing a radio ship or a transmitting antenna, all of the stations faced financial disaster at any given moment. Although plenty of money could be had in the offshore radio business, many entrepreneurs eventually lost their shirts.

Radio Nord and Radio Syd both pre-date Radio Caroline and the whole British and Dutch experience of the 1960s and early 1970s. The origins of Radio Syd go back further than Radio Nord, although the latter was first on the air. In 1958, Radio Syd began broadcasting as the Swedish service of Radio Mercur, which beamed its transmissions toward Sweden several hours per day. A special Swedish staff ran these programs and Britt Wadner ran the advertising department. After several years, Wadner bought another radio ship and essentially took over for Radio Mercur. Radio Syd continued despite laws being instituted in Sweden against offshore broadcasters and those companies who advertise on such stations. Wadner wound up in prison in two separate incidences and a number of companies were prosecuted for advertising. The station also had one ship sink, a collapsed mast, several broken anchors, and was ice bound several times, but the broadcasting continued. Finally, five years after the anti-offshore laws were instituted, the *Cheeta II* sailed to England for repairs, then to then to the tiny West African nation of Gambia. Radio Syd could easily dismiss any

Fig. 5-4 *Radio 270 was one of the stations that broadcast up until the last evening before the Marine Offences Broadcasting Act took hold.*

rumors of being a fly-by-night station; it received a license to broadcast from the land in Gambia, where it is still operated by the Wadner family, more than 30 years later. The *Cheeta II* was used as a floating restaurant and disco, but it later sunk.

Radio Nord was almost an anti-Syd. The station worked through many physical disasters while still in the planning stages in 1960 and 1961. But soon after taking the airwaves, Radio Nord claimed an audience of nearly 25% of the homes in Sweden, an amazing share. However, Radio Nord closed immediately after the Swedish anti-pirate laws were instituted in 1962.

Another oldie is Radio Veronica, which goes all of the way back to 1960. In a sense, you could say that Veronica is the Dutch version of Radio Caroline, but some might even consider it to be a bit of an insult. Radio Veronica is older than Radio Caroline . . . and is still in existance today. Like most of the other pirates, Veronica was one of the pop and rock music stations, and was extremely popular as the only Dutch offshore station for years. Most of the history of Radio Veronica represented a best-case scenario for offshore radio stations. It didn't have many of the serious and costly problems on the high seas that befell most of the stations that were anchored off the coast of England and Scandinavia (although it wasn't entirely safe from these dangers, running aground on a Dutch beach on one occasion).

Throughout its career, Radio Veronica made money and was known as an "above-board" station—especially in comparison to the free-wheeling and rather desperate operations that broadcast from the forts. The Veronica organization even filed appropriate taxes and other paperwork. However, some incidents tarnished the station's clean-cut image in the early 1970s, when competition arrived in the form of Radio North Sea International, which broadcast in Dutch and English from the *Mebo II*.

Radio North Sea International operated from a psychedelically painted state-of-the-art broadcasting vessel (Fig. 5-5), as opposed to some of the old ill-prepared, makeshift ships that were converted for radio use. RNI began broadcasting to Holland with the standard format of pop and rock music with veteran off-shore DJs in spring

Fig. 5-5 *A beautiful QSL card from Radio Nordsee International, showing the* Mebo II *in its full glory. Unlike the many mediumwave-only offshore pirates, RNI also broadcast on shortwave, so it was widely heard in North America (as well as the other continents).*

1970. In early Summer, the station relocated to the British coast and changed its name to *Radio Caroline* and broadcast anti-Labour Party (UK) rhetoric. The Labour Party lost the election and the *Mebo II* returned to the Dutch coast with Radio North Sea International programming.

At one point, the *Mebo II* was attacked by a boarding party, and the spray from a water cannon was used to knock the shortwave transmissions off the air. The staff continued broadcasting through the attack, announcing the names of the ships, and making for rather exciting radio. The boarding party was foiled, despite warnings that they would return. It turned out that the boarding party was a neglected part owner in the station, who evidently had not polished his skills of negotiation.

Before long, RNI went silent as it was announced that the ship had been sold to an African company and the vessel would be moved off the European continent. Later, it was disclosed that Radio Veronica paid a large sum of money to prevent competition. But the monetary alliance alone was not enough to prevent more confict. Before long, the *Mebo II* (which, by that time had a Radio Veronica crew in place) was boarded by armed RNI staff, which took over the ship. RNI then voided the contract with Veronica and began broadcasting their old pop/rock format in Dutch and English. Weeks later, RNI and Veronica went to court, which ruled in favor of RNI.

On May 15, 1971, a team of frogmen placed a bomb in the *Mebo II*'s engine room, which rocked the ship and set everything ablaze. Station announcer Alan West continued the broadcasting, calling Mayday and stating that they had been bombed. The announcer only left when the lifeboats were called in, and the transmitters went dead. The entire episode was broadcast to millions of astounded listeners, and made enough of an impression that the story was picked up by North American media (such as *Newsweek*). The plot became more intricate when it was announced that the frogmen had connections to the beloved Radio Veronica. Another court case ensued, and several members of the Radio Veronica board were sent to jail, although other board members were cleared of any involvement in the bombing.

The Veronica/RNI feud didn't just involve the participants; a station that was just starting during the pirate feud, Capital Radio, was also a victim of the violence. Capital Radio was only a week into regular broadcasting when the two anchors on its ship the *King David* became entangled. One was cut free, the rudder snapped, and the antenna system collapsed. While the ship was in port being repaired, someone attempted arson by filling the engine room with a foot of diesel fuel. But Capital Radio pulled the funding together to repair the ship. Just a month after the anchor had been replaced, the new anchor chain snapped in a storm, setting the vessel adrift. Although attempts were made to pull the ship back to sea, it grounded itself on the resort beaches of Noordwijk, Holland, hundreds of feet from the Palace Hotel. While beached, someone "lubricated" the diesel engine's moving parts with diesel fuel so that it would catch fire if anyone attempted to start the engines. The *King David* was pulled out of the sand, but despite efforts to receive extra funding, the station's organization went bankrupt.

After the initial pirate war claimed Capital Radio, Radio Veronica continued until late 1974. It received a license to broadcast independent radio from The Netherlands and returned in 1975. Today, it is a Dutch pop media giant, owning radio and TV stations, and producing a magazine, called (not ironically) Veronica. Radio North Sea International continued broadcasting via its powerful transmitters on AM and shortwave. For some listeners and future pirates in the United States and Canada, Radio North Sea International was the first pirate they had ever heard. In fact, it would not be surprising if Radio Clandestine (which started in 1973) was modeled after RNI. As Dutch offshore legislation pressured RNI, the station pulled the plug in 1974. Rumors abounded that the station was moving to the coast of Libya to broadcast for Gen. Mummar Quadaffi.

These pirate battles are featured in the almost-impossible-to-find book *To Be a Pirate King* by Paul Harris. Harris was one of the founders of Capital Radio, and was best-known for his earlier pioneering book, *When Pirates Ruled the Waves,* which is also difficult to find in North America. Obviously, Harris was biased for Capital Radio, but he presented some fascinating theories about RNI's ownership that don't seem too far-fetched. Harris believed that the owners of RNI were connected with the East German government and that they were in the process of building a number of radio ships for the Soviet Union. These would be used to broadcast propaganda and to provide leverage for the Russians to have Radio Free Europe and Radio Liberty (which the USSR despised) removed. Harris also claimed that the shortwave outlets of RNI occasionally broadcast coded messages. The basis for these theories were RNI's known connections to an East German group; that the station was jammed by a number of countries in Europe (including England's one-million watt transmitter), which was an unprecedented move against a pirate station; and because anyone with connections to RNI was questioned at length by federal agents.

If these theories were true, either in part or in whole, this might have been why Radio Veronica became the only offshore station to be licensed in Europe. The Dutch government might have felt that licensing Veronica would shift the power in Veronica's favor and cause RNI to fail or look for greener pastures. If this was the government's course of action, then it evidently worked. But if those Soviet propaganda ships were in the construction process or were purely myth is anyone's guess.

One station that certainly deserves a mention in this book is Radio London, which broadcast from off the coast of England for only two and a half years, from late December 1964 until August 1967. The station was another of the pop music ships, but this one was very high powered and popular. In fact, Radio London operated with a 50-kW transmitter, was pulling in between 70,000 and 75,000 pounds in advertising per month, and had an estimated 8,000,000+ listeners during its peak in mid 1966. At a discussion of the second book at a book store in New Jersey, I met one of the engineers of Radio London, who said that it was very difficult to run a 50-kW transmitter on an all-steel ship in water. "RF got into everything," he said. Unlike some of the operations that broadcast from offshore, Radio London's owners made out quite well, bringing in hundreds of thousands of British pounds without the serious incidents that plagued the Dutch stations of the early '70s, then quietly closing shop when the Marine Offences Act was passed in 1967.

Even to the fan of unlicensed radio, it seems as though the British government was forced to do something about these stations. When Radio London gradually increased its power to 50 kW, Radio 270 upped theirs to 75 kW, and Radio Caroline went from 10 to 50 kW. The British government started receiving complaints of interference from deep in southern and eastern Europe. And with the high powers used to cover such a small territory, the pirate signals were actually interfering with each other in some areas. Also, plans for more broadcasting vessels were underway. What would happen if five more stations joined the seven British offshore stations and participated in the power struggle (both in terms of transmitter output power and violence)?

The Marine Broadcasting Offences Act of 1967 made it illegal for any British citizens to aid or work on an offshore radio station and for any British companies to advertise on such stations. Even before the act passed, the fort-based stations were being prosecuted for unlicensed radio operations from British territory. The owner of Radio 390 had been jailed and was fighting its case in court along with Radio City.

The Marine Broadcasting Offences Act of 1967 effectively cut the financial and supply lines to the offshore stations and six of the seven were off the air by August 17, 1967. On the evening of August 16, the remaining pirates were giving their tearful goodbyes on such stations as Radio London, Radio 270, and Radio Scotland (Fig. 5-6). Considering the combined tens of millions of listeners for these three sta-

Fig. 5-6 *Radio Scotland, anchored off the northern coast of the UK, also held out until the end in August 1967.*

tions, it really was an emotionally charged evening for listeners and disk jockeys alike. The listeners to the offshore stations identified more with the DJs than in today's broadcasting. Most offshore DJs were paid very little and were risking life and livelihood (offshore broadcasting could be hazardous and the operators were always under the threat of prosecution). Also, the listeners heard DJs getting seasick on the air, when storms beat against the ship, and when emergency help was necessary. On top of the "rock star" image that the DJs had, the listeners also felt other very strong bonds. So, after the stations closed down at midnight, literally tens of thousands of fans turned out to greet the announcers when they came ashore. Parties for Radio Scotland and Radio London drew thousands of people, and riots occurred in some areas. At 12:00 A.M. on August 17, Radio Caroline, now the only independent, commercial radio station in the UK, greeted their new listeners.

To many, the biggest problem with the Marine Offences Broadcasting Act wasn't that it closed down most of the pirates, but rather that the British government failed to replace the service. For approximately five years, independent radio was still off-limits in the UK, and the closest replacement was a strictly regimented BBC service.

Tony Greaves, engineer of the British land-based pirate Radio Aquarius, summed up the licensed radio situation as the reasons for broadcasting: "Many of the offshore DJ's joined the BBC's new Radio One pop music station, but the teenagers of Britain felt cheated; their favorite stations had been silenced and replaced with an inferior service, sharing its airtime with middle-aged Radio Two, and run by more or less the same bunch of people who just the day before had been in charge at the 'Light.'

"To add insult to injury, the BBC used the very same jingles that Radio London had used, with the wording changed from 'Wonderful Radio London' to 'Wonderful Radio One.' The ex-pirate deejays did their best to make the shows sound as fresh and exciting as they had on board the ships, but with the tight needletime restrictions, inflexible playlists, and scripts that had to be okayed before going on the air, it wasn't surprising that the programmes sounded dull and forced. If the average teenager was disappointed, the young radio enthusiast was inconsolable. How could a so-called 'democratically elected' government disenfranchise the youth of Great Britain by outlawing such an innocuous activity which brought enjoyment and entertainment to millions of people? Was it because they couldn't control it?"

The inflexibility of the governments in England and other European countries ultimately led to an explosion of land-based pirates in the late 1960s and 1970s.

The Last European Offshore Station

One of the last offshore stations, Laser 558 (later known as *Laser 576* and *Laser Hot Hits*) took the same route as many of Radio Caroline's colleagues. Started as an American alternative to European radio in a commercially viable area, Laser began broadcasting from the Communicator on May 24, 1984, on 558 kHz. Immediately after hitting the airwaves, Laser 558 became the most popular station in the British Isles, amassing an estimated 6,6000,000 listeners. One English free radio magazine even suggested that Laser was the most popular offshore radio station since the 1960s.

Laser's popularity was due to its use of American disk jockeys with a nonstop popular music format. With Europeans craving American-style radio for decades, it is no wonder that an essentially American radio station transplanted in the North Sea would become successful. In fact, the announcers became celebrities. Many listeners took the time to send letters with comments and music requests to their favorite air personalities.

Sheer success with listeners is no guarantee of a lasting radio operation, however. Laser's problems stemmed from technical difficulties and off the air staff tension. Toward the end of 1985, Laser's broadcasting vessel, the *Communicator*, broke away from its moorings. The ship was unable to backtrack, lost all power, and was simply floating in the North Sea. After narrowly missing a ferry, news services reported that the crew panicked and called the British authorities for help. Soon a government vessel arrived and towed the *Communicator* to port in the United Kingdom. With Laser Radio in port, the government used anti-offshore broadcasting laws to hold the ship there. For each day off the air, the Laser investors lost money, and to further complicate matters, the old 558 kHz frequency was taken over by Radio Caroline.

After over a year of absence from the airwaves, Laser returned to a new frequency, 576 kHz, in December 1986, with strong signals reported across the British Isles. However, the station went through various funding changes, and before long was in need of money. Apparently, the investors felt that pumping more funding into the Laser project would be unprofitable. Once again, the station fell silent—this time for good. As is often the case with the Europirates, the station name lives on through the current land-based British pirate, Laser Hot Hits. Laser Hot Hits even uses some of the old jingles from Laser 558 (which at times used a "Laser Hot Hits" slogan).

European governments are universally opposed to the concept of pirate radio, especially in the form of offshore stations. Although government pressure has reduced the flow of offshore pirate radio, it has not entirely stopped it. More than anything, the cost of maintaining a large radio ship with a full-time crew in the rough waters of the North Atlantic have prevented more offshore pirates from entering the shortwave world and becoming successful. In fact, as mentioned earlier, the rising costs and changes in technology have at least temporarily eliminated all actual shipboard broadcasts from Radio Caroline. Presently, the station is heard in Europe via satellite and occasionally via one of the radioprivates (explained later in this chapter) shortwave transmitter in Ireland. Although some of the Caroline programming is still produced on the ship, it is used more as a fund-raising novelty than as a functional broadcast studio.

With the increase of independent licensed radio stations and the sudden increase of satellite radio, it appears that the era of European offshore radio has passed. Still, Radio Caroline still owns the *Ross Revenge* and some people in the world still have hands-on offshore radio ship construction and engineering experience.

RSL Radio: Offshore Pirates since 1990

The British Radio Authority has instituted a means by which individuals can place temporary low-power radio stations on the air. These stations, called *RSLs (Restricted Service Licenses)*, enable a number of different voices get out on the air for the purpose of testing a particular type of service or providing temporary broadcasting (such as broadcasting a sporting event). Nearly all of the long-term RSL licenses are for hospitals, military bases, or schools. Some of the offshore stations have also returned via RSL, including Radio London and Radio Caroline. Radio London (Figs. 5-7 and 5-8) last broadcasted a reunion with old station DJs and some younger people from Clacton Pier in Essex in August 2000. Radio Caroline's RSLs have been from their current ship, the *Ross Revenge*, anchored in the Sheerness area. These irregular broadcasts help complement Caroline's regular satellite service, but it's still ironic for a licensed, low-power station to be broadcasting from a pirate ship.

As noted in Chapter 7, although the old offshore pirates will surely continue broadcasting, RSLs, and possibly via other means (such as on satellite radio or licensed shortwave stations that offer commercial airtime), it is doubtful that any regular offshore pirates will broadcast from the coast of Europe in the near future. In order for offshore radio to be worth the effort and the risks of being anchored in the North Sea, it's probably necessary for independent radio to be more restricted than

Fig. 5-7 *Radio London's ship,* M.V. Galaxy, *being raised for the last time, after spending seven years at the bottom of the harbor.* John S. Platt

Fig. 5-8 *Another shot of the* M.V. Galaxy *after being raised in 1986. The ship was melted down soon afterwards and sadly, to the best of my knowledge, the logo was not saved by a radio museum.* John S. Platt

it is at the present. Currently, the licensed independent stations in England command a great enough listening audience that Radio Caroline hasn't taken any chances with hauling the *Ross Revenge* back out into the ocean.

American Offshore Pirates

Most unlicensed broadcasters love the notion of "piracy" as much as kids adore Captain Hook and Peter Pan. "Pirate ship" imagery brings a mythical, fairy-tale-like aspect to the free radio hobby. The nostalgia of the raiding ships from the early years of the New World gives identity to the radio hobbyists of today. Radio pirates are sometimes thought to be a modern version of the ship-sailing thieves, with transmitters and audio equipment substituted for swords and cannons, commercials for the treasure chest, and the FCC for the Navy.

Many announcers on unlicensed stations frequently exploit this connection in their programming. For example, WDX's main announcers called themselves Pirate Jack and Pirate Mike. A favorite KPRC segment was "Yo Ho Ho and a Bottle of Rum" played on a cart (a cart machine is a cartridge tape deck that is used for playing relatively brief audio bits at radio stations). Radio North Coast International claimed to pirate from the "good ship *Sphincter*" in the Great Lakes. Radio Clandestine often ran promos with pirates announcing their address, and dozens of stations have used the skull and crossbones in the designs of their QSL cards.

Even with this hype, most broadcasters who actually operate from offshore are usually associated with Europe; only a few existed from the United States. The first radio pirate to operate from international waters operated off the coast of California, in the early 1930s. just two others from the United States operated since the 1930s, and both of these only lasted for several days.

One reason that offshore broadcasting has never thrived–even survived–in North America is because most organizations that have the money to buy and equip an ocean-going ship would rather buy a legal station. The hazards of broadcasting from a vessel, plus the problem of finding a staff willing to spend weeks away from their families, quickly sober the would-be offshore pirate. Add to this the thousands of local commercial stations already broadcasting from the United States and Canada, and it is clear why offshore broadcasting is not feasible in North America. Only in a few unusual situations have international waters stations been successful.

RXKR: The First Offshore Pirate

RXKR is still one of the few stations in the United States (or the world) to have actually pirated the airwaves. Although many pirates absorb the symbols of piracy, RXKR was a true pirate in every sense of the word. As covered in excellent articles by Tom Kneitel in *Electronics Illustrated* and *Popular Communications*, the station operated from the *S. S. City of Panama*. Although both the station and the ship were licensed in Panama, neither followed the stipulations contained in the legal contracts. *The City of Panama* was supposed to be a floating showboat to display the glories of tourism in Panama to Californians. Instead, the owners of the ship left it as it was before the new registry: a floating speakeasy and casino.

Likewise, RXKR was a sham. The station was licensed on 815 kHz with 500 to 1000 watts under the RXKR callsign with experimental, noncommercial programming. Instead, it pumped out a beefy 5000 watts of popular music and commercials. RXKR actually tricked the Panamanian government into believing that the station would be used to promote tourism and industry within the country, but RXKR really wanted to turn a fast buck.

Even before pirating the airwaves, though, RXKR fell into legal troubles. Worries that the station would destroy the signals from other broadcasters across the United States plagued businessmen and the U. S. government alike. A demand was cabled to the Panamanian government from the State Department to cancel the registration of the *City of Panama* and the license of RXKR. The Panamanian consul to the United States became angry at the demand and claimed that the U. S. government had no right to interfere with vessels in international waters. In closing, the consul also noted that the operators would abide by the rules of the Federal Radio Commission.

Shortly after these incidents in May 1933, the station began its operations from off the southern California coast. The owners supported the popular music format with funds raised from commercial spots sold to advertisers in southern California. An office was opened in Los Angeles to receive advertising inquiries. In an article commemorating the fiftieth anniversary of RXKR by Tom Kneitel in

Popular Communications, it was reported that some companies signed up for as much as $1500 per month in advertising.

Companies obviously flocked to advertise on RXKR because of its high power (for that time) and clear channel frequency; they also worked with the station because there were so few other legal broadcasters in the West Coast area. High power and the clear channel frequency might have been beneficial to RXKR, but the other stations and radio listeners were angry. RXKR's signal from 815 kHz wiped out stations on 810 kHz and 820 kHz throughout the West. The pirate ship was heard as far away as the East Coast, Hawaii, and northeast Canada, with fair signals.

The affected stations and their listeners mailed complaints to the Federal Radio Commission, which was quickly swamped by mail. After realizing that the FRC had given up on the offshore pirate, some stations wrote directly to RXKR, requesting that the operation move from 815 kHz. They were horrified and angered when RXKR replied that it would change frequency for a payment amounting to thousands of dollars.

This Mafia-style "frequency protection racket" upset the State Department once again. They cabled the Panamanian government and requested that RXKR have its broadcasting and sailing licenses revoked immediately in June of 1933. Panama either realized that the United States was serious or discovered that their floating public relations tool was only a floating speakeasy and extortion instrument. Within a few days, the *City of Panama* officially lost its registry, and RXKR lost its license. Despite this, the station continued broadcasting from off the California coast.

Somehow, government officials from the United States and Panama must have been confused about the station, its illegality, and the fact that it no longer was licensed. RXKR remained on the air as if nothing had happened for several weeks; the two countries were discussing the steps required to dismantle the station. Finally, the State Department received the information it needed to remove RXKR, and in August 1933, the *City of Panama* was towed into a Los Angeles harbor. Neither the ship nor the station owners were associated with radio again.

RXKR never truly set any trends for broadcasting. No one attempted to use a similar site of operation until a quarter-century later when Radio Mercur began broadcasting from international waters. Due to the interval between the two operations, it is unlikely that RXKR directly influenced the Dutch pirate. Whether or not RXKR directly influenced other pirates is unimportant; the noteworthy characteristic of the station is that it tried new methods of broadcasting.

Fortunately, no pirates have ever tried to extort money from other stations since 1933. While RXKR operated, the U. S. government worried that if nothing was done a fleet of pirate ships would sail near the Atlantic and Pacific Coasts and jam legal stations. If the government remained defenseless against these pirates, the entire radio spectrum could have been held at will by criminals and scam artists. Some suggest that the radio spectrum was controlled by criminals and scam artists anyway, but the government's reaction in this instance is understandable; the measures they took against RXKR had a great impact on future broadcasting laws. In fact, international radio laws were written later that decade to prevent further attempts.

Radio Free America

Throughout the 1950s and 1960s, when offshore broadcasting became a popular and commercially viable trend in Europe, the North American coasts remained silent. After RXKR was silenced, international broadcasting laws effectively dampened the plans of hopefuls in the United States who wanted to imitate them. Besides, in the 1960s, the FM band was beginning to open up across North America. Some low power and college radio stations were already experimenting with alternative new music formats in a much less commercial style than the European pirates.

Politically, the broadcasting situation here was not especially ripe for offshore activity as it was in Europe. Nonetheless, an unexpected operation anchored off of Cape May, New Jersey, began sending signals in 1973. Active pirate broadcasters are generally in their teens or twenties, and in the early 1970s they were also characterized by liberal politics. But in a bit of historic irony, 1973 saw the debut of Radio Free America, a religious pirate operated by a paunchy 68-year-old right-wing radio preacher (Fig. 6-1).

In a huge departure from other pirates, Reverend Carl McIntire operated Radio Free America not as an instrument of progressive free radio or for self-satisfaction, but to return to the days when he owned a legal station, WXUR. WXUR broadcast under McIntire's ownership from 1965 to 1970. Problems cropped up when the FCC revoked the station's license under the Fairness Doctrine, which then required all broadcasters to allow those with opposing viewpoints equal time to express themselves. One program, "The Closet," was a telephone call-in show moderated by an announcer

Radio Free America

Radio Free America, the good ship "Columbus," symbolizes the current struggle for the free exercise of religion, free speech, and a free press in the United States of America. The ship, carrying a 10,000-watt radio transmitter, is stationed in international waters off the coast of the USA near Cape May, N. J. A historic "Manifesto of Freeom" was read on deck during its dedication, Labor Day, September 3, 1973. The ship is being supplied from the Christian Admiral, Cape May Bible Conference and Freedom Center. Christians are asked to pray for this Gospel ship that it may be used for Revival '76 and to deliver the radio and television stations of the nation from government repression and control. The ship has the backing of the 20th Century Reformation Hour, 756 Haddon Ave., Collingswood, N. J., and all gifts for its support may be marked Radio Free America. It may be heard on the AM band at 1160 kc.

Fig. 6-1 *Promotional literature from Radio Free America in 1973. The ACE*

accused of cutting off and insulting listeners who did not share his views. Since this was deemed a violation of the Fairness Doctrine, the FCC pulled the plug on WXUR.

McIntire caused controversies throughout his lifetime and did everything with grand gestures. In 1936, the United Presbyterian church publicly denounced his actions before its general assembly because the board felt that he was too conservative. He left the United Presbyterians to form his own Bible Presbyterian denomination. He preached to the 1200 members of his Collingswood, New Jersey, church from inside a tent until a new building could be constructed. Within several years, he helped establish the American Council of Churches and the International Council of Christian Churches. McIntire served as president of the latter, leading over 200 fundamentalist denominations from 73 countries around the world.

While amassing large numbers of followers, McIntire also built a huge empire of organizations and properties. Among his enterprises were a college, a hotel, a convention center, three apartment buildings, and 280 acres of undeveloped land near Cape Canaveral, Florida. He owned a college in Philadelphia; in Cape May, New Jersey, he possessed the four largest hotels. *The Christian Beacon*, a religious weekly newspaper with a circulation of 145,000 in 1973, was also owned by McIntire. Certainly, his enterprises greatly surpassed the resources of the average pirate station.

Controversies surrounding McIntire and his beliefs stemmed not from his conservative social or religious practices, but rather from his political rhetoric. Although he told the *Washington Post* in 1973 that "the body of man belongs to God, not to the state," his preaching featured a collage of politics and Christianity that were either praised or despised by listeners. Most controversial of all was McIntire's devotion to the destruction of any policy that even slightly resembled socialism or communism.

His anti-left sentiments might have won McIntire popularity throughout the conservative 1940s and 1950s, but the social revolution in the 1960s began to strip his power away. Despite the sometimes violent protests of Vietnam demonstrators, McIntire opposed the negotiated peace in that country. In fact, during 1970 he led several groups through Washington, DC, protesting the expected withdrawal from Vietnam. Later that year, McIntire even flew to Saigon and Paris to try to convince Nguyen Cao Ky, then the vice president of South Vietnam, to speak at the pro-war rally. Ky declined the invitation.

WXUR was only the flagship for the 610 stations that carried

McIntire's "Twentieth Century Reformation Hour" program. Many of these legal stations suffered a confrontation with the FCC because of the program and subsequently dropped it to protect their own licenses. This caused McIntire to lose money. Funding for the broadcasting empire came from listeners who (by McIntire's estimate) donated $100,000 per year, so the FCC's license denial came as a severe blow to the organization.

The monetary problems continued to plaque McIntire's communications kingdom. His Faith Theological Seminary was mortgaged in 1965 for $425,000 to pay for WXUR. By 1973, the Seminary found the payments so difficult to cover that his newspaper, *The Christian Beacon,* urged readers to make the Seminary a beneficiary in their wills. With his empire suddenly crashing down around him, McIntire was forced to take drastic measures to prevent a total collapse.

His drastic measures, of course, meant entering the obscure world of pirate radio, still a barely recognized, heard, or organized territory in 1973. And McIntire's revenge did cast the eyes of the public in his direction for a brief time. In September, *The Columbus,* a rusty World War II minesweeper, was anchored off the coast of Cape May, New Jersey, in international waters.

On board was a 10,000-watt AM transmitter, a crew for navigational and broadcasting purposes, and a cache of rifles to protect against potential raiders of the station.

Radio Free America touted itself as revolutionary just before firing up the transmitter, but no violence erupted despite the FCC threats to close the station and McIntire's threats to use his rifles. No boisterous incidents were caused by McIntire's presence in international waters, but Radio Free America did pound 10,000 watts of conservative religious and political programming across the Eastern Seaboard for a very short time. The already interference-heavy frequency (McIntire used 1160 kHz in the middle of the AM band) confined the signal to the East Coast of North America, but it was heard by many nonetheless (Fig. 6-1).

Broadcasting was hampered during the first day of operations on September 19, 1973, when the powerful, heat-emitting transmitter caught the deck of the aging minesweeper on fire. Output power from the transmitter was then lowered considerably until Radio Free America could barely be heard outside of New Jersey. After ten hours of broadcasting, the station closed down for the evening and never returned to the air. The excursion evidently discouraged McIntire, whose empire crumbled within a few years. *The Columbus* was sold soon after the incident.

Federal agents were prepared to get a court order that would force McIntire off the air. McIntire, in turn, pledged to stay on the air and, with rifles or lawyers, defend his right to broadcast. But Radio Free America's own equipment and vessel saw to it that the offshore broadcasts subsided. The conflict never came, and whether or not offshore broadcasting is technically legal under the United States' laws was never determined by the Supreme Court at that time. No FCC action was taken against McIntire, but it would have been interesting if Radio Free America would have remained on the air for a few more days.

Radio Newyork International
Round 1

Continuing its policy (that it is illegal to broadcast from off the coast of the United States without a license), the FCC effectively discouraged everyone considering this type of operation for years after the demise of Radio Free America. The federal agents said they would close down any station's attempts, fine and possibly prosecute the operators, and confiscate the equipment. No one had the nerve to see if the FCC was bluffing. Besides, court costs alone in a case such as this would rapidly diminish the funds of anyone but a large company.

In the mid-1980s, a group of loyal radio enthusiasts decided that the FCC was bluffing. Even if the government was not, the enthusiasts felt it was their right to have freedom of speech. Anything less than the ownership of a radio station was considered by the enthusiasts an infringement of rights because it was their desire to operate one, and furthermore, all had experience on legal broadcast stations.

Radio Newyork International (RNI) was created as a reaction to what was considered a stagnant state of rock-and-roll radio in New York City. A group of 20 radio enthusiasts shared the new station. Despite working for a legal license for 16 years, none of the group could pull the millions of dollars required for a station in the New York market. So the group found a loophole in the system: only stations based in the United States' territory are required to be under the control of the FCC (Fig. 6-2).

With this in mind, the broadcast enthusiasts bought a 160-foot Japanese fishing vessel (rechristened *The Sarah*), bought and modified FM, AM, shortwave, and longwave transmitters, and registered *The Sarah* under the Honduran flag. After one-and-a-half years of fitting the ship with broadcasting equipment, they took it to sea and

R.N.I.
496 LaGuardia Place
Suite 451
New York, N.Y. 10012

RADIO
NEWYORK
INTERNATIONAL

Andrew Yoder,
This is to confirm your reception of RNI from the MV. SARAH on July 28, 1987 at 0211-0339 utc and Oct 15, 1988 at 0259-0418 utc.
Sorry about no QSL til new. Some former RNI personal didn't do their jobs! Thanks for your reports
Randi Steele

Fig. 6-2 *A QSL from the shipboard broadcasts of Radio Newyork International in the 1980s.*

anchored four-and-a-half miles off the coast of Long Island, New York, in international waters. On July 23, 1987, RNI began broadcasting rock music, comedy, and social programs to listeners as far away as Oregon and Europe.

The operators planned to begin commercial broadcasting on August 1; they stated the format would not be playlisted to keep from becoming "stale." Also, 10% of all advertising proceeds were to be donated to social programs that help the homeless in New York City. Despite receiving hundreds of favorable letters from listeners, the station did not please the FCC.

On July 27, the FCC visited *The Sarah* and warned the crew of the "illegal nature" of the broadcasts. The next day, the U.S. Coast Guard and the FCC raided the station. They hauled Allan Weiner (owner), Ivan Rothstein (station manager), and R. J. Smith (a reporter from the *Village Voice* newspaper) away in handcuffs. Then the Coast Guard and the FCC tore cabinets of equipment out from the control boards and dumped them in the center of the floor. Weiner and Rothstein were later charged with conspiracy to impede the Federal Communications Commission (a felony) and operating a broadcast station off the shore of the United States (a misdemeanor). Each faced a maximum prison term of five years and $250,000 in fines from the felony charges.

However, on August 28, 1987, federal authorities dropped all charges against RNI, Weiner and Rothstein, as long as they promised to never broadcast illegally again. Andrew J. Maloney, the U. S. Attorney for the Eastern District of New York, said, in the *New York Times*, that "no further government purpose could be served by pursuing criminal charges. By shutting down the illegal station, the FCC achieved what it set out to accomplish."

If the story had ended quietly, maybe little controversy would have arisen on the subject. But immediately following the court case, Rothstein and Weiner vowed to return to the airwaves. In fact, the pirates described in depth the manner in which they would return to the air during an interview on WNBC radio in New York City and in an article written by Weiner in the April 1988 issue of *Popular Communications*.

The FCC and other government officials remain firm on their position. "If they are dumb enough to resume broadcasting," said Assistant U. S. Attorney Matthew Fishbein to the *Washington Post*, "then we would prosecute—not only for future violations, but for past violations. We have clearly established that what they have done is clearly against the law. If they resume broadcasting at this point, we have given them every opportunity and will prosecute."

Rothstein and Weiner claim that their equipment was vandalized. Bill Martin, lawyer and publisher of *The ACE* at that time, called the FCC's media representative on August 28, 1987, the date that the charges were dropped against the operators. Martin described his call in the September 1987 *ACE* bulletin.

"When I asked about allegations that the FCC engineers had damaged equipment and cut cables at RNI, the FCC representative advised me that the station did not continue its unlicensed broadcasts." If these statements regarding the legality of vandalism and intentional destruction of another's property in the name of free airwaves are correct, the government may have exceeded its authority to regulate radio.

In the original raid of RNI, R. J. Smith, a reporter for the *Village Voice*, was also arrested. Martin Gottlieb, editor of the *Village Voice*, said that Smith was handcuffed and prevented from producing his press credentials. Smith was held for seven hours; the FCC confiscated his camera and other personal belongings. Apparently, the Coast Guard would not listen to Smith's claims. Regardless, it was not an impressive initial presentation of FCC policies on offshore broadcasting to the media.

It is possible that FCC officials took their personal revenge on RNI's crew for past experiences. Nearly every member (maybe all) of the crew had broadcasting experience on illegal land-based radio stations in the past from New York City. The earliest broadcasts came from Weiner and Ferraro's affiliated stations WKOV, WXMN, WFSR, and

WSEX in 1971. Other illegal stations with probable connections to RNI included KPRC, KPF-941, WGOR, WGUT, WHOT, Stereo Nine, KSUN, and others that transmitted on AM, shortwave, and/or FM.

Round 2

It was clear that the staff intended to put RNI back on the air at all costs. In fact, the station proceeded to anchor off the coast of Long Island, New York, and wait for permission to be on the air. The radio enthusiasts' desire to return to broadcasting, combined with the FCC's harsh threats to anyone that dared to attempt offshore broadcasting, formed a volatile situation.

"When push came to shove and the day in court came," said J. P. Ferraro, they dropped the charges because they couldn't make them stick. The government had no jurisdiction—even though they said they did." Fueled with this fire of enthusiasm and social activism, the RNI crew refitted *The Sarah* in Boston's harbor in late 1987 and early 1988.

While the ship was being prepared for operation, the RNI staff whipped the mass media into a frenzy. During this time, various RNI staff members were featured on MTV, others participated in guest broadcasts and interviews on local commercial radio stations, and *Rolling Stone* magazine even voted RNI as the best radio station of 1987. This all followed the large-scale media attention that was given while RNI was on the air in the Summer of 1987, when the station was featured on the front page of the *New York Times*, *Time*, and *Newsweek*.

Just over a year after the first broadcasts of 1987, RNI returned to the airwaves on October 15, 1988, on 1620 kHz. The difference was tremendous. In 1987, RNI operated around the clock on 1620 kHz, 6240 kHz, and 103.1 MHz FM with slick jingles and a full staff of professional radio announcers. When the station returned in October 1988, only the 1620 kHz transmitter was running, the operations were limited to a maximum of three hours, no jingles or prerecorded announcements were used, only one announcer (who was a much less than professional air personality) spoke, and the programming was riddled with technical difficulties, errors, and glitches. The resulting sound was less entertaining and less professional than the typical rookie pirate station. The FCC raided the station again on October 17, 1988.

As a result of the poor quality radio programming and the lack of consistent air time, Radio Newyork International was silenced with no newspaper or magazine reports, awards, or guest appearances. The publicity from 1987 helped RNI to be able to fight onward into 1988, but the poor planning of the second RNI installment killed the project.

Between the two rounds of RNI vs. the FCC, several radio crew members became disillusioned with the project and dropped out. Several of those were key members from the former New York station WHOT, which was one of the most professional-sounding pirates ever to operate in North America. Although the lack of the WHOT announcers put a dent in the station programming, the remaining air personalities could have produced entertaining programs. Why they didn't is a mystery.

Round 3

Less than two months after the last RNI bust in mid-October 1988, a new shortwave pirate appeared on the frequency of 6240 kHz with the name of Allan Weiner's late 1960s pioneer pirate station, Falling Star Radio (Fig. 6-3). To add more to that connection, Falling Star Radio often relayed live sociopolitical programming from KPRC.

Like RNI, the transmitter of Falling Star Radio packed a wallop and was heard with *extremely* good signals across much of North America and the broadcasts often lasted several hours. One change from the RNI programming was that the an-

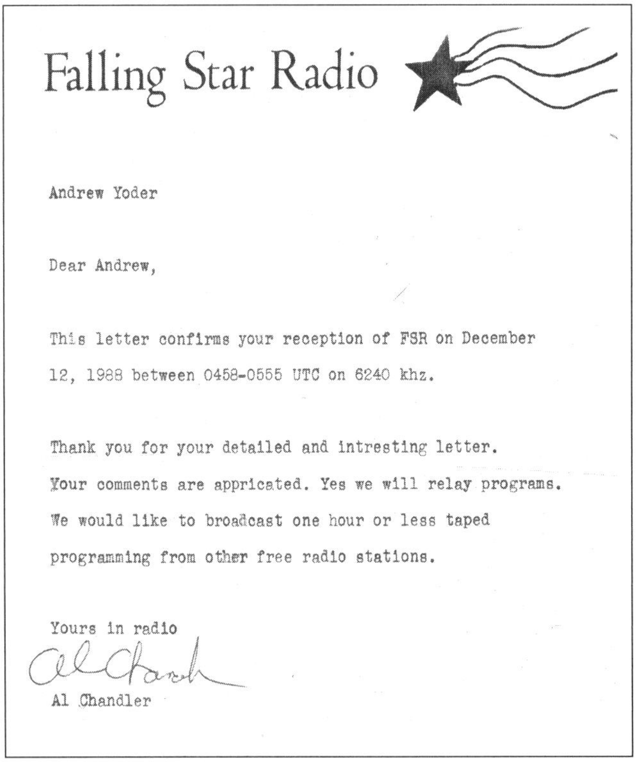

Falling Star Radio

Andrew Yoder

Dear Andrew,

This letter confirms your reception of FSR on December 12, 1988 between 0458-0555 UTC on 6240 khz.

Thank you for your detailed and intresting letter. Your comments are appricated. Yes we will relay programs. We would like to broadcast one hour or less taped programming from other free radio stations.

Yours in radio

Al Chandler

Fig. 7-3 *A QSL from Falling Star Radio, which operated in the brief time between the last ship broadcasts from RNI and the first syndicated programs from RNI on WWCR.*

nouncer regularly requested a $5 donation to be mailed in with reception reports for QSLs. Later, the station informed the listeners that although the donations were desired, they were not necessary for the QSL sheet.

In February 1989, Falling Star Radio disappeared unexpectedly from the air, even though the station was requesting cassette programs from other stations for relay. After the station retired from broadcasting, it was rumored that these broadcasts were emanating from the old 1000-watt RNI shortwave transmitter installed in southeastern Canada. However, these rumors have not been verified by anyone.

Round 4

After an official two-year silence, Radio Newyork International returned yet again in September 1990. This time, the crew seemed resigned to the fact that broadcasting from offshore the United States would be impossible. During this time, Weiner, Ferraro, and newcomers Dan Lewis, Johnny Lightning, Steve Cole, and Julie Weiner made an unprecedented move; they began broadcasting on Sundays from 9:00 P.M. to midnight Eastern time via the licensed transmitters of

Fig. 6-4 *A color photocopied QSL showing* The Sarah, *from the days of RNI's syndication on WWCR, Nashville.*

WWCR in Nashville, Tennessee, on 7435 kHz (Fig. 6-4).

The RNI programming was divided into lighter programming early in the evening with Dan Lewis on the mailbag show, Johnny Lightning with oldies and occasional humor, and "Big Steve" Cole with "Crossband RNI," the RNI DX program. The rest of the evening was filled with heavier programming. Pirate Joe (and occasionally Julie Weiner) discussed a variety of social and political topics and received telephone calls via a toll-free number.

The early programming tended to be fast and loose with a talk-based free speech emphasis. The programming from Pirate Joe was serious and liberal from a pro-Democrat platform. In fact, the station even took a hardline "Don't Vote Republican" stance during the 1992 Presidential election. These political slants drew flak from a number of critics and split the pirate listening ranks. In addition, because of the difference in programming styles on the station, RNI began to fall apart on and off the air.

First, Dan Lewis participated in an on-air argument with some of the other members of the station. On one of Lewis's last broadcasts, he spent his airshift complaining about how his cow lawn-ornament had been stolen. Steve Cole evidently became disillusioned with RNI and left the station, taking Crossband elsewhere. Julie Weiner also disappeared after the early days of RNI; in one of her last airshifts she spent approximately 15 minutes attempting to make one of the other announcers apologize for a comment that he made while he was on the air earlier.

By the end, Johnny Lightning, Pirate Joe, and Allan Weiner were the surviving members of the air staff, with Weiner contributing infrequently to the broadcasts. At this time, it appeared to listeners that the station members spent less and less time preparing the programs each week. The funds to keep the station on the air were running short and the staff had to spend more and more time asking listeners for contributions, which allowed them less time to broadcast the programming that they were asking people to support. By 1993, RNI could no longer pay its bills to WWCR, and the former offshore broadcaster fell silent once again.

Although the preceding paragraphs in this section cover the problems that plagued and eventually destroyed RNI during its WWCR days, the station did have a number of positive attributes. Unlike many call-in shows, the RNI call-ins were laid back, and they were often thought-provoking, unlike many of the rude, shock-oriented talk show hosts on commercial AM stations. The hobby radio news, letterbag programs, and call-ins supported the shortwave-listening hobby.

Unfortunately, the station appears to have been poorly planned. No matter whose fault it was, many people seemed to have had greatly differing opinions about what RNI was to be. Also, some of the former staff members acted less than professionally by fighting on the air and later by venting their complaints and frustrations over computer on-line services, such as the Internet. Although this fighting makes for a good story, it did not make for a long-lasting broadcast station.

Voyager Broadcast Services
Round 5

By early 1994, it almost seemed as if the RNI project had died. Nothing had been heard from the station in nearly a year and the rusted, listing *Sarah*, the vehicle of the original offshore RNI broadcasts from 1987 and 1988, was sold to MGM for the final explosive scene in the movie *Blown Away*, starring Jeff Bridges and Tommy Lee Jones. As was stated in the article "An Old Pirate is Blown Away," by Christopher Jones in *Monitoring Times*, *The Sarah* had one final kick. "The ensuing four explosions at five-second intervals sent a fireball billowing over three hundred feet into the air, dwarfing the Boston skyline. . . . Hundreds of windows were blown out by the explosions and further inspections revealed cracked walls and foundations, as well as a few collapsed ceilings."

After *The Sarah* was blown up, the wreck was bought by another company that was planning to sell it either to a Central American company or to a scrap metal yard. Although *The Sarah* is no longer a symbol of American pirate radio, it is immortalized in *Blown Away*.

As the charred, rusted hull of *The Sarah* was being dismantled in the Charleston Naval Yard, a new broadcasting vessel was being fitted for operations. Unlike *The Sarah*, which had always looked as though it had been abandoned for several decades, the new ship, the *Fury*, was a sharp, well-maintained 140-foot fishing trawler.

In late 1993, *The Fury* (Fig. 6-5) was being repainted, and huge generators and transmitters were being installed. It was rumored that two 40,000-watt and two 10,000-watt transmitters would begin broadcasting from the Caribbean country of Belize in the upcoming months.

But the oddity of the Voyager Broadcast Services was not the well-kept ship or the number of transmitters; it was the combination of the forces that created the group. VBS was not RNI reborn; instead

Fig. 6-5 *The impressive* Fury, *bearing the insignia of Becker Broadcast Systems, anchored in the harbor.* Martin Van Der Ven

it was a peculiar coalition between the Overcomer Ministries and "other investors, engineered by none other than Allan Weiner. According to *Monitoring Times*, three of the transmitters would be leased and the fourth would be operated by R. G. "Brother" Stair of the Overcomer Ministries.

This coalition seemed unlikely because the Overcomer Ministry is a conservative Christian fringe group and RNI was a liberal, often antireligious broadcaster. Although RNI was never implicated with Voyager Broadcast Services, VBS investor Scott Becker had provided satellite links for RNI while it was on WWCR. Tensions were occasionally evident, such as when Becker told the *Charleston Post and Courier,* "The only way I can describe this group is bizarre. . . . It gives me the creeps. . . . They walk around like zombies. He's a David Koresh waiting to happen." Likewise, Stair was critical of the swearing from and long hair of the VBS crew on his regular syndicated AM and shortwave programs.

Stair has received quite a bit of attention over the past decade as a result of his prophesies and his simple, but radio-based, communal lifestyle. In 1978, he purchased an old motel in South Carolina and encouraged people from around the United States to sell their possessions, donate their money to the Overcomer Ministries, and move into the center. Otherwise, except for a Christian Science-type belief that doctors should not interfere with people's lives and that the world will soon end, the Overcomer Ministries function somewhat similarly to a typical conservative Christian group.

After purchasing the motel, Stair has received much attention for predicting the end of the world several times. Because of some missed predictions and quite a bit of negative publicity, Stair now refuses to grant interviews to reporters, so little direct information (opinions, beliefs, etc.) about Voyager Broadcast Services is available.

As *The Fury* was nearing completion, the ship was raided by FCC agents, gun-toting U. S. Federal marshals, and the U. S. Coast Guard in Charleston harbor. The charge? Unlicensed broadcasting. The next day, FCC-hired electricians removed the large transmitters from *The Fury* with cutting torches. Other equipment, including tubes, spare parts, test equipment, etc., was also taken.

According to the FCC, agents from the Commission traced unlicensed signals on 7415 kHz to *The Fury*. These broadcasts consisted of test tones and Radio Newyork International programs from Johnny Lightning. Indeed, these programs, complete with identifications stating that the signal came from international waters, were widely heard in December 1993. Some reception reporters, such as Kirk Trummel in Missouri, received the RNI programs with tremendous signal strength and speculated as to why they would be foolish enough to broadcast from their own ship while still docked in the United States. Others, such as myself, received the signal very poorly and wondered if some pirate station was rebroadcasting the programs from another location.

According to the FCC, the transmissions were being emitted from a small amateur transceiver that was on board the ship. However, Allan Weiner has contended through various press releases that no transmissions originated from the vessel and that the transmitters were not yet even capable of broadcasting. If that was the case, why had those mystery RNI transmissions been specifically recorded for the Caribbean broadcasting from *The Fury*? Surely those tapes hadn't fallen into the hands of others.

Even if those tapes were rebroadcast by other people, why did the Voyager Broadcast Services team announce their plans in advance? They knew as well as anyone else in the world that the FCC rabidly opposed offshore broadcasting and that their vessel would be a lame duck while it was docked in U. S. territory. The fact that Stair was announcing the broadcasting plans regularly over his AM and shortwave outlets doomed the project from the beginning.

Everyone who keeps abreast with pirate radio in the United

States knows that the FCC is willing to play "hardball" when challenged. And they did. Although the four broadcast transmitters were not capable of emitting the unlicensed signals that were claimed to be from *The Fury* (the broadcasts that the listeners heard used single sideband, SSB, modulation while the large broadcast transmitters on board were only capable of transmitting amplitude modulation, AM), the FCC agents had these transmitters "arrested" anyway. Not only were these rigs taken, but they were cut out, causing considerable damage to the ship. In effect, the FCC agents caused tens of thousands of dollars of vandalism for an administrative violation. Considering the penalties of the day, VBS should have received a Notice of Apparent Liability for anywhere from $1000 to $20,000. Instead, the FCC destroyed a large amount of private property without a trial, knowing that the VBS didn't have the resources to fight the agency and win back the equipment.

The FCC's press release, which provided the background information on the bust, even failed to correlate with their actions. The release stated "The seizure was accomplished pursuant to 47 U.S.C. 5 10 which authorizes a civil seizure of radio station equipment used to violate 47 U.S.C. 301, which, in turn, prohibits the operation of a radio station without a license within the jurisdiction of the United States." However, the costly blow from the FCC was that they took equipment that was not used to violate any FCC rules; this equipment is clearly not covered in this ruling. Also, the release states, "When the group made known its plans to resume broadcasting from the high seas without a license, the U. S. Attorney's Office in Boston and the FCC sought the above injunction from the U. S. District Court." Actually, *The Fury* was to, as Weiner put it, ". . . broadcast from a safe anchorage in Belize territory," which inferred that VBS was to be approved or licensed by the government telecommunications authority in Belize.

After this incident, Voyager Broadcast Services died and there were no apparent plans for any of the participants to attempt offshore broadcasting. Stair has continued with his AM broadcast and shortwave ministry, although it was rumored that he might purchase the former Voice of America transmitter site in Belize. Weiner then disappeared from the radio scene for several years to, as we would later discover, work on more shortwave radio projects.

Lawrence Clance, the FCC's bureau chief for law, said in an article in the *Charleston Post and Courier* that the FCC considered criminal prosecution, "but it was concerned that it would make Weiner a martyr." However, the FCC had no leads on who made the broadcasts, so it seems more likely that the FCC decided against criminal prosecution because they realized that they couldn't make the charges stick. As Weiner said, "Why was an entirely legal radio station destroyed without a hearing and due process?" If the FCC didn't want Weiner to become a martyr for free speech, they should have licensed him to operate a small, remote radio station; by banning him from ever broadcasting, they were merely asking him to try again.

Round 6

Unlike the prior broadcasting attempts, very little information leaked out about the mini-broadcaster, the *MV Electra* (Fig. 6-6). According to Dr. Martin van der Ven, *The Electra* was an 88-foot ocean-going tug boat. It was readied for operation in 1998 and 1999, with an intended anchorage somewhere in the Carribbean. In December 1999, Glenn Hauser announced on "World of Radio" that *The Electra* had sunk off the coast of Rhode Island.

Fig. 6-6 *The little, ocean tugboat* Electra, *in better days, showing off the base of its extensive transmitting tower.* Martin Van Der Ven

WBCQ

And try again, he did. In 1998, the near-impossible occurred; the FCC granted a shortwave broadcasting license to Allan Weiner. Was someone asleep at the wheel or did the FCC administration finally decide that they could rid themselves of numerous future headaches? Regardless, The Dream became reality and WBCQ ("The Planet") began broadcasting with 50,000 watts from upstate Maine in late 1998. Perhaps as a tribute to the pirate roots of WBCQ, the station broadcasted on the long-time pirate frequency of 7415 kHz, which had been abandoned years earlier because of interference from the Voice of America (Botswana relay) on 7415 kHz and WEWN on 7425 kHz.

WBCQ is of particular interest to pirate listeners because numerous pirate programs, such as W.D.C.D. and the Voice of Juliet, can also be heard via the powerful transmitters. In addition, some people who would otherwise be encouraged to take less-than-legal avenues to air their programming now appear on WBCQ, thanks to the strong signals, low airtime costs, and pursuit of alternative programming. Thanks to the success of the station, WBCQ has also installed new transmitters and antennas, so the station's total airtime has increased significantly in the past year.

Round 7

The 2001 SWL Winterfest in Kulpsville, Pennsylvania, is the annual Mecca for a few hundred hardcore shortwave listeners. Weiner, who had typically avoided these types of radio gatherings in the past, attended, apparently primarily as a PR piece for WBCQ. Some of the group that also attended included Scott Becker and WBCQ engineer (and longtime AM ham) Tim Smith (WA1HLR) (Fig. 6-7).

Of course, the Winterfest organizers made the most of having an independent shortwave broadcaster (and hobbyist) in attendance, and Weiner spoke at several forums (staff members from the Voice of America, China Radio International, HCJB, and YLE

Fig. 6-7 *WBCQ chief engineer Tim Smith modifying the original 50-kW station transmitter.*

Finland also spoke at various forums).

As fascinating as it might be to hear about the difficulties of shortwave broadcasting and stories from the past, the truly shocking news was not WBCQ related (well, not entirely). Allan Weiner announced at a forum that another broadcasting ship was in the works—and this one was intended to operate from somewhere in South or Central America. Adding validity to the words, a final deal for powerful diesel generators was completed later that evening.

In late Spring 2001, Allan Weiner updated his initial statements with more specifics about the project. At this point, the ship project is on hold and has tentatively been scrapped for true offshore broadcasting. Originally, one of the WBCQ airtime purchasers wanted to broadcast from the ship—and it would have been outfitted and operated by that company. Weiner's $20,000 to $25,000 estimate to totally outfit the ship, although costly for many individuals, was quite reasonable for such ship work. However, when it was time to begin work on the ship, the company ignored requests for the money to commence work.

Months later, the motor sailor *Katie* remains anchored in Boston Harbor, and despite the lack of a backer, offshore broadcasting is still in the future plans for the ship. Weiner has no plans to broadcast his own programming because he has everything he needs with WBCQ. So, the new plan is to make legal remote broadcasts from the *Katie* as a high-profile promotion for WBCQ and for shortwave radio in general. "We're not going to do it to make money; it's just for fun to promote shortwave and get more people listening," said Weiner. After *The Katie* has been readied for broadcasting, it can be sailed to different ports, where it can publicly produce programming that will be connected to the WBCQ transmitters via a studio-to-transmitter (STL) link. In this way, it will be operating legally from a ship, in much that way that Radio Caroline currently broadcasts.

But Weiner cautioned that a remote broadcasting vessel is far from the top priority of WBCQ. "Next we need to pour a concrete slab for another antenna and then we need to modify another transmitter [for the land-based WBCQ]," said Weiner. "It will happen, but I'm not sure when."

The Future Of U.S. Offshore Broadcasting

Offshore broadcasting in the United States seems to be an impossibility at the present time. Apparently, if it is to resume, it will be in another, smaller country, such as one in Central America, where the radio laws are less restrictive. Then again, why should a broadcasting group sail a broadcasting vessel to a Central American country when they could just as easily be granted permission to operate from within that country? A case in point would be Radio for Peace International, which is an American-backed, licensed station that operates from Costa Rica.

Another example is Radio Copan International, which was operated by Jeff White from Honduras until he received his FCC license to operate from Florida with 50 kW on shortwave. And WBCQ has actively sought programming from current and former pirates, some of which have switched exclusively to legal syndicated broadcasting. Legal avenues for broadcasting, such as this, will put an end to offshore broadcasting much quicker than heavy FCC enforcement. As it stands at the present, the combination of access to commercial shortwave airtime, the heavy FCC enforcement, and the extensive costs to construct and maintain an offshore radio station will prevent the phenomena from occurring again in North America in the near future.

International Pirate Radio

My 1986 Volkswagen Golf is parked at the edge of a baseball field a little more than a mile away from home. It's 2:15 A.M. on a Saturday night and I'm shivering in the back seat while curled up in a sleeping bag. Nearly a foot of snow has already blanketed the field and even though my breath has frozen across the insides of the windows, I can still see the shadows of more flakes landing on the rear hatch. But the shadowy precipitation and snowy isolation are only atmosphere; my attention is focused on the electronics and notebook in front of me.

Placed on the rear deck of the Golf is an old cassette deck and a well-worn Kenwood R-5000 communications receiver. It has enough knobs and buttons on the front panel to keep even the most savvy remote-control expert couch potato busy for hours. I keep my hands in the sleeping bag, under my legs to keep my fingers from going numb, only pulling my hands out to spin the large tuning knob, jot a few details in the notebook, or change a cassette. The tape deck "pops," alerting me that the auto stop has shut off the cassette that was recording. I dig another out of the bag, pop it in the deck and take a spin through the band. I quickly pass through the loud two-way communications that are probably just the shipboard radio operators contacting the shore. I've been out long enough to turn down the volume because the digital station around 6238 kHz would soon make me deaf before I'm 35.

Sure enough, it's there and it's loud. I pass it and turn up the volume and begin looking for my friend, Peter Verbruggen, who said that he'd be on tonight. Maybe "friend" is stretching it a little bit. I've never met Peter, in part, because he's from The Netherlands, and although I would love to travel through Europe and meet everyone that I've written to over the years, I never have either the time or money to make such plans. I first made my aquaintance with Peter about 11 years ago when I started writing a radio column for his newsletter, *FRS Goes DX*. I wrote for the newsletter about five years until the workload became too much for Peter and the rest of the gang. Since that time, I have occasionally exchanged friendly e-mails with Peter, but I can't say that we've ever eaten bad Mexican food, shot basketball, or changed a tire together.

The band is almost dead until I get up to 6289, where I hear several signals mixing together. As I tune across, I notice some audio on one, although I'm having a tough time tuning anything in. Before you can say, "You've got a friend in Pennsylvania," I've figured out which signal has the audio and that the station is rapidly drifting upward in frequency. I start tracking it and, lucky for me, the signal is strong enough that I can pick out "Free Radio Service Holland" from one prerecorded promo ID. Bingo! Peter plays a few songs and mentions that he hopes some listeners in North America are tuning in. At 2:28 A.M., he announces that FRS Holland is shutting down for the morning and he wishes everyone good morning before abruptly turning off the transmitter. In addition to taking notes in my logbook, I've also captured the past 15 minutes on cassette. After hoping to hear the irregular FRS Holland for more than 10 years, finally tuning in was so easy that it was almost anti-climatic (Fig. 7-1).

With my main goal realized, I pull out some paperwork from my workplace so that I can at least sort of justify skipping half a night's sleep to freeze in a car and risk turning my runny nose and deep chest cough into pneumonia. Of course, this also makes me

Fig. 7-1 *A QSL card from Peter Verbruggen's Free Radio Service Holland.*

wonder if the snowfall will bury the car deep enough to stick the Golf at the baseball field until the next thaw. Let's see, I'd be desperately ill, terribly inconvenienced, and somewhat embarassed. So, I pull out my digital camera, so I have something to remember my experience at the field (Fig. 7-2).

Fig. 7-2 *The cassette deck and Kenwood R-5000 receiver sitting on the rear deck of the VW Golf. Notice the inches of snow packed on the rear window.*

Time to spin through the dial and pop in another cassette. I note another carrier squeal around 6300. It goes dead a little above 6301, and the audio pitch is about right at 6301.5 kHz. This one is weaker than FRS Holland and although I can pick out a few details, I can't log the name of the station. Too bad, but at least I've got it on tape.

By 3:30 A.M., I can't find any more stations, so I decide to pack it up before my little car is buried like a stegasaurus fossil in the bottom of an ancient lake bed. Overall, it was a great night—five new stations. The one station that I had already heard a few weeks earlier was coming in so strong this morning that I could easily understand every word. Because the signal was so strong, I included it on the CD that is included with this book.

I hopped out of the car to discover that the snow was still dumping and the Golf's deep tracks were almost buried and erased like an Etch-O-Sketch picture in a coin-operated massage bed. I tried to get a photo or two of the car in the snow with the digital camera, but it wasn't turning on (afflicted by the cold?). So, I threw the camera in and started winding up the 100-foot electrical cord and the 500+ foot antenna before my hands could go totally numb. Too late. Everything went numb. Somehow I managed to wrap the wires, despite the big masses of largely unfeeling tissue at the ends of my wrists.

I threw everything into the car and started it as quick as I could. My bed called and I was all ears. You probably think that I got stuck and had to walk home. Nothing quite so dramatic happened. I pulled right out through the deep snow in the field, drove home on the snowpacked, unplowed road and even hopped in bed without waking my wife. As much as I'd like to think that it was skill, I guess I should probably just attribute everything to dumb luck.

But before I hit the sack, I zipped off some e-mails to stations and listeners in Europe, both to answer some questions about what I heard and to get the information out so fast that everyone would know that my reports were real and not detail-skimmed from listener reports in Europe.

At this point, most everyone that I know seems to think that I'm either crazy or stupid. My mother-in-law tells me I'm crazy, my mom says that I need my sleep and should take care of myself or I'll get sick, my wife knows better than to argue with me, and my kids look at me strange. Even other DXers tell me that they wouldn't get up at 1 A.M. to go out in a snowstorm.

Most people seem more interested in the end-results of the listening than in the actual activity of listening attentively through the static for IDs and other pertinent programming information. My nearby co-workers seem to appreciate when I receive detailed e-mails from people in Europe, although those in broken English go over best. My QSL albums garner wide interest at radio conventions and other functions where I set them out. And even my mother-in-law was interested by the postcards from Holland and the nice personal letters.

But, is this really European pirate radio? No, I can't say that it is. It certainly is one aspect of listening to Europirates, but it is clouded by a sort of tunnel vision. All of the work that is required for me to just catch a few bits of audio causes my focus to be on my preparation and effort to hear the stations, rather than on the signals and programming from the pirates who are broadcasting.

Land-Based Stations

Just like in the United States, pirate radio stations in Europe and across the world seem to date back to some of the earliest days of radio broadcasting. However, because of the turmoil and economic depressions that occurred throughout the Western world from 1914 to 1945, these countries didn't have the sheer numbers of pirates as the United States or Canada. Or, as in the Middle East and many parts of Asia, because of the political oppression, there any unlicensed broadcasting is focused on the government (clandestine radio).

Although there were some exceptions, such as Sweden's Radio Black Peter in the 1950s, the organized "pirate radio scene" didn't start in Europe until after the offshore stations disappeared with the advent of anti-pirate laws between the late 1950s and early 1970s. The radio hobbyists who were fans of the offshore stations began constructing land-based operations. The new pirates were usually radio and electronics experimenters who built their own transmitting equipment from spare and surplus parts and contacted aspiring announcers to record programs for the station. Other stations featured a crew of announcers and technical engineers working together to produce popular programming. Generally, these stations survived for a considerable time period. Unlike today's pirates, which broadcast for radio hobbyists and the counterculture of people who are disenchanted with commercial broadcasting, the 1960s and 1970s, land-based pirates were targetting the millions of listeners who were left without many of their favorite stations in August 1967.

The land-based pirates were essentially a small-scale migration from the ocean. Most of the new broadcasters followed the offshore example by imitating United States top-40 radio. But unlike the offshore stations, the land pirates operated only a few hours per week to avoid detection by the equivalent of the FCC in many European countries. Many land pirates operated in shortwave bands, which were apparently not as thoroughly policed in European countries as the AM and FM bands. And because less power was needed to cover a wide area on shortwave, AM and FM land pirates were rare in Europe.

American and European free radio developed quite differently. The original commercial offshore pirates became the base from which the European free radio movement grew. North American pirates, on the other hand, grew from a series of hoaxes, shortwave listening, and alternative cultures. Most American pirates exist for the entertainment and enjoyment of the operators, with a minor effort directed toward community service (with the exception of some of the post-1990 FM "microbroadcasters," such as Free Radio Berkeley). As a result, American pirate radio is filled with creative but erratic and sometimes irresponsible stations. In contrast, most European free radio stations operate with a regular schedule and attempt to please their listeners.

One of the reasons for this desire to please is that popular pirate radio personalities from the 1950s and 1960s were commonly offered jobs with legal government stations. This desire has mostly dissipated today and very few stations are broadcasting with the hope to be legalized. Because of the legalization of some independent stations in England a decade ago, some pirates had hoped that by broadcasting on a regular schedule with a professional sound, they would eventually be granted a license. However, after the closures of Radio Fax, Radio Merlin International, and the Belgian International Relay Service in the early 1990s, few British pirates can be hopeful that they will ever become licensed.

A Little History

Interestingly, the offshore era caused several other land-based stations, but many of these were actually offshore (or pre-offshore) stations. For example, Radio Sutch and Radio Veronica both started broadcasting from land, claiming to be from the sea, in order to attract advertising money and financial investors before they were technically ready for regular operations.

Some of the earliest post-offshore pirates go back to 1968, when Radio Free London (Fig. 7-3) and the Helen Network took to the airwaves. There is some debate as to which group was first on the airwaves, but most people consider RFL to be first and Helen second. Two of the major difficulties with pirate radio from England in the 1960s and 1970s are that most listeners were on AM (medium wave)

Fig. 7-3
The Radio Free London on-air studio in the 1970s.
Radio Free London

and the post office was determined to close down any stations that took to the airwaves. AM is a real problem because the wavelength is so long that a very long and/or tall antenna is necessary for a good signal. These two requirements limit the possible locations for a station, which in turn makes it much easier for the active post office agents.

According to the Landbased Pirates of the '60s and '70s web page, Radio Free London's first major broadcast was in August 1968, at a gathering of free radio listeners who were celebrating the first anniversary of the Marine Offenses Broadcasting Act. RFL was raided by late afternoon. But the station returned over and over again, and eventually broke into a little Radio Free London network. The network included Radio Free London North, Radio Free London South, and occasionally Radio Free London Southeast (alluding to the Radio Caroline North and South services). The format was rock music, which by 1968 had once again become difficult to find on the radio. At that time, rock radio was limited to Radio Caroline and nighttime slots on Radio Luxemburg. Evidently, the station had some very professional moments because some of the staff members later moved on to jobs with offshore pirates.

The Helen Radio Network consisted of a loose confederation of stations that coordinated very brief broadcasts on the same frequency to avoid direction-finding and raids. The individual stations in the network operated with either Helen callsigns (such as Helen 3, Helen 4, and Helen 5) or Helen-related slogans (such as Radio Free Helen, Helen Radio International, etc.). The network, successful as it was, was more important as a training ground for a number of stations that would continue on as major pirates in the 1970s. Some of the former Helen Network stations included Radio Spectrum, Radio Telstar, and Radio Jackie (the latter became extremely popular with general listeners in the late 1970s and 1980s).

Radio Free London disappeared by the early 1970s as the Dutch offshore stations took to the airwaves as well as dozens of English AM (medium wave) pirates. One fascinating station, Radio Aquarius (Fig. 7-4), started in 1973 and operated every weekend for approximately two years. The motivation and dedication of Radio Aquarius was unbelievable. With a staff of only three-to-six radio hobbyists, the station managed to make new programs and broadcast from various locations (mostly outdoors) every weekend. Even when the station was raided and those caught were prosecuted and fined, Radio Aquarius would reappear the next weekend with a new transmitter that had been built by the station engineer in just a few days.

Fig. 7-4 *Radio Aquarius in the 1970s:Tim in the studio (left) and the DTI catch two members of Radio Aquarius in a roadblock at the Heald Green railway station.* Andrew Howlett

Amazingly, the station was funded by the staff and had no outside sources of revenue.

Tony Greaves from Aquarius described the end of one of the broadcasts: "Near the end of the transmission, when the last tape was playing, a group of cross-country runners came gasping by. Bob had seen them coming a mile off, but then a shout of 'NOW!' rang out, and all hell broke loose. PO operatives, policemen, and dogs appeared out of nowhere having completely surrounded the hilltop, and panic set in. Trevor and Bob ran for about half a mile with dogs at their heels, and Pete Welton ran so fast that one of his shoes flew off. After his capture, he was not allowed to return to the site in order to find it.

"The transmitter was lost, along with the cassette player and the aerial, and in a small-minded display of victory, the PO men smashed up the car batteries to prevent the Aquarius team retrieving them for future use. A fairly successful broadcast had ended with the loss of a complete set of equipment and more summonses in the pipeline."

Despite facing the real possibility of such full-scale attacks during any given broadcast, such stations as Radio Aquarius, Radio Free London, Radio Kaleidoscope, Radio Jackie, Radio London Underground, London Music Radio, and many others operated with regularity (sometimes even with set schedules).

In the mid-1970s, shortwave pirate radio began to take off in

Europe. The origins of the 1970s European shortwave pirates are a bit sketchy, but the first major station was World Music Radio, which operated on Sundays from Holland and Denmark, starting in 1967. Although other early stations existed, such as Radio Lancashire in 1972, it seems that most of the mid-1970s pirates were inspired by either World Music Radio or Radio Viking (as it was later known). WMR and Radio Viking were heard all over Europe for several hours every Sunday morning with taped block programming from different DJs. The format was generally rock music, but some other programming was also featured. The station was raided several times, and finally closed in 1979 after a court prosecution. However, it returned in the early 1980s as World Music Radio, purchasing airtime on the private stations Radio Andorra, Radio Milano International, and Radio Dublin. WMR disappeared when those stations closed on shortwave, but the World Music Radio concept remains alive, as can be seen on their web page: http://www.wmr.dk/.

To avoid possible attacks from the government, European pirates place a much greater emphasis on alternating transmitting locations and other avoidance techniques (Fig. 7-5) that the North Americans generally have not caught onto yet. The most commonly used European technique is to take a transmitter, car battery, power

Fig. 7-5 *Pat of Radio Kaleidescope tuning up the transmitter in the early 1970s. The two black boxes are batteries, the chassis connected to the battery is the power supply, and the last open chassis is the transmitter. Notice that the makeshift antenna is a barbed-wire fence!* Nick Catford

inverter, cassette player, and antenna into a forest. Once the equipment is arranged and connected, the transmitter is turned on. The staff member then leaves and returns only when the program has ended. This method has been effective, but it has mostly failed to interest the North Americans.

A more complicated technique is sometimes used by shortwave stations, but it has become especially useful to regular broadcasters on AM or FM. The system is the same as the others, except that the programming is transmitted from a studio with a low-powered UHF, VHF, or even infrared link. This signal is then received at the site of the broadcasting transmitter and fed through the system. With this method, even if government agents do find the main transmitter, they will still have to track the radio link back to the studio. If the main transmitter has been removed by the authorities, the pirates can turn off their radio link so that the signal cannot be traced. Should the authorities enthusiastically track a station, this method can become expensive; for many stations, the security is worth it.

Differences among Europirates

It's long been said that America is a "melting pot" of cultures, where peoples from many lands arrive and become "American." American culture isn't a composite of the cultures that arrive here, but is something almost entirely different. Much the same effect has simultaneously occurred in Canada, where the culture is more closely related to American culture than to any other. As a result, someone can travel a vast area in North America and find very few regional cultural differences. Of course, there are some variances, but these are generally minor.

Europe is a completely different spool of wire. One huge barrier is language. Although millions of Europeans are multilingual, the Continent contains so many languages that it's not possible to broadcast in a single language that everyone understands. And those listening in a second language won't necessarily get the jokes or other subtleties of a broadcast. As noted by Norwegian DXer Kai Salvesen, "Pirate radio is probably not so funny when you have to broadcast in a foreign langauge."

Numerous American and Canadian pirates have been frustrated when having their creatively programmed shows aired by Europirates. Often, the stations will receive reception reports that either say "music, then man talking, music," etc., or "I think the program was funny, but I don't comprehend English." Exactly the same thing happens in Europe, where listeners might have problems understanding a broadcast from England, Germany, or Finland, for example.

Salvesen also stated, "One important point about European pirate radio is that the pirate communities of the different countries are a bit cut off from each other. This has something to do with the language barrier, of course. For example, I don't think that Subterranean Sounds (who emphasizes speech rather than music) has more than a (tiny) handful of listeners outside Britain. Many British stations broadcast primarily for Britain and Ireland and don't really want to be heard on the continent."

In addition to the language barriers, different countries have their own personalities as far as pirate radio goes. For example, the main pirate radio countries in Europe on shortwave are Ireland, United Kingdom, Holland, Germany, and Italy (in no particular order). Denmark, Norway, Sweden, Finland, France, Belgium, and Switzerland occasionally have stations. Shortwave pirates very rarely (if ever) appear in Spain, Austria, Portugual, Poland, the Czech Republic, Slovakia, Hungary, Yugoslavia, Greece, etc.

For some time, I have been fascinated with how pirate radio scenes develop. For example, why Germany or Holland and not Austria or Switzerland? Economically, the countries are all similar, the people groups are Germanic, and the cultures vary a little. Enforcement is strict in Austria and Switzerland, but such is also the case in Germany. Why don't Austria and Switzerland have more pirates? I still don't have a good reason.

I've heard very little about why the English (and also the Scots) and Germans are inclined toward shortwave pirating, but I've heard a few comments about The Netherlands, which is absolutely saturated with pirates. One listener from England theorized that the Dutch have lived in the ominous shadow of Germany for so long that they have developed a rebellious national psyche toward all authority. The longtime (24 years) Dutch pirate Radio Torenvalk (Fig. 7-6) agrees to at least some of these statements. "The reason why there are so many pirates in Holland has got something to do with the Dutch attitude toward some things. Most Dutch people are to some extent 'outlaws.' For example, practically nobody over here will stick to the speed limits, unlike other countries, like Sweden, Scotland, or the USA, where most people drive much more relaxed."

To generalize the pirates in Europe: The UK stations are more professional, with a leaning toward regular music services. Great examples of UK pirates are Radio Free London, Laser Hot Hits, and Radio Nova International, which sound like commercial rock stations—

Fig. 7-6 *Is anti-authoritarianism a Dutch national trait? A sticker from long-time Dutch pirate Radio Torenvalk.*

plenty of music, jingles, and professional DJ patter from several different announcers. The broadcasts from the UK stations often last for several hours, but occasionally run for several days at a stretch via autoreverse cassette decks (Fig. 7-7). The British phenomena is completely understandable. Most of the UK pirates were either directly or indirectly inspired by the many offshore stations that operated from just off the country's coast as recently as 15 years ago.

The Dutch stations are one-man operations by the technically inclined. Most Dutch pirates build their own transmitters and simply enjoy broadcasting. As a whole, they are extemely friendly and broadcast music live. At times, it appears that some of the stations wouldn't care what was being broadcast so long as they could be transmitting. One of the only exceptions to the Dutch pirate image is the Free Radio Service Holland (FRSH), which occasionally broadcasts with professional blocks of music and radio news, more in the UK pirate vein.

The Dutch way is to let some friends know that a broadcast is coming up, put on favorite CDs, drink a few beers, and turn on the transmitter. After taking a few telephone calls or checking e-mail, the announcer will sign off and maybe chat on the frequency with a few other Dutch pirates. But not everyone enjoys hearing the carefree Dutch participate in their favorite hobby. Some listeners, particularly those in the UK, complain about the Dutchies "playing polka and reading out greetings to Aunt Gulda and Uncle Willem." It's common

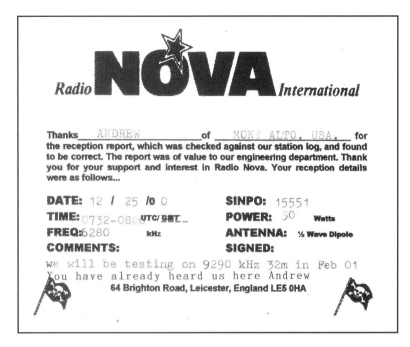

Radio **NOVA** *International*

Thanks ANDREW of MONT ALTO, USA, for the reception report, which was checked against our station log, and found to be correct. The report was of value to our engineering department. Thank you for your support and interest in Radio Nova. Your reception details were as follows...

DATE: 12 / 25 /0 0   **SINPO:** 15551
TIME: 0732-080 UTC/GMT   **POWER:** 30 Watts
FREQ: 6280 kHz   **ANTENNA:** ½ Wave Dipole
COMMENTS: **SIGNED:**

we will be testing on 9290 kHz 32m in Feb 01
You have already heard us here Andrew
64 Brighton Road, Leicester, England LE5 0HA

Fig. 7-7 *With a sound like a professional rock station, Radio Nova International from the UK has broadcasted for more than a week straight.*

for some listeners to complain about hearing "barrel-organ music" (the derogatory term of choice for Dutch folk music).

Another reason that some listeners dislike the Dutch pirates is their high power. Because the enforcement against Dutch shortwave stations is among the lowest in Europe, the pirates have the luxury of operating from their own houses from AC power. Instead of operating with small, homemade battery-powered transmitters, nearly all of the Dutchies are running anywhere from 100 watts up to a scorching 2000 watts with military-suplus or home-brew equipment. So the English listeners and pirates sometimes get angry when a 10-watt signal from the UK (lovingly put on the air by someone who strung up the antenna in the fog and is hiding in the woods) is flattened by a 1500-watt Dutch flamethrower whose operator couldn't hear the other station.

German pirate radio is much more varied, with brief programs ranging from simple music and IDs to techno music and heavy production. In this sense, German pirate radio is more like that which is broadcast in North America. German DXer Martin Schoech agrees, "Germans are indeed a 'mix' and always quarrel about the quality of the programmes. They often quarrel about 'using a relay' or 'building an own transmitter'"; the former is also a common argument in North America. Unlike the Dutch stations, which often broadcast in English or broken English, the German stations often strictly use the German language to reach the large audience within their country.

To illustrate a bit of the German program diversity, Radio Marabu broadcasts an *extremely professional* alternative music (techno, industrial, punk, etc.) mix on shortwave, FM, and legally via

Fig. 7-8 *The country music sounds of the German cowboys can be heard via Radio Perfekt.*

satellite; Radio Lollipop is a children's station; Radio Perfekt (Fig. 7-8) only plays German country and Western music; and Radio Dr. Tim offers music and DX news.

Italy and Ireland mostly just relay programming from other stations. The only regularly active stations from either country that broadcasts its own programming are the country music stations Jolly Roger Radio (Ireland) and Ozone Radio (Ireland). And even both of these stations spend much of their time relaying programs from other operators.

Another anomaly is Russia, which had a number of local "radio hooligans" throughout its communist years in the Soviet Union. These stations were local or regional operations, with two-way conversations and occasional songs aired in the 1600- to 2000-kHz range, just above the AM broadcast band. In essence, these were bootleg hams, not pirates. Two true shortwave broadcasters began operating from the

Moscow area in early 1991, just after the USSR began to break up. Radio Without Borders International (RWBI), which was featured in the second edition of this book, operated much like a typical Europirate, using the same frequency ranges, broadcasting regular programming, and issuing QSLs through a maildrop. The RWBI transmitter produced in excess of 100 watts and was heard from Eastern Europe to the British Isles and even in Canada. The other operation, Romantic Space Radio, was a tape station with close ties to RWBI. It was heard via RWBI and transmitters in France, England, The Netherlands, and the United States. RWBI was raided in 1993, but successfully pleaded its case in court. RWBI was raided again in 1994 and the station disappeared, the maildrop closed, and all e-mail addresses for RSR and RWBI bounced. Russia's once-promising pirate scene has remained silent since 1994.

The leading pirate country in Scandinavia has fallen largely silent, but for slightly different reasons than the Russian pirates. Finland has had an active shortwave pirate scene since the mid-1970s, with numerous regional pirates broadcasting in Finnish and only a minor emphasis on the English language and reaching countries outside of Scandinavia. In the early 1990s, such Finnish pirates as Radio Matilda, Radio Meteor, Radio Mayday, Radio Milliwatti, Radio

Fig. 7-9 *Radio Diablo was one of a group of Finnish radio pirates that was active in the early 1990s.*

Diablo (Fig. 7-9), and the Voice of Free Radio were active most every weekend, with a range that covered little past Scandinavia and northern continental Europe. In 1994, Finnish authorities raided the maildrop in Jyväskylä that forwarded mail for nearly all of the shortwave pirates in Finland. With the contact information for the Finnish pirates in hand, the stations were quickly closed. As a result, only Radio Milliwatti remains from this original group. After a relatively silent period throughout the mid-1990s, Finnish pirates such as the Scandinavian Lighthouse, Northern Music Radio, Radio Old Time, and Scandinavian Beer Drinkers have begun to creep back onto the airwaves. Still, the scene is more than an 807 tube short of its early 1990s glory days, and Finnish pirates have rarely reported by European DXers in 2000 and 2001.

Recent Activity

Europe is still filled with hundreds of active pirates on AM, FM, and shortwave. Strong government threats successfully deter enthusiasts from broadcasting in some European nations, but pirating has been thriving elsewhere. The British Isles, Sweden, Germany, The Netherlands, and Belgium are saturated with activity (including AM, FM, and shortwave stations), but even the less-active countries, such as Finland and Switzerland contain low-powered community pirates. Despite the fact that most of these stations operate for the community on AM or FM, many pirates are widely heard across the continent on shortwave. In fact, it's common for listeners to log more than 20 different stations on one Sunday.

For the entire existence of pirate radio in Europe, the time to listen for these stations has been Sunday morning. Likewise, the frequency range has been 48 meters (6200 to 6300 kHz), although several other bands have been used to a much lesser extent. But as the sunspot cycle began to bottom out at the beginning of the 1990s, some pirates shifted operations to the European 80-meter broadcast band (known as the 76-meter pirate band) during Saturday evenings.

By the late 1990s, most of the stations left on the 3900- to 3950-kHz area were Dutch pirates that recently moved up from

Fig. 7-10 *A sticker from Holland's Radio Black Power--a long-time AM pirate that occasionally broadcasts on shortwave.*

broadcasting just above the AM broadcast band, as had been the case for decades. These old AM stations typically broadcast on a night-time schedule at different times throughout the week or weekend (Fig. 7-

10). The range between 3000 and 4000 kHz is short enough that the pirates who moved up could continue broadcasting during the hours of darkness without having problems with signals skipping over most of the potential listening audience. In addition to the Dutch, Laser Hot Hits set up a 24/7 outlet on 3945 kHz. Other British stations, such as Live Wire Radio and Subterranean Sounds, plus Russia's RWBI broadcast regularly in the evenings.

For a time in the mid-1990s, 76 meters was the international pirate band for listeners across Europe. Today, about all that's left in the band are Laser Hot Hits and the occasional Dutch AM pirate, most of which don't broadcast on any other shortwave frequencies.

The 3900- to 3950-kHz range is also a section of the 80-meter amateur band in the United States and Canada. Most of these stations are running anywhere from 100 to more than 1000 watts. This is also a very popular and crowded band in North America. For a time, it seemed unlikely that anyone could hear a 50-watt broadcast from several thousand miles away, in the middle of a crowded band with powerful local interference. However, Laser Hot Hits was noted by several people in this range, running only low power. I started checking and heard Live Wire Radio (UK) with a strong signal, Radio Spaceman (The Netherlands), and Radio Korak (The Netherlands), in addition to Laser Hot Hits.

Also in the early and mid-1990s, the 41-meter band (7300-7500 kHz) was somewhat actively used by European pirates. Regular users were Britain Radio International on 7365 kHz, Radio Stella International on 7444 kHz, Radio Waves International on 7473 kHz, and Radio Marabu on 7484 kHz. As the 41-meter band fell into disuse in the United States, Europe followed course. Today, 41 meters only contains a few pirates, most notably Laser Hot Hits on 7415 kHz, Ozone Radio (Ireland) near 7445 kHz and Radio Benelux (Germany) and Crazy Wave Radio (Germany), which coordinate broadcasts on 7485 kHz.

The radio band of choice remains the unofficial European 48-meter pirate band, which ranges from 6200 to 6320 kHz. Any frequency in this area is used, depending mostly on the amount of interference from the licensed ship-to-shore stations that transmit in the area. In 2001, plenty of these digital data stations were operating in the 6240- to 6265-kHz area, so this range was rarely used by pirates.

The currently active experimental broadcasting range is the 19-meter broadcasting band. Most of the activity occurs at the low end of the band, near 15070 kHz, which is occupied nearly every weekend by Alfa Lima Interational (The Netherlands). ALI has spearheaded European pirate testing to North America and the rest of the world by starting a pirate listserve and chat room, and broadcasting on high frequencies to regions beyond Europe. Some of the other Benelux-area pirates that have operated in the 19-meter band include the powerful Radio Borderhunter (Belgium), Radio Black Arrow (The Netherlands), and Radio Sunflower (Holland).

The odd experimental bands used by European pirates are 31 meters and 13 meters. Most of the recent 31-meter pirates are using the low end of the band, from 9290 up to 9330 kHz. Free Radio Service Holland has made a few successful tests in the area, but the primary users of this band are Mike Radio (The Netherlands) (Fig. 7-11) and Radio Nova International (UK), the latter of which broadcast for more than a week straight in the Spring of 2001. In late 2000 and early 2001, ALI tested

Fig. 7-11 *A QSL from Mike Radio, featuring its hefty German military-surplus Rohde & Schwartz transmitter. Although Mike often uses 48 meters, this transmitter is being stationed on 9290 kHz in the 31-meter band throughout winter 2001-2002.*

frequently on 21490 kHz in the 13-meter broadcast band. The results were very positive and I heard Alfred with a powerful, local-strength signal. Before many other stations could move up to join him, ALI moved down to 15070 kHz, where it appears that he will remain for some time.

Current European Pirate List

The following list contains basic frequency/time information for the more common Europirates. This list will quickly change. For more information, see the *SRS News* webpage at http://www.srs.pp.se.

Freq.	Station	Country	Notes
3925	Radio Korak	HOL	Often on late evenings
3935	Laser Hot Hits	UK	24/7 Dance music
5805	Radio. Free London	UK	Long broadcasts, often 24 hours
6210	Mike Radio	HOL	Pop music
6210	Radio Nova Int.	UK	Long broadcasts, often 24 hours
6220	Laser Hot Hits	UK	24/7 Dance music
6238.6	Jolly Roger Radio	IRE	Country music and station relays
6261.2	Radio Dr. Tim	GER	Music and pirate radio news
6266.3	U.K. Radio	UK	Pop music
6276	Swinging Radio England	UK	New station with long programs
6285	WKNR	UK	1980s pirate, testing for a return in 2001
6306	Farmers from Holland	HOL	Dutch folk music
6306	Tower Radio	HOL	Pop music and Dutch folk music
6373.6	Radio Brigitte	HOL	Dutch folk music
6399.6	Westcoast Radio	HOL	Pop music
7306U	Radio Europe	IT	Mostly relays of Europirates
7460	Laser Hot Hits	UK	24/7 Dance music
7480	Radio Benelux	GER	
9290	Mike Radio	HOL	Winter testing frequency
12256	Wreckin R.Int.	UK	Hours of pop music every Sunday
15070	Alfa Lima Int.	HOL	Regular tests and relays

Regular 48-meter stations with no set frequencies (all stations are from Holland, except where noted): Radio Borderhunter, Radio Black Arrow, Radio Foxfire, Radio Likedeeler, Mike Radio, Radio Perfekt (Germany), Radio Blue Star, Eastside Radio (Germany), Radio Torenvalk, Voice of the Netherlands, and Radio East Coast Holland.

The Italy Effect

Obviously, the free radio movement created many problems and politically difficult situations for the different governments in Europe. Feeling they had lost control of the situation, the Italian Supreme Court ruled that the people of the country owned the airwaves and had the right to broadcast as they wanted with some government oversight. Pirate control by the government has been limited to forcing stations into the international broadcast bands. In spite of this slight limitation, Italy has become a true haven for pirates.

Although the Italian legislation gave pirate stations the freedom to broadcast safely, the broadcasters were still neither licensed nor legal; they were simply no longer in danger of prosecution. For this reason, unlicensed Italian broadcasters are usually called "privates" rather than pirates in reference to their private ownership (and not government ownership). Some of the private stations that were thought to be unlicensed by the government actually were licensed. While researching an extensive article for *The ACE*, Kirk Trummel discovered that peculiar loopholes continue to plague the Italian broadcast authority. Several of the regular Italian stations are actually covered in a loophole, but a few are still true pirates (the government just has not attempted a raid as of yet).

The most widely heard Italian shortwave privates (such as IRRS), operate much like large American commercial stations. Each station has large, expensive studios, powerful transmitters on AM and FM (as well as shortwave), a staff of professionals, and commercial airtime for sale. All three of these stations have been heard in North America in the 1990s, and their signal strength in Europe is quite good.

One Italian commercial station that is much like the old Radio Milano International from the 1980s is the IRRS, also from Milano, Italy. SWR Switzerland, a non-profit radio group, bought chunks of airtime on the IRRS 30,000-watt 7125-kHz transmitter for the specific goal of having pirate programs aired. *FRS Goes DX* reported that IRRS president Alfredo Cotroneo said, "We want to continue to be a real international community radio, which maintains a very tight and personal relation with all European listeners, and this is a very splendid opportunity for us to fulfill our aims." The relays have indeed been popular and a number of stations, including Radio Marabu, Radio Francis Drake, Southern Music Radio, Radio Joystick, and others have been heard via this outlet.

Uncontrollable pirate situations plagued countries other than Italy. Irish legislators passed bills like those passed in Italy. In 1978, the

Irish officials allowed pirate stations to register as radio broadcasting companies with full legal status. Although the registered stations were still not licensed as such, the government no longer had the authority to make raids on them. As a result, hundreds of commercial and hobby stations frequently operate on shortwave, AM, FM, and even television bands.

Radio Dublin is one of the founding Irish pirates (now a private) and is possibly the oldest and most widely heard unlicensed broadcaster in the world. Created in 1966 as an alternative to Radio Caroline, Radio Dublin began with a meager 10 watts or less on AM frequencies. For 11 years, the station only used low power on weekend afternoons. But after extensive testing in December 1977, Radio Dublin switched to a 24-hour music format–the first Irish pirate to attempt such operations. The hobby-style programming was then traded for a more cost-efficient commercialized format. Additionally, all broadcasts were moved to a single studio instead of multiple transmitter sites which had changed each week.

Frequent Department of Posts & Telecommunications (DPT, the Irish equivalent to the FCC) raids caused slight monetary damage to the station, but outdated legislation regarding pirate radio allowed Radio Dublin to continue. In fact, only one successful prosecution of the station was recorded, which resulted in a small fine. This antagonizing situation with Radio Dublin was the vehicle that eventually drove the government to allow pirates to register as private broadcasters in 1978. Finally achieving a semilegal status, Radio Dublin moved to a new location, bought new studio and transmitting equipment, and opened one new outlet on both the AM and FM band.

Always a leader in the Irish pirate and private radio scene, the 1978 legislation allowed Radio Dublin to explore the possibilities of an international audience via shortwave broadcast stations. But instead of merely broadcasting on shortwave as many Irish pirates had done in the past, Radio Dublin had the opportunity to become the nation's voice because Ireland has no authorized shortwave broadcast stations. This dream was fulfilled in 1980 when the station began operations on 6315 kHz with 300 watts on a 24-hour schedule. Transmissions on this frequency caused interference with a British government station, so Radio Dublin eventually moved to 6910 kHz and boosted the output power to 900 watts.

With the increased power, Radio Dublin became a relatively easy catch for listeners in eastern North America and has been heard throughout many parts of the world. Radio Dublin realizes the service it provides and is quick to promote and manipulate the situation. They actively promote tourism in Ireland by playing advertisements, sending information guides, and offering contests. Radio Dublin also heavily promotes within Ireland by sponsoring contests to raise money for hospitals and community centers, donating decorated floats for parades, and providing free music for charity events. Whether motives are innocent or not, the station must at least know that these charitable activities will offer protection against future government actions. Legislation expected in 1985 would have established a commercial radio network throughout Ireland, forcing the private stations off the air. This legislation was never passed and many questions remain about the present broadcasting situation in Ireland.

Rumors spread that some privates would become part of the network. Obviously, a responsible community-oriented broadcaster would be given the first opportunity for licensed operations. The government legislation never was passed and the privates have not been closed, but these possibilities still remained at the time this book was written.

Another colorful front-runner of the Irish privates with a hobby twist is Westside Radio International, now known as Ozone Radio International (Fig. 7-12). Founded in the mid-1970s by announcer Dr. Don and Prince Terry, Ozone Radio International is still operated for

Many thanks for writing to
Radio Ozone
International

Name .... Andrew Yoder ....
Date .... 20.2.2000 .... Time (UTC) .... 08.06 ....
Frequency (Khz) .. 6.200 .. Tx Power .. 165 watts
SINPO .... 34454 .... Signed .. Mr Ozone

Fig. 7-12 *Radio Ozone International has been broadcasting for nearly 25 years from Ireland. For the past few years, the station has played its own rock and metal programs and relayed Britain's Better Music Station.*

several hours every Sunday morning on shortwave. Listeners across Europe have been treated to a consistent helping of oldies and heavy metal music (a rare format in Europe) and relays from other free radio stations around the world. The station was silenced for a year, beginning in mid-1976, but since its return it has rarely relinquished its 6280-kHz frequency.

As can be expected, some pirates have migrated their operations to Ireland and Italy, although Ireland seems to be the choice site. Radio Fax, which was funded by a British private radio group, installed a transmitter site in Ireland for its 24-hour per day broadcasts. After it was being heard by thousands of people around the world, the Irish government threatened to close the station, and Radio Fax complied by leaving the air. However, Reflections Europe, which was relayed by Radio Fax, is still heard via the old transmitters.

Another somewhat similar situation occurred when an Irish local FM pirate known as DLR-106 (Dun Laoghaire Local Radio) bought a shortwave transmitter and took to the airwaves in 1991. For a number of months, DLR was heard on weekends; later, it was heard 24-hours per day with strong signals across much of Europe. During this time, DLR featured volunteer DJs who played blocks of music, much like American college radio stations. By August 1993, the DLR shortwave station fell silent, and it was rumored to have been threatened off the air by the Irish authorities.

As has been the case for the past 20 years, no one knows exactly what will happen to international shortwave broadcasting from Ireland. For years, pirate radio from Ireland has appeared to be both on the verge of being closed down and yet just about to be legalized at the same time.

The situation in Greece is very similar to that of Italy. Unlicensed radio was out of control, so the government allows all unlicensed, noncommercial radio. The key word is noncommercial. Any unlicensed stations that choose to air commercials suddenly cross the line and will be raided by the government. Even those that run very high power, yet remain noncommercial, are within the boundaries of the law in Greece. The big difference between Greece and the operations in Ireland and Italy is that the Greek stations broadcast only for a local audience and have no interest in broadcasting on shortwave. Of the thousands of unlicensed Greek broadcasters that have operated over the years, none are known to have used shortwave.

Inside A European Pirate

Unfortunately, there is not enough space to cover all of the interesting European stations in as detailed a manner as the North Americans in Chapter 3. Because hundreds (maybe thousands) of different stations have operated from Europe in the past 15 years, full coverage of each broadcaster would be impossible (furthermore, extensive coverage of each broadcaster would be extremely repetitious because most station histories are similar on many points.) Several books that briefly describe pirate broadcasting in Europe are listed in the bibliography.

A number of high-profile pirates operate at any given time from Europe. In this edition of *Pirate Radio Stations*, I have chosen Alfa Lima International to show "behind the scenes" because Alfred has spearheaded the movement to test on the high frequencies and to close the space between shortwave listeners around the globe.

The story of Alfred's involvement in hobby radio stretches 18 years prior to the beginning of Alfa Lima International in 1980. A much younger Alfred visited his uncle several times while he was broadcasting as a medium-wave pirate. Although Alfred found the experience interesting, it didn't immediately make a long-term impact. But not long afterward, while running through a junkyard with a teenage friend, Alfred spotted something that looked much like his uncle's transmitter. Indeed it was, and it sparked an interest in hobby radio.

Before long, Alfred borrowed a mixer from his uncle and bought a five-watt FM transmitter, operating on 103.5 MHz as Edison (Dutch local pirates often go by a single name, rather than with a callsign or as "Radio ——"). "At that time, there were maybe only 20 pirates active in the area and we had a lot of listeners every Saturday," said Alfred. Of course, 20 pirates in an area would be a phenomenal number almost anywhere, except in Holland.

A few years later, Alfred started experimenting with CB, in particular, on 11 meters in the range just above CB's 40th channel. Unlicensed stations from around the world, often called *outbanders*, congregate in this part of the spectrum. These stations are often identified by several phonetic letters, and Alfred was known as 19 Alfa Lima 401 on 11 meters. 19 Alfa Lima 401 operated with 650 watts into a three-element yagi antenna, guaranteeing strong signals around the world.

In 1998, Alfred built a 100-watt shortwave transmitter from scratch and tested it on March 14, as Alfa Lima International, with excellent signals around Europe. Because of his DXing background, he wasn't content only to broadcast to Europe; Al wanted to reach the USA and other parts of the world. So, he began testing in the evenings and almost immediately received responses from the United States. In order to reach more distant regions of the world with stronger signals, Alfred found a German military surplus Rohde & Schwartz SK-050 transmitter (Fig. 7-13), which was not only much more powerful, but could also operate on other frequency bands.

Fig. 7-13 *The huge and powerful Rohde & Schwartz SK-050 transmitter of Alfa Lima International. When the RCD dismantled Alfred's original SK-050, he bought another.*

With the large cabinet-mounted transmitter operational, ALI began making regular tests on 15030 kHz (just below the old 15050-kHz pirate frequency), as well as 21890 (13 meters) and 11480 kHz (25 meters).

On Sunday, August 29, 1999, there were a few rings at the doorbell during the show. Alfred opened the door and saw two policemen and another man from the RCD (Dutch radio enforcement) asking if they could come in. They followed Alfred into the attic studio where the Rohde & Schwartz transmitter was running with 450 watts AM modulation. After a little talk, it was soon announced that ALI had been transmitting on a frequency (15030 kHz) used by the UK's R. A. F. (Royal Air Force). This interference is why they decided to visit. After that little talk, the transmitter was shut down and parts of it were moved to the van outdoors. A few sections of the transmitter were too heavy and were smashed with a hammer.

Alfred was a bit angry about the raid and vowed to return. "If they had such interference, why didn't the R. A. F. give us a call? (A telephone number is regularly announced in every ALI program.) We didn't know the frequency was used by them; if we had known, we never would have used that frequency." Immediately, Alfred paid his $500 (U. S. equivalent) fine and spent $1000 (U. S. equivalent) on another Rohde & Schwartz SK-050 transmitter.

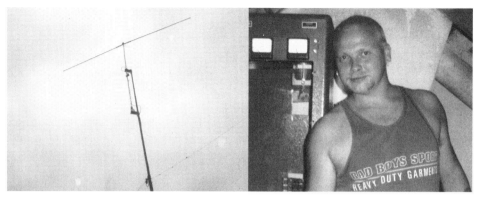

Fig. 7-14 *At left, the Alfa Lima rotatable dipole for 19 meters. At right, Alfred standing beside his SK-050 transmitter.*

ALI has bloomed since the acquisition of the new transmitter. Alfred set up and moderates a shortwave pirate radio listserv that now has a subscriber base of 400 and he has created a chat room with organized chats of European pirates and listeners. In addition, he broadcasts most every weekend for hours and announces the transmissions in advance in several places, such as the FRN board.

The main inspiration for Alfred (Fig. 7-14) is the listeners. "Of

Fig. 7-15 *The Alfa Lima studio, including a Yaesu FRG-7700 receiver in the center, CD players at top left, and MD player/recorder at lower right.*

course for me, the friends and contacts we make are the best thing of our hobby. I hope that lots of people keep on turning those dials for SW pirates." Although Al has not yet met any of his listeners in North America or Australia, he's met many of them from Europe (Fig. 7-15). "It is always a honor and pleasure to meet all of them, as well as my other fellow broadcasters."

Since 1998, Alfa Lima International has received hundreds of reception reports from 37 different countries, including such diverse locations as Latvia, Iran, Chile, and the Phillipines. He has also received reports from 23 different states in the U. S. Although ALI is currently the easiest European pirate to hear in North America at this writing, pirate radio can change in an instant. Although Alfred's interest in broadcasting is as strong as ever, the Dutch authorities could eventually become a problem. "We do think that it should be soon that they will catch us again. Many times, I think, 'Not this week,' but when the time is there, we are on again."

Pirate Pranks

Not all pirates follow the status quo as closely as the privates or the top-40 stations. In fact, a few European pirates have created ingenious satirical broadcasts that have stunned authorities and listeners alike. The hoaxes are especially effective in Europe where pirate broadcasting is, on the whole, much more honest than that of North America; thus, European listeners are more apt to believe the claims of a pirate.

One of the most interesting pirate operations that ever operated from anywhere in the world only broadcast for 35 minutes on FM in 1976. Just over half an hour on FM might seem rather dull, but the programs were aired by several BBC Radio One network transmitters throughout the south and southwest of England! The prank made newspaper headlines across the country and later forced the BBC to employ new systems of remote program transmission to replace the manipulated system.

The story behind this scam is very interesting. Unlike North America, where FM radio service is provided by a large number of independent commercial and educational stations, the United Kingdom and some other European countries operate with several country-wide networks. In order for the network system to operate, all stations must be receiving the same programming simultaneously. Although most of the stations received the programming direct via land lines, several sites tuned to another BBC radio frequency and fed the audio into their transmitter. The method was severely outdated, but it provided sufficient service until the government could replace it with a land line.

Several BBC transmitters in southern England and the Channel Islands were fed from a station in Wrotham, Kent. In order for a takeover of BBC Radio One transmitters in southern England to succeed, a high-quality signal had to be inserted over the same frequency of the Wrotharn station. The pirate equipment had to be exactly on frequency (otherwise it would be rejected by the BBC off-air receivers) and near enough to the BBC Rowridge, Isle of Wight transmitter that the Wrotham signal would be overpowered.

So the pirates constructed a stable 30-watt stereo FM transmitter, connected it to a cassette player, and hid it in a hedgerow less than a mile away from the Rowridge station. At exactly 11:00 P.M. (local time), as announcer John Peel was about to open a pop music program, his initial words were suddenly drowned out in a hail of machine gunfire. Appropriately, "Substitute" by The Who became the first musical selection following the gunfire. Afterward, a bizarre program of records banned by the BBC and fake announcements continued until government engineers regained control at 11:35. Among the announcements was a brief talk from "Idi Amin, the new chairman of the BBC" and a public service message service from the "Metrification Board," warning that anyone who continued using nonmetric units might find their "houses mysteriously demolished by bulldozers during the night."

Unprepared for such a situation and unable to locate the illegal transmitter, BBC engineers were forced to return with programming fed through regular telephone lines. As a result, the BBC audio was distorted, brassy, and in mono, in contrast to the high fidelity stereo signals inserted by the pirate. Just before the regular sign-off of the BBC at 12:07 A.M., the presentation announcer merely apologized to "listeners in southern England who may have been listening to the wrong program." Nothing more was said about the situation, but security measures were immediately taken to ensure that the BBC transmitters could never again be overtaken.

The BBC network was presumed safe after the installation of the security systems, but the pirate group wanted to try it again. Although the transmitters were protected with a security system that made it virtually impossible for anyone to hijack the equipment while it was in operation, the hobbyists found that the system was turned off every evening with the transmitters as the broadcasts ended. The transmitters were then turned on every morning at 6:00 A.M. with a 19-kHz signal. The pirates chose August 14, 1977, the tenth anniversary of the Marine Offenses Act that closed virtually all offshore broadcasting in England, as the day to take over the Rowridge transmitter again.

After the sign-off of BBC Radio Two and Three on August 14, the 19-kHz stereo pilot tone was transmitted, returning the carrier to the frequency. With transmitters back in operation, it was then possible for the station to be pirated in much the same manner as had occurred the year before. Because this broadcast took place after midnight, no BBC workers were monitoring the frequencies.

This time, instead of claiming to be the BBC, the pirates called themselves K-SAT, a new legal operation that was to offer commercial programming nightly via the government transmitters. Programming featured pop music, frequent jingles, and commercials similar in format to private stations, such as Radio Dublin or Radio Milano International. The hoax was more believable than the previous one because no overtly humorous segments were aired and the station diligently attempted to copy a private broadcaster. Some actual commercials were aired, although others were odd fakes produced by the hobbyists.

The post-midnight broadcast was planned out of necessity, and although fewer people tuned in, this program lasted for three hours on Radio Two and three-and-a-half-hours on Radio Three. After realizing that their transmitters were being pirated, a station engineer rushed to the Rowridge site. The next morning, BBC officials had little more to say about the situation than they did the last time; this time they noted that the broadcast was "probably heard by a few sheep." Despite this lighthearted response, the BBC diligently revised the network again. Their transmitters have not been taken over since.

Although BBC transmitters are no longer pirated, radio hoaxes still occur. The more recent broadcasts are usually easily identified as unlicensed. Most hoaxes in Europe operate on holidays for a number of hours, imitating private or offshore stations. Radio Galaxie, Radio Bouvet, and the Global American Network are virtually the only North American pirates to operate in this manner. Many North Americans do make false claims, but few pirates attempt to produce programs that are exactly like the imitated stations.

Tuning in the Europirates

In the past, most European pirates operated with low power, and even the larger stations do not expect an audience off the Continent. Many of the UK and German stations generally use transmitters with about 20 watts of output on shortwave, compared to an average output power of 100 to 150 watts SSB for a North American pirate. But with the recent influx of high-powered (often running from 100 to 1200 watts) Dutch pirates on shortwave, tuning in these stations on continents beyond Europe has become much easier.

European pirates can be heard in eastern North America if the listener is patient and informed. One problem with listening for pirates is that signals are absorbed by land masses. The lower the frequency, the greater the absorption. If the amateur radio interference in the 3900- to 3950-kHz (76 meters) range is not stifling enough, the land mass of North America soaks up the signals and prevents them from straying too far West. The typical 6200-kHz (48 meters) pirate area is less affected, but afflicted nonetheless. The listeners who receive the strongest signals live along the coastal regions, such as New England and Virginia Beach. The signals are still there, back into Pennsylvania, Ontario, and West Virginia. Some had thought that this was close to the limit of these signals, but David Hodgson of Nashville, Tennessee, heard a number of 48-meter Europirates in early 2001. Evidently, listeners at least as far West as Ohio, Indiana, Michigan, and Illinois have a chance to hear some of these signals.

Recent high-frequency tests from the Dutch stations have allowed listeners deep in North America to hear Europirates. Alfa Lima International, Radio Borderhunter, and Radio Black Arrow (Fig. 7-16) have all been heard with good signals by John Sedlacek in Omaha, Nebraska (near the center of the U. S.), during their test broadcasts on 15 MHz, which is much less affected by land mass absorption than 76 or 48 meters. With such signals, these stations should be

Fig. 7-16 *A QSL from Radio Black Arrow, which has recently been heard deep into North America on 19 meters.*

capable a penetrating further West on this band.

To hear many European pirates, it's essential to:

* **Have a decent receiver**   Although you can hear some Euros with a portable receiver, many more are audible with a table-top receiver, such as the Drake R8B, AOR 7030, Grundig Satellit 800 Millennium (the latter, a semi-portable), Kenwood R-5000, JRC NRD-545, Icom IC-R75, etc.

* **Listen during regular broadcast times when signals are capable of propagating to the U. S.**   Easier said than done. Propagation at these distances on 48 meters relies on a path of darkness between the transmitter and receiver. Most Europirates broadcasts on Sunday mornings, so by 1000 UTC (5 A.M. EST) in the Winter, everything has or is fading away. The best times for listening are 0700-1000 UTC in the winter and 0400-0630 UTC in the summer. The British Isles stations do best later in this time period; the continental stations are best received earlier. Also, the window for hearing Europirates is larger in December (when the days are shorter than at times when the days are longer).

* **Have a good antenna**   Those who hear the most Europirates use several hundred feet of wire pointed approximately NE. Antenna height doesn't matter; in fact, I unroll my antenna directly on the ground or snow. I'm sure that people could have fine results with other antennas, but the simple longwire has provided the best results among DXers.

* **Find a good location**   I can't stress this one enough. I haven't discovered all of the factors that make one site great and another mediocre. However, one very real problem at most locations is interference. Outdoor lights, televisions, computers, and other electronic equipment can radiate radio-frequency interference that can seriously degrade radio reception. At my own house, I have never heard a Europirate—the noise threshold is simply too high. The alternative is to find a quiet location.

But not all quiet locations are equal. I've listened at several very quiet locations, but have had far better success at a local base-ball field than at several others. I have no idea why. Is it the shape of the surrounding mountains, the ground conductivity, or is it just my Field of Dreams? This is why it's best to experiment with listening locations.

DXing Europirates on 76 or 48 meters requires quite a bit of effort, so it's not like spinning through the dial, kicking back, and enjoying your favorite music. Instead, it's struggling through static, whistles, and interference from other stations, deciphering bits of words and playing Name That Tune with the 15 clear seconds of music that you heard. It's a lot of work, but it's a challenging and fun way to combine science and geography.

Some of the additional challenges when tuning in Europirates are images and other stations. Images are false signals, caused by strong signals into a receiver that has weak image rejection. In the 1980s, when I had a lesser receiver, I often mistook images from Radio Canada International and the BBC on 6275 kHz for real pirates. It was extremely frustrating to frantically take notes for an hour, only to hear Big Ben toll at the top of the hour. Other problem signals are licensed broadcasters in the range. A few years ago, a Latin American station was broadcasting with a faint signal on 6299 kHz. Two other longtime "deceptive" stations are Vatican Radio on 6245 kHz and Zambia's National Broadcasting Corporation on 6265 kHz.

The easiest way to find Europirates is to make an initial sweep through the band in one of the SSB modes (I prefer USB). When you tune across an AM signal while in SSB mode, you will hear a tone that will vary in frequency as you tune across it. This is incredibly handy because you can hear very faint signals that you would otherwise miss in AM. One interesting phenomena of AM modulation (particularly if a station's audio is weak) is that the carrier is audible before the audio. For example, sometimes I will tune in a signal around 2300 UTC and I will only hear a weak carrier with no audio. By 0000 UTC, I will start to hear a few bits of audio, and by 0015, I might hear continuous audio.

In my initial band sweep, I write down where I hear dead carriers and signals with audio. Some of the dead carriers are licensed data stations (military or commercial ship-to-shore) that leave the transmitters on between transmitting data streams. I take note of the strength of the stations that are regular, such as Laser Hot Hits and UK Radio. For example, on the morning that I could hear LHH and Radio Borderhunter (Fig. 7-17) with three feet of wire as an antenna, I knew it was going to be a good morning and that I should try for some of the tough-to-hear stations. Then, I go through and listen for the stations that I think are from the Continent (they will fade out first) and go for the British Isles stations next. I take notes while recording the signals for reception reports.

Of course, if you live in a continent outside of North America, you can apply these techniques to DXing other stations. However, it works best with the Europeans because it's rare for more than two

Fig. 7-17 *Belgium's Radio Borderhunter, which has been widely heard in North America. Most interesting have been its low-power tests, where it was heard by several DXers with only 30 milliwatts of power!*

North American pirates to be broadcasting at the same time. The South American pirates are a little more DX-friendly, using high frequencies to reach distant locations. I doubt that many DXers outside of North America would care to drive out into remote locations to listen unless a number of stations were planning to test or it was a holiday that usually brings good listening results, such as Halloween.

A list of stations heard by Dave Valko and Andrew Yoder in January and February 2001.

3935	Laser Hot Hits	6210	Union Radio
5805.4	R.F. London	6220	Laser Hot Hits
6206.2	Staunder Radio	6235.3	Radio Black Arrow
6209.9	Radio Marabu	6235.3	Radio Foxfire
6210	R. Borderhunter	6240.5	Radio Brigitte

A list of stations heard (continued).

6241.5	Radio Black Power	6298.6	R.. East Coast Holland
6243.8	XTC	6299.8	VOTN/RECH
6266.1	R. Dr. Tim	6301.6	Radio Perfekt
6266.6	U.K. R. Int'l	6305	R. Shadowman
6267	Westcoast Radio	6306	R. Northlight
6270	Delta Lima Radio	6310	Radio Wonderful
6271.2	R. Dr. Tim	6399.9	Astoria Radio
6273.9	Astoria Radio	7459.5	Laser Hot Hits
6275	UK Radio Int'l	7474.2	BBMS/Ozone
6279.7	Radio Likedeeler	7480.1	R. Benelux
6285	VOTN/RECH	7484.9	VOTN/RECH
6289.4	Free Wave Hit R.	9290.1	R. Nova Int'l
6289.9	Alfa Lima Int'l	12256.5	Wrekin R. Int'l
6290.9	Radio Nova Int'l	21884.9	Alfa Lima Int'l
6291v	FRS Holland		

Pirates in South America and Oceania

For some reason, the pirate hobby radio has not taken off in regions of the world other than Europe and North America. One theory suggests that pirates are present all over the world, but most do not broadcast in English and so are ignored by much of the world. However, many of these stations are not hobby pirates as such, but instead are merely unlicensed commercial broadcasters.

The difference is that they closely imitate the other commercial stations that are on the air from their particular country and are not any part of a pirate scene, such as those in North America and Europe. This is not an indictment of those stations, but a reason why a station such as this from a foreign country would be very difficult to identify as a pirate. Peru is one example of a nation where the difference between a pirate and an unlicensed commercial station is difficult to determine.

Two more factors that affect whether or not a given area is likely to have pirate radio is the general political climate and the wealth of the region. For example, many countries are relatively poor and a radio hobbyist would probably not be able to afford the necessary radio transmitting and audio equipment to broadcast. On the other hand, some countries, such as China, have plenty of access to high technology, but the present political climate in the country rules out hobby broadcasting on shortwave.

Presently, one of the ripest relatively new area for pirate broadcasting is Oceania, where several wealthy Westernized countries are located. Two of the countries, Australia and New Zealand, sport shortwave pirate stations.

Radio G'Day from Australia announced its tests to the world over Radio Netherlands' Media Network program in 1992. After some test broadcasts, Radio G'Day was heard throughout Oceania and even in western Canada! Likewise, KIWI Radio (Fig. 7-18) from New Zealand ran tests in 1992. Although the broadcasts were heard well throughout the south Pacific, they weren't reported in North America. However, all of this changed in 1994, when KIWI called a number of listeners in North America and arranged specific tests on 7445 kHz USB. These tests were heard across all of North America, even into New England with occasionally fair to good signals. In fact, in spite of using only 350 watts of power, KIWI probably sent more QSLs to North America in 1994 and 1995 than all of the Europirates put together in the 1990s.

Although many pirates last for a handful of broadcasts, KIWI is a true long-term operation. The earliest transmissions from this

Fig. 7-18 *Graham Barclay of KIWI Radio (New Zealand) sitting in the studio.*

operator date back to 1977, but it wasn't until 1980 that the station was well heard with new, higher-powered equipment for AM, FM stereo, and shortwave transmissions. On December 4, 1980, the station, then known as "Radio Freedom," was raided by the New Zealand Post Office. But 11 days later, Radio Freedom returned to the AM band, although it was not well heard throughout 1981. As the situation was improving in 1982, the station was raided again by the New Zealand Post Office and the police. This time, the officials interrupted the broadcast while the song "They're Coming To Take Me Away!" was being broadcast! However, Radio Freedom returned in 1982 from a new location and reports were received from as far away as the Canary Islands. On July 6, 1985, Radio Freedom was raided once again and station operator, Graham Barclay, began to receive a great deal of local newspaper coverage. Of course, it was not long before the station, now known as KIWI Radio (Fig. 7-19), returned to the air.

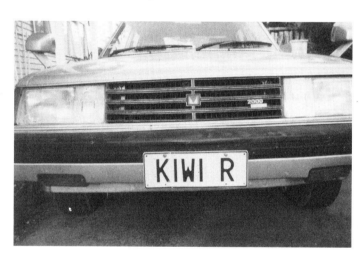

Fig. 7-19
Always brash, Graham Barclay even had a custom KIWI license plate made for the front of his car!

In the 1990s, KIWI was heard across the world, receiving telephone calls and reports from across the United States and Canada, Korea, Australia, Iceland, Germany, India, and more. In spite of very strict warnings from the New Zealand Post Office to cease operations, Graham kept broadcasting. It seemed that nothing could stop him from pirating. Well, broadcasting no, but pirating yes. In 1997, KIWI fell silent, not the victim of a government raid, but of better opportunities. Unstoppable Barclay, received a low-power FM license for a pop music station called Soundwave FM on 88.8 and 102.2 MHz. Although there have been rumors that KIWI would return to the air, even the freshly updated web page says "returning soon," nothing has been heard from the station since 1997.

Since the departure of KIWI, shortwave pirate radio in the South Atlantic has essentially fallen silent. The only station that has been heard is Radio Eureka International, an '80s rock Australian pirate that made several publicized tests to the world in 2001.

South America

In part because of the language and cultural barriers, it seemed unlikely that shortwave pirate radio would ever take off in Latin America. This is not to say that Latin America is devoid of unlicensed broadcasters. Particularly in the Andes Mountains of northern South America and the Amazon region of Brazil, shortwave broadcasters proliferate. And these broadcasters are far from the government-controlled voices from Germany, for example. Nearly all of the dozens of stations are tiny commercial community radio outlets run on shoe-string budgets.

Low-power shortwave community and hobbyist radio dates back to literally the first Latin American shortwave station. TI4NRH started in 1928, using only 7.5 watts of power from a homemade transmitter in Heredia, Costa Rica. Operated by Don Amando, the station was dedicated to international friendship and ran more as a hobby than as a business. Unlike the QSL policies of United Patriot Radio, TI4NRH responded to every letter, sending out more than 110,000 QSLs over the years, the very first of which was a full-sized chair, made from Costa Rican tropical woods! Amando, who had lived in the United States for several years, frequently traveled to meet people and broadcast in both English and Spanish to reach a broader audience. The legendary TI4NRH was frequently mentioned in radio and electronics magazines in the 1930s, and Don Moore wrote a fascinating article about the station in the March 1993 issue of *Monitoring Times*, after a visit to the original transmitter building and a long discussion with Don Amando's daughter.

Contemporary stations from the Andes are local operations, not quite following in the footsteps of the first Latin American shortwaver/international goodwill ambassador. These fiercely independent broadcasters often operate from a shack or an apartment with an ancient or homemade transmitter that might break down at any time. The stations transmit local music programs and ads, and the more remote stations include "communicados" (personal messages to those who live in the mountains, where no telephone service is available). The Bolivian stations even include numerous outlets operated by mining unions, which considering governmental instability, are often unlicensed and literally fighting to stay on the air. Unlicensed stations frequently turn up in Ecuador, Columbia, and Venezuela. In Peru and Bolivia, the situation is so chaotic that few people ever know which stations are licensed and which aren't.

As a result, the largest barrier to shortwave pirate radio in Latin America is neither, in most countries, the risk of operating without a license nor transmitting on shortwave. It's being connected to the international pirate scene. The first contemporary South American pirate was Radio Pirana International (Fig. 7-20), which was an active Europirate in the early 1990s prior to relocating. At the

Fig. 7-20 *A QSL card from Radio Pirana International, the first "well-connected" South American shortwave pirate. Pirana broadcasted for years from Europe before moving to South America.*

time, RPI was regularly active and using as much as 500 watts into a directional antenna array. While on vacation in South America, Jorge Garcia made a number of broadcasts from a homemade transmitter in 1993. When Garcia returned to his native land in 1995, he assembled a new transmitter and began testing again. In the six years since Garcia's return to South America, RPI has rarely been active, typically only with scheduled tests. However, the station has reactivated and has broadcast more in the Summer of 2001 than it has since 1996.

If Radio Pirana International was the first pirate, then Radio Cochiguaz (Fig. 7-21) jumpstarted the scene. In February 1997, Cachito

Fig. 7-21
*From the
Andes, Radio
Cochiguaz has
been widely
heard across
the world.*

Mamani, who, unlike Jorge Garcia had not formerly lived in North America or Europe, created a truly international, yet absolutely South American station. For example, the first broadcasts from Cochiguaz didn't just appear on the tropical bands (such as 60 or 120 meters), only to be found by American DXers. Instead, advance scheduling for the long test broadcasts on typical pirate frequencies was announced to numerous pirate listeners in North America and Europe. Soon, it had its own web page and was regularly relaying programs from pirates in Europe (primarily) and future pirates in South America. Mamani said he was inspired by "the pirate hobby stations, and I know about this through the foreign bulletins and web sites." Despite all of the European relays (more than two dozen different stations relayed to date), Radio Cochiguaz maintains a strong South American flavor by only airing Andean folk music in its own programs. Their web page describes the origins of the station: "When he and his friend Paco Jeréz got together, they suddenly came upon the name "Cochiguaz," which is of indigenous Quechua origin, it had an appealing phonetical sound to it, naturally born from its Andean roots, which was the main essence of the broadcaster in itself."

One of the first stations to be relayed via Cochiguaz was Radio Blandengue, which was a regular feature for RC's first two years. Blandengue operator Raul Gonzalez was a South American listener Blandengue (Fig. 7-22) took the relay process one step further by sending programs to Europirates for relays. Inspired by the successes

Fig. 7-22 *Radio Blandengue's high-frequency tests have provided listeners from around the world with solid signals from South America.*

of the relays, Blandengue purchased three transmitters, including one old tropical band rig, before they found one that worked properly. Before long, Gonzalez had the transmitter running and was operating in a similar style to Cochiguaz: e-mailed schedules, occasional long relays of other stations, and used a high frequency in the afternoon, then jumping to a lower frequency for night broadcasts.

Another station followed the lead of Cochiguaz by offering a similar service: the Andino Relay Service (Fig. 7-23), located "in a secret location, near to the beautiful and majestic Andes mountain range" began broadcasting in 1998, with the intention of relaying pirate stations around the world in South America. Despite sticking to frequencies in the 6900-kHz range, the Andino Relay Service was heard simultaneously in Europe and Africa during one broadcast in June 1998. Despite receiving letters from around the world, Pachakuti says that most of his listeners are in Chile and the bordering countries. Also, he said that their aging transmitter is not currently functional, but that ARS should soon be on the air.

Regardless, more stations are on the way. Zamba Radio features Brazilian rock, samba (the origin of the station name), pagode, axé, and toadas. To date, their programs have been aired via Radio Cochiguaz, but the operators are attempting to find a transmitter of their own. Radio Marabunta, another of the relayed programs, is

Fig. 7-23 *Like the other major South American shortwave pirates, the Andino Relay Service occasionally makes serveral-hour-long broadcasts with numerous relays of other stations from around the world.*

looking for a transmitter. Some of the other South American pirate-relayed programs include: Emisora Z del Dragon, Radio Bosque, Dark Pampa Radio, Radio Exotica, Radio Pasteur, R Sin Fronteras, and FLARS.

Except for Radio Pasteur and Zamba Radio, all of the South American stations are quite guarded about their broadcast locations. By contrast, all of the stations from Europe announce their country of origin and many of the UK and Dutch stations even name their town! None of the active South American pirates announce their country and Radio Cochiguaz has stated that they have transmitted from several locations in different countries. The only tip about any of these stations is that Cochiguaz has stated, "We must pay much attention to the Subtel (Chilean radio enforcement) and the security for us is the point number one." Of course, this comment might be a subtle obfuscation, rather than a tip of the hand.

The South American pirates are fascinating in part because they offer material not heard anywhere else. Sure, someone in the U. S. commercial broadcasting hierarchy might complain that it's irresponsible for someone to pirate when they should be transmitting on the Internet. But some of these stations might be in locations with limited (or no) Internet access. For that matter, many rural towns in the Andes only have electricity during certain hours of the day. Whether any of these stations are broadcasting from such areas is unknown, but their signals certainly do cover people in such regions. It's fascinating to imagine that in some mountain village, a teenage boy with almost no access to the mass media might be tuning in, listening to the peculiar combination of Andean folk music and German pirate radio.

The South American stations are all receivable in North America, and to a lesser extent in Europe and Oceania, depending on propagation and the frequency. Most of these can only be heard with a faint signal in North America, but I've heard some, such as the 14565-kHz broadcasts from Radio Blandengue with good, steady signals.

Frequencies of South American Pirates

6880U	Andino Relay Service
6925	Radio Pirana International
6925U	Radio Blandengue
6935U	Andino Relay Service
6950U	Radio Blandengue
6950L	Radio Cochiguaz
6965U	Andino Relay Service
11420	Radio Pirana International
11440U	Radio Cochiguaz
15465L	Radio Blandengue

An Outlook

Unlicensed radio activity should certainly continue at a high rate throughout at least the next five years in Europe. Although the Irish stations have essentially disappeared and the British shortwave pirates have declined, the recent influx of Dutch AM stations has perked up the scene. In fact, activity in 2001 is considerably higher than it has been at times throughout the 1990s. Also, many radio enthusiasts in countries such as the UK and Sweden abandoned pirate radio for independent commercial radio. Some of these people have returned, noting that the commercial radio became much more bland than they had imagined. I suspect that in the upcoming years, more will rejoin the ranks of the pirates.

For years, the pirate radio situation has been "out of control" in the United Kingdom, where dozens of stations are active. Nearly 15 daily FM pirates broadcast from London alone; some of these operate 24 hours per day. Clearly, the British authorities are not able to control the airwaves. This is a blessing for the British shortwave pirates because the authorities always focus on closing the FM stations. This enforcement technique allows the shortwave pirates to broadcast almost free from harm.

As noted previously, the older South American pirates are continuing to broadcast and some of the relayed stations are working on purchasing their own transmitters. Thus, it appears that the South American scene has not yet peaked. Imagine the success of these stations if they would only broadcast with AM modulation, rather than in SSB, for which a small percentage of South Americans have receivers.

As a result of these developments and the increased communications with listeners and pirates in North America (via the Internet), it appears that shortwave pirates from outside of North America will continue to make for plenty of interesting listening, whether you live here or abroad.

Political Pirates

Western society generally visualizes politics as a method or manner by which national governments are operated.

Although this definition is correct, it is not complete. Politics is an important aspect of everyday life, from the largest government to a social group consisting of a few members. Any time that several individuals convene as a group and make decisions together, they are engaging in political communications.

Furthermore, every time a radio enthusiast breaks the law by broadcasting without a license, he or she is making a political statement. Even if the pirate station considers its actions to be no more than a fun pastime, the operator has still chosen to break the laws of the country. By such standards, every pirate could be considered political by nature and would be included in this chapter. However, this chapter will deal only with pirates whose programming or raison d'etre is explicitly political in nature.

Politics is the motivating force behind most shortwave broadcasters, especially those operated by governments. From Moscow to Washington and from Pyongyang to Addis Ababa, billions of dollars every year are funneled into broadcasting political rhetoric to convince listeners to support certain causes. Many countries operate expensive local and international radio stations and use them to maintain control despite the unrest in their countries (widespread hunger, unemployment, crime, pollution, etc.). Broadcasting is obviously an important and powerful tool of these governments.

Broadcasting is also a tool for those not directly involved in government, such as civilians, military officers and troops, and leaders waiting to alter or overthrow the establishment. One of the first signs of power within a revolutionary organization is ownership of a broadcasting station. It takes a dedicated staff to produce programming and keep the equipment in operation, a powerful or sneaky enough outfit to not get caught by government troops, and the money to finance it. Only large revolutionary groups can do it all, and few ever fail to operate radio stations as soon as the resources are available. Such hidden, extralegal stations with political aims are known as clandestine broadcasters. In North America, several clandestine stations have operated from El Salvador and Nicaragua in recent years during periods of civil war and insurgency.

Many shortwave listeners enjoy tuning in to clandestine radio stations because of the unpredictability of their operations and the first-hand information that is sometimes revealed days later by the news media in the United States. But all too often, clandestines are little more than outlets for shallow, emotion-based programming. Complaints against the status quo are reduced to name-calling and occasional fake news reports concerning the recent victories by revolutionary troops are common. The government radio stations retaliate by occasionally faking news reports to suggest superior power. The result of the propaganda war is confusion and rather boring attempts at credibility.

Shortwave pirates in the United States, on the other hand, initially rebelled against the established programming by avoiding all politics. Most just stuck with playing music, reading listener letters, and taking telephone calls. Some stations such as Ozone Radio International, from politically volatile Ireland, have public policies against the mention of any politics on the air. WHYP, Psycho Radio, KIPM, Voice of the Angry Bastard, Ground Level Radio, Take It Easy Radio, and many others take no particular political stance. The most popular programming of many pirate stations has been that which ridiculed all politics without taking sides or delving into the issues.

However, a few years after shortwave pirate radio took off in North America, a small departure from the nonpolitical started occurring. One pirate, the Voice of Tomorrow, is now considered by many to have been a clandestine. Stations such as KNBS, Tangerine Radio, WGOP, WEED, the Voice of the Real World, and a few others have become known as political pirate stations. Unlike Europe, where most pirates are mainly emulators of commercial broadcasting in the United States, North American stations have been moving toward alternative programming and picking up a few political radicals along the way.

Before the North American pirates began showing their political colors, shortwave listeners often complained that the stations lacked intelligence and were nothing more than clones of the legal FM rock stations that existed throughout the country. Then, for a time, some listeners complained that American pirates were too political and had lost the carefree nature that made them fun. Now, it appears that the pirate scene has balanced out again and listeners should be prepared to expect anything! With the specialization and addition of the counterculture into free radio, more political stations will surely appear, although it is safe to assume that at least as many nonpolitical pirates will crop up as well.

Current Political Pirate Activity

Truly political or semi-clandestine stations are rare in North America and especially in Europe. Over the past 30 years, only a handful of European pirates were even slightly political. As a result, this chapter is more of a history of different political-based pirates that have broadcast since the late 1960s. You can't expect to turn on the radio and hear Nazi programming from The Voice of Tomorrow, socialist talks from the Crystal Ship, or anarchy from Tangerine Radio or Defiance! 90. They're all gone.

However, as of this writing (July 2001), the closest to a true North American clandestine station ever is very active. Because it is the only clandestine that you can hope to tune in right now, I've separated this section and placed it at the beginning of the chapter.

The first broadcast from Kentucky State Militia Radio was noted on March 10, 2001, on the unusual frequency of 3260 kHz, well below the 80-meter amateur radio band in range where some small Latin American stations broadcast. Before long, dozens of radio listeners were tuning down to 3260 kHz to hear the nightly 10 P.M. broadcasts from KSMR. The station programming has been peculiar from the start, consisting almost entirely of satellite downloads of other patriot-movement radio programs mixed with live, often long, commentaries about the current state of politics in the United States and the rest of the world.

Major Steve Anderson (a major in the Kentucky State Militia) hosts all of the programs. The general line of thought from KSMR is that the current government in the United States is socialist and will soon destroy the average person. Some of the issues supported by the station include a total states rights government (a federation of states, rather than a republic), an end to all U. S. participation in such foreign organizations as the U. N., an end to all government agencies (such as the IRS, CIA, FBI, FCC, etc.), and a currency that is 100% backed with gold or silver. Also, Anderson frequently displays some racist philosophies, most significantly the Nazi-like belief that there's a Jewish conspiracy that controls the world.

United Patriot Radio Mission Statement

http://www.posse-comitatus.org/unitedpatriotradio/

As the Jew World Order owns and controls all the media, except for a few Patriot stations here and there, the "good" shows have been all but silenced by removal from air time. United Patriot Radio is dedicated to keeping the "endangered species" of the Patriot shows ON THE AIR! Freedom of speech must include those that are considered too controversial by licensed commercial stations. By being licensed, they place themselves under the rule of the destroyers of freedom. As they are commercial, they may not risk their income by angering their sponsors, or present information that would bring down the wrath of the likes of the ADL, in the form of financial punishment. UPR is not licensed, or commercial in nature.

•It is strictly a free, First Amendment station. There is no income to lose, or jurisdiction by any agency by a license. We exist on listener donation ONLY. We don't charge for air time, and advertise for Patriots FREE! As you can imagine, this causes considerable consternation among the money whores. We care nothing for the railing of the likes of the commie ADL, and their minions. To my knowledge, at this time, we are the only one carrying "The Intelligence Report," and Pastor Wickstroms' show. My show, "The Militia Hour" is also exclusive to UPR.

•As I am Identity myself, I feel very blessed to be able to get our message out through Pastor Wickstrom and the Sheriff's Posse Comitatus website of his associate Pastor Kreis. We have letters of support coming in in overwhelming numbers, and the negatives are extremely few. To the fans of "The Intelligence Report," we have been a lifesaver, and are also thankful for that blessing. I say to any detractors, that we realize that there are controversial views expressed on the shows, and if you dislike one show, but like another, listen only to what you like. Free speech must include the controversial, as well as the "mainstream". The "mainstream" shows out there are not cut off from air time, and we don't feel the need to carry un-endangered broadcasts, commonly heard on many stations. It is our mission to provide a free voice to those wrongly censored for their "politically incorrect" views. The First Amendment applies to all, including us. We fully intend to use it.

Anderson has operated the station nightly and also sometimes in the mornings with hundreds or thousands of watts of power and announced his general location. Obviously, Anderson is not worried about an FCC raid, and some listeners have commented that he must be anticipating some sort of Waco scenario.

By July 2001, KSMR's name was changed to United Patriot Radio (UPR) and the frequency was moved to 6900 kHz, not far from the 43-meter pirate band. Radio Bingo followed UPR down to 6900 kHz with some bingo programs, announcing that Steve Anderson had won the big $1000 prize. No doubt, some of this programming was the result of UPR's nightly broadcasting, the general ideology, and also the general distain for radio hobbyists. Despite the regular broadcasts, UPR refused to QSL any reports. "We are not a ham station, or commercial, and our target audience is not into that stuff. They are militia," said Anderson. Of course, this contradicts the dozens or even hundreds of clandestine stations that have QSLed over the years. For example, it's difficult to imagine the Voice of the Tigre Revolution, the voice of the armies that overthrew the communist government in Ethiopia, as not being "serious enough" because they regularly QSLed reception reports. Despite the potential positive PR that could be gained by issuing QSLs, Anderson holds steadfastly to the issue, "We are not here for a hobby."

In addition to not QSLing, Anderson refuses to describe his transmitter and will not release any photos or other hard information about the station, aside from the issues that he supports. As a result, a few (apparently incorrect) rumors have spread about the station. First was the inamicable split between the Kentucky State Militia and UPR, which was mentioned by some on the Internet. Anderson said that there was no falling out. In late July, he stated, "As a matter of fact, Charlie Puckett (of the Kentucky State Militia) will be the host here tonight on the Militia Hour. Internal operations of the Militia beyond that are classified."

The shortwave DX program "World of Radio" reported that United Patriot Radio had been raided by the FCC on June 27, at 0500 UTC. Evidently, this proclamation was premature because UPR never missed a beat, continuing the nightly schedule with the same high-powered transmitter.

Although Anderson claims that UPR is not a pirate station and that the FCC does not have the jurisdiction to impose any regulations on the broadcaster, UPR obviously raises questions related to each group. Whether you call it "pirate radio," "freedom radio," or "chop suey," UPR matches the broad criteria of stations that are covered in this book—even if the programming is different. UPR has already inspired the United Militia Bingo Radio parody stations. Will the station inspire other patriot radio stations to form a sort of militia radio network on shortwave? Will UPR inspire the younger, left-leaning FM pirate organizations to put a counter-UPR station on the air?

And just because Anderson says that the FCC has no jurisdiction for taking action against UPR, doesn't mean that agents won't be sent out to close the station. Traditionally, the FCC has never allowed an unlicensed shortwave station to broadcast on a nightly schedule for more than about two weeks or so. After months of broadcasting, what is the FCC waiting for? Are they hoping that the station will either lose interest or fall victim to a transmitter failure? Is this a dormant period for FCC activity and they simply haven't noticed that UPR has been broadcasting? Are they waiting for approval to use an armed force from other government agencies to close United Patriot Radio? If and when UPR is closed by the FCC, will the station present an armed resistance and force a government shootout, all broadcast live on shortwave radio? Stay tuned!

Right-Wing Pirates

Conservative pirates have not been numerous because popular politics in the country have only recently swung over from the left. During the 1960s, some liberal, radical, and revolutionary groups owned radio stations. In the early 1980s, such liberal groups faded and were replaced by increasingly powerful right-wingers.

The new American radically conservative organizations pull their ideological views from the largely "traditional" views of racism and xenophobia found in the Ku Klux Klan (KKK) and the National Socialist Worker's Party (Nazi). The KKK has grown in size; the sects of Nazism, the National Aryan League, and a loose federation of skinheads, have received wide media coverage for attempting to create a homeland in the Northwest and for scattered acts of violence.

When a new station, Radio Vanguard International, briefly appeared in the Summer of 1983, with a professional format of pop music, few listeners showed more than a light interest in the broadcaster. Then again, it was barely heard by anyone and the loggings of the station that did appear in the hobby media were inconspicuous. But "the soldiers poured out of the Trojan horse" the next month when listeners across the country received photo postcards announcing the initial broadcasts from the Voice of Tomorrow, formerly Radio Vanguard International. The card listed a test schedule of six broadcasts that would be transmitted on 6240 kHz and 7410 kHz during a weekend in mid-June.

To the surprise of radio listeners, the Voice of Tomorrow (Fig. 8-1) did arrive on the pre-announced frequencies at the proper times. Until this time, very few of the pirates that had scheduled broadcasts in hobby publications ever showed up. The Voice of Revolutionary Vinco, Radio Fluffernut, Radio North East Michigan, CHHH, Radio Ohio International, and Radio Prophylactic International were

the

Voice of To-morrow

transmitter Baltimore, MD 2 kW

studios located in Providence, RI

21728 Campfer und Silvaplana

Fig. 8-1 *A placid-looking QSL from the Voice of Tomorrow, a neo-nazi pirate that operated occasionally on shortwave and above the AM band from 1983 until 1990.*

just a handful of stations that had never met their published operating schedules. On the other hand, few pirates were ever creative enough to send a professionally printed postcard to potential listeners around the country. (WUMS was one of the few stations to try this method before the DXers found that the production values of the programming were as professional as the printed "camfer and silvaplana" photo-card schedules they received in the mail. But this was only a small consolation for the hateful blend of racist commentaries and Nazi music. The hobby radio press universally denounced the station for the programs. Some bulletins and reporters refused to print loggings for the Voice of Tomorrow, especially after it announced that it would only remain on the air if it was actively supported with listener letters.

Hot debates followed in the hobby press about reporting the Voice of Tomorrow. Was it ethical to report such a station in the hobby bulletins? Many claimed that giving the operators publicity would only satisfy their need for attention and if DXers stopped logging the station, it would cease broadcasting. On the other side were the "free flow-ers" who believed that all opinions and information should be voiced; anything less would be censorship and an infringement on human rights. Besides, they argued, ignoring the station would not make it go away because the operator appeared to be in it for more than notoriety. Perhaps the best argument for reporting the station came from Robert Horvitz, ANARC Executive Secretary, who said, "Ignoring such people doesn't make them go away and quoting them doesn't necessarily help them."

The Voice of Tomorrow had a prominent and easily identifiable signal. Of the four frequencies used with their transmitters, only 7410 kHz was commonly used by other pirates. Additionally, their transmitter power was great enough for the station to be heard across the eastern half of North America with very strong signals. They claimed to be running 2000 watts, which could have been true. The equipment had been modified for high-fidelity audio response and sounded as clear and crisp as a legal broadcaster.

Their programming was also conspicuous. The sound of a howling wolf was repeated several times as an interval signal and "Tomorrow Belongs to Me," the pre-World War II Nazi youth song, was the theme. Announcer Phillip Carey often read commentaries against ethnic groups, with light pop and rock songs separating the segments. Some programs were merely the same commentaries with different songs inserted in the spaces. But in the mid- to late-1980s, some programming appeared to originate from outside sources. In one of its last years on the air, the Voice of Tomorrow aired a critique and biography of American poet Ezra Pound and an interview with a spokesman for CIFTRA, a white separatist group in Spain, among other things.

Programming from outside sources and the frequent change in maildrop locations suggested that at least a small organization operated the Voice of Tomorrow. It could even be that Phillip Carey merely had access to Nazi magazines or newsletters or sent for cassettes of information that could be edited down to suitable program segments. Maildrops for the Voice of Tomorrow were located in Bristol, Virginia, Ferndale, Michigan, and Clackamas, Oregon. One *ACE* member reported that the Oregon address was owned by *The Patriot Review*, a conservative interest tabloid, so the station might have had ties to much larger organizations.

The Voice of Tomorrow disappeared after approximately eight years of irregular broadcasting. By the end of its career in the early 1990s, the station was limiting its output to less than five broadcasts per year. Like the Nazi skinheads that have achieved great amounts of publicity in comparison to the actual limited threat involved, the Voice of Tomorrow was often exclusively mentioned in mainstream newspaper and magazine articles about pirate radio, and it was the only North American pirate mentioned in the editions of *Passport to World Band Radio.*

The truth of the matter is that the Voice of Tomorrow made fewer broadcasts in eight years of operating than many lesser-known stations, such as One Voice Radio, made in one year. Many of the late 1980s and early 1990s listeners who heard the station for the first time after reading about the Voice of Tomorrow for years stated that the station was actually more boring than offensive (or enlightening).

After all of the publicity and the years of irregular operations, the Voice of Tomorrow quietly closed its doors, but not because of an FCC raid or lack of interest. It appears that the Voice of Tomorrow aspired to reach for bigger dreams and audiences. In the early 1990s, a new radio program, National Vanguard Radio, appeared on 7355 kHz over licensed broadcaster WRNO and via a handful of commercially licensed AM broadcasters. The programs were very similar in content and production as those from the Voice of Tomorrow. At the time the second edition of this book was written, National Vanguard Radio had changed its name to American Dissident Voices. It is just another paid program on the international shortwave bands and it appears that it will stay that way.

Since the absence of the Voice of Tomorrow, only one other ultraconservative shortwave pirate has appeared on the bands. In 1992, WCCC (Consolidated Conservative Confederacy) appeared on the shortwave bands for several broadcasts. Listeners said that WCCC (Fig.

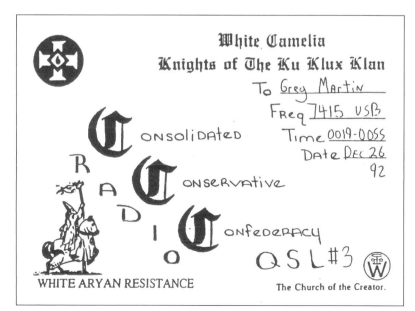

Fig. 8-2
Little more than a one-shot pirate, KKK-related WCCC operated near Christmas 1992.

8-2) sounded like it was operated by the delinquent son of the Voice of Tomorrow–a very young male announcer with an amateurish production of white power rhetoric. After several broadcasts, the station disappeared.

Right-wing pirates, such as the Voice of Tomorrow and WCCC, have been rare in North America but several have existed throughout the rest of the world. England experienced Radio Enoch in the late 1970s, a mouthpiece for People Against Marxism (PAM), a populist right-wing group. Two announcers presented the programs, which contained news, commentaries, features, letters, and music. The organization denounced immigration into England, supported the policies of South Africa, and argued for a strong defense to guard against the "communist threat." As in the case of the Voice of Tomorrow, the pirate hobbyists unanimously denounced the operations of Radio Enoch. In fact, English pirates in opposition to Radio Enoch began jamming its signal.

The station's programming might have been tolerated by the other pirates if it were not for the obnoxious attitudes of the announcers. For example, when Radio Enoch was initially jammed, the announcers responded by saying that the jamming was done by "pathetic souls with flea-powered transmitters." The jamming did, in fact, destroy the major portion of the signal, rendering it virtually indistinguishable to the listener. Radio Enoch eventually faded out with the beginning of the 1980s. Maybe the operators saw that their radio voice was accomplishing little when it was audible and jamming made it accomplish even less.

Not all political pirates are as serious and hate-oriented as the Voice of Tomorrow or Radio Enoch. In fact, few are. Most are merely radio hobbyists with a sociopolitical conscience. One odd exception to this rule was the Voice of Free Long Island, an average shortwave hobby pirate from the late 1980s who expressed his conservative Republican political views in hope that they would protect his station from an FCC raid. The main operator, The General, believed that because his views corresponded with those of the Reagan and Bush Administrations, the FCC would not waste its time chasing a colleague. According to the FCC, this theory is bunk; the FCC will raid any operation that violates the broadcasting rules regardless of the program content.

Even though these opinions have been expressed, some more middle-of-the-road conservative politics have appeared across the pirate bands. One of the longer-lived conservative pirates that frequently delves into politics is Omega Radio. Omega Radio is a conservative Christian station. Although station operator Dick Tator often plays rock music or discusses various aspects of Christianity, he also occasionally airs his conservative political views, sometimes with fake advertisements from the Rush Limbaugh radio program.

A contemporary of Omega Radio, WGOP, broadcast several brief pro-Republican programs throughout 1992, the year of the Presidential election. WGOP appeared to be a counter against the "Don't Vote Republican" sloganeers from Radio Newyork International (then being commercially relayed via WWCR) and the pirate Radio DC. Strangely, WGOP disappeared after Bill Clinton was elected President. Considering the rapid decline in Clinton's popularity early in his first term, it was surprising that WGOP didn't continue its pro-Republican broadcasts.

From studies taken of free radio operators and from their written opinions in shortwave bulletins, a high percentage apparently support the small Libertarian political party. Although very conservative, the Libertarian Party leans neither towards fascism nor racism. The ideological difference between Libertarian thought and strong conservative Republicanism is that the former believes in a weak central government with little interference into local and private affairs, while the latter group normally follows the opposite. The center of the ideology of Libertarianism is the absolute freedom of the individual without government interference. These certainly are politics made for someone who broadcasts without a license.

Although the station did not announce itself as being related to the Libertarian Party in any way, Free Radio One (Fig. 8-3) was one of the most aggressively Libertarian pirates that has operated in North America. From its start in the Spring of 1989, Free Radio One broadcast talk radio programming about "Christian patriots" and "God-given rights." Announcer Dr. David Richardson despised the federal government and often featured interviews with people who had fought the law (sometimes literally). Free Radio One was extremely active throughout 1989 and announced plans to create a network of stations across the United States to free the country from the liberal media. The station disappeared toward the end of the year (possibly as a result of an FCC raid?), but Free Radio One appears to have been much like the Voice of Tomorrow—an operation that almost qualifies as a clandestine from the United States.

Left-Wing Pirates

The creation of liberal, socialist, or communist pirate stations has partially resulted from the fact that the U. S. had conservative Repub-

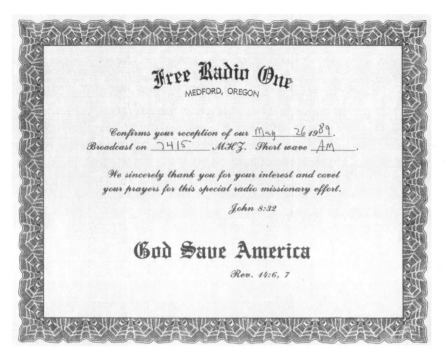

Fig. 8-3 *Dr. David Richardson offered a sort of 1850s-style conservative isolationist politcal forum on Free Radio One that foreshadowed popular patriot and militia programming of the last half of the 1990s.*

lican control of the government through most of the 1980s. Furthermore, many liberals still exist and the left-wing dedication to being outspoken is still strong. Although the mass media has been more liberal than the government throughout the 1980s, those leaning toward socialism feel that they have few voices and are in the minority. So, they turn to pirating and, as a result, left-wing pirates are numerous.

Of course, the Falling Star Network, the Voice of the Purple Pumpkin, WGHP, and a few other late 1960s pirates, pioneered political revolution of free radio. As was mentioned before, most pirates before the 1960s were either young radio hobbyists with a flair for the technical or young broadcasting hopefuls. The sociopolitical revolution changed this by replacing the hobbyists with young people searching for an outlet to present their opinions and music. Anti-establishment attitudes from the hippie culture gave listeners an interesting break from music and technically oriented pirates.

But with the passing of the 1970s, the popular liberal ideology also faded away. No truly political stations operated after the pirate explosion of 1976 and throughout the rest of the decade. Finally in 1982, after six years of partying (Voice of the Voyager, Radio Liberation), music (KVHF), telephone calls (WCPR, WGOR, WFAT), DX programming (Voice of Syncom), and comedy (Radio Clandestine, Radio Confusion), left-wing radio returned in the form of two stations, one on shortwave and the other on the AM broadcasting band.

First on the air in early 1982 was KPRC ("Pirate Radio Central," not to be confused with a legal Texas broadcaster using those call letters), a phone-in AM pirate that pumped strong signals out across the Eastern half of North America. Unlike most pirates, KPRCs style, technical quality, and programming on 1616 kHz never varied from its first evening on the air.

Main announcer Pirate Joe single-handedly led an unusually tight, but relaxed, format of 1960s music, commentaries, and telephone calls. Commentaries on KPRC ranged across almost every topic covered by the Democratic Party, but the favorites were war and nuclear weapons (in opposition to both). Seeing that the main topic appearing on KPRC was politics, a number of listeners dubbed it a clandestine. Pirate Joe was less violent and serious than someone like Phillip Carey, so KPRC could never truly be considered "clandestine" in nature. In fact, callers often changed the station's topics to discussions about pirate radio, and sometimes listeners would announce pirate tips. (More information about KPRC and the other affiliated stations, such as the Falling Star Radio network and Radio Newyork International, are contained in Chapters 2 and 6.)

The easy identification and political programming might have made for interesting listening, but it also created a target for others to tamper with. Although KPRC was rarely jammed, several times stations transmitted comments about the programming over the station's signal.

Originally, Pirate Joe had announced for another New York City station, KSUN on FM, in the early 1980s until KPRC was created in 1982. At this point, the KSUN transmitter apparently was used for KPRC on 91.5 MHz as well. Later on in 1983, the group ran a transmitter on 6275 kHz shortwave that was heard across the entire country. KPRC always "stacked" these transmitters in parallel, making a predictable arrangement and schedule.

The 91.5-MHz transmitter was used at least once per week. Sometimes, usually once or twice per month, the AM transmitter was thrown into action. From the time that the shortwave equipment was first used, it was always run in parallel with the other two transmitters. But the shortwave transmitter was never in operation without the AM, and likewise, the AM never ran unless the FM was on.

The FCC searched desperately for KPRC as they would for any operation heard across the country (and especially New York City) on AM, FM, and shortwave for as long as eight to ten hours per transmission. No one knows where the other transmissions emanated from, but the Belfast, Maine, FCC field office tracked the broadcasts on 1616 kHz to a location in northern Maine, not New York City. Surprisingly, these broadcasts were then traced to a building housing the antenna of WOZW, a legal FM station, according to *Billboard* magazine. The pieces began to fit when it was found that the owner of WOZI/WOZW from Presque Isle was Allan Weiner.

Weiner operated the Falling Star Network along with J. P. Ferraro in the early 1970s (see Chapter 1) and ran into many difficulties when attempting to work with the FCC in the past. After years of work in the radio industry, Weiner eventually pulled himself through the established system and bought the two legal stations in Maine. When the FCC traced KPRCs signal to WOZW transmitting building, they claimed that Weiner refused the agents entrance to inspect the facilities. Although the FCC charged that Weiner had broken the law by broadcasting KPRC, he claimed to have no knowledge of the incident. Regardless, all broadcasting of KPRC ended at this point.

In the Autumn of 1984, several months after the silencing of KPRC, Weiner and Ferraro teamed up again on the KPF-941 project. KPF-941 was the callsign of a legal remote broadcast unit licensed to WOZI/WOZW on 1622 kHz. A 100-watt transmitter and a vertical antenna were installed in Yonkers, New York, the former location of the Falling Star Network.

Experimental broadcasting from KPF-941 began in late 1984, with mostly 1960s album rock music hosted informally by one of the staff members. KPF-941 never became a political voice; however, its similarities to KPRC make it a sort of ending to the story. The 100 watts of output power from a Western Electric broadcast transmitter covered the New York City area and the rest of the Northeast faintly, but it certainly was not comparable to the powerhouse KPRC was in terms of range.

Although KPF-941 claimed to be legally operating as a production tool for WOZI/WOZW in Maine, the FCC was critical of the transmissions. Instead of just offering a remote facility for WOZI/WOZW, KPF-941 played music and appeared to be broadcasting to the general public, especially since a telephone number was given for listeners to call. By the end of November, after broadcasting for several weeks on 1622 kHz, the station received a telegram notifying the operators of a discrepancy between the station's transmissions and the broadcasting laws of the United States. The next day, KPF-941 was closed by the FCC.

Weiner and Ferraro still claimed that KPF-941 existed within the rules by which it was licensed, but the FCC felt differently. Because of the legal conflicts caused by the operators, the FCC took action against Weiner and reviewed his qualifications for owning radio stations. Because of their past confrontations with Weiner, the FCC revoked all of his broadcasting licenses. Since he lacked the funding to take the decision to court, he was forced to sell the radio station at a 25% loss to a minority group in exchange for the renewal of his licenses. Still searching for a loophole to bring their style of radio to the New York City area, Weiner, Ferraro, and a group of radio enthusiasts and pirates pooled their funds and created Radio Newyork International two years later (see Chapter 6).

Also beginning in 1982 was The Crystal Ship (Fig. 8-4), a widely heard but less notable left-wing shortwave hobby pirate that operated on 41 meters. Throughout its broadcasts in 1982 and 1983, little of the political programming was audible. Besides having a limited frequency range on the audio, the transmitter also popped, squealed, and coughed out fits of frequency modulation (FM) while operating in the amplitude modulation (AM) mode.

After it had caused interference to other stations and had accomplished little itself, the operators of the Allied Knight-Kit T-150 transmitter tore it apart and repaired it. The staff found that whoever built the kit 20 years earlier made between 12 and 20 very big mistakes in wiring and quite a few parts needed replacement. An RF feedback problem was also solved with a new audio

The Crystal Ship

QSL

Shortwave

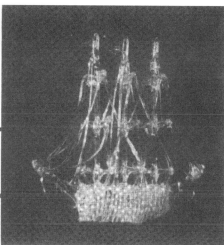

Fig. 8-4
With more of a "junior socialist" platform, The Crystal Ship broadcasted from 1982 to 1984.

board and rewiring.

With the audio and modulation problems finally corrected, listeners could understand what The Crystal Ship had to say. Much of their ideology, slogans, and the station name were derived from the 1960s rock group, The Doors. From a sign-on featuring segments of Doors' songs, including "We Want the World and We Want It Now" and a poem that pieced together a smattering of The Doors' and the operators' ideals, the overall philosophies were often music-related. Even pirate ideals were condensed in a theme song, "We Want the Airwaves" by The Ramones.

When The Crystal Ship was not playing music or music-related segments, announcers The Poet and The Radical discussed a variety of current events on the air. Usually the two denounced the U. S. foreign policy in Central America and other nations around the world, but topics including censorship in the music industry, FCC limitations on broadcasting, the draft, and the Reagan Administration were also aired. Although The Crystal Ship took a liberal Democratic stand by endorsing Jesse Jackson for President in 1984, programming was often described by both listeners and station personnel as socialist in nature.

Just as The Crystal Ship finally began to offer coherent, well-organized programming through a properly functioning transmitter, it suddenly became less active by August and then disappeared. It is odd that a station as active as The Crystal Ship, which had overcome many problems (technical and otherwise), would disappear without a trace.

Another political music station that suddenly disappeared was Rebel Music Radio, a 1983 holiday pirate from the New York City area. Like KPRC, the main focus of programming from Rebel Music Radio was centered on the issues of nuclear weapons and war. Songs such as "Masters of War," "You're in the Army Now," and "The Draft Registers" rounded out their "golden protest weekends." The station opened on 1616 kHz like KPRC, but with a much weaker signal that was barely audible farther than 100 miles outside of the city. Rebel Music Radio was first heard on May Day 1983; it returned on Christmas Eve and New Year's Eve, but has not been heard from since.

Nonviolent political radicalism reached a peak in 1984 when Tangerine Radio began broadcasting on shortwave. Raunchy Rick, announcer and owner, faithfully preached a steady strain of anarchist values and beliefs with practical applications to everyday life (Fig. 8-5). Unlike the common definition of anarchy–a temporary chaos or lack of control–Tangerine Radio was among those who

IMAGINE THERE'S NO COUNTRIES
IT ISN'T HARD TO DO ~
NOTHING TO KILL OR DIE FOR
AND NO RELIGION, TOO ~
IMAGINE ALL THE PEOPLE
LIVING LIFE IN PEACE ~

IMAGINE NO POSSESSIONS
I WONDER IF YOU CAN ~
NO NEED FOR GREED OR HUNGER
A BROTHERHOOD OF MAN ~
IMAGINE ALL THE PEOPLE
SHARING ALL THE WORLD ~

YOU MAY SAY I'M A DREAMER
BUT I'M NOT THE ONLY ONE.
I HOPE SOMEDAY YOU'LL JOIN US
AND THE WORLD WILL
    LIVE AS ONE.
            —LENNON

QSL

to _ANDREW YODER_
GMT date _OCT 7 '84_
time _0305-0422_
freq _7421_
power _75 W USB/20 W AM_
antenna _V-BEAM_
(X) from our xmtr
( ) relayed

Thanks for your
report.

x _Laundry Dick_

TANGERINE RADIO

Fig. 8-5 *Tangerine Radio, the most creative and serious of the anarchist shortwave pirates, primarily broadcasted in 1984 and 1985.*

believed in anarchy as a political system. The nonviolent anarchists believe in the destruction of all established forms of institutions and management with a peaceful new society based on individualism arising from the remains.

Anarchism today receives its worst stigma from the violent segment of the original political movement of over 70 years ago. During that time, anarchism was divided between the violent and nonviolent sects. The violent half was the most extreme left-wing political group of the time, committing many murders of government leaders, including President McKinley in the early 1900s. This violent sect eventually became absorbed into the communist party, and today the only organized portion of anarchy left is run by the nonviolent. However, hardcore punk music is often synonymous with anarchy and sometimes supports violent overthrow of the establishment.

Just because today's anarchist movement largely condemns violent actions, it cannot be considered nonaggressive or restricted to the laws of any country. Since anarchists do not consider themselves to be living under the law of any country, "Robin Hood-style" robberies and like actions are encouraged. In every program, Tangerine Radio featured segments suggesting ways the common anarchist can destroy the establishment, such as tampering with time cards to receive extra money at work, methods of wreaking havoc on the

nation's food supply, how to start your own pirate radio station, *etc.*
Although no widespread destruction of property was noted across the
shortwave listening community following Tangerine Radio broadcasts,
they might have drawn more attention from the CIA or FBI (as well
as the FCC) if they had been more frequent.

Tangerine Radio programming, like most other political pirates, was well-organized and well-produced. Every program featured
a different theme that is carried with music, commentaries, and
sometimes fake commercials. Antiwork and antipolice programs were
aired, along with their perspective of the Vietnam War, a Tangerine
Radio telethon, and "Adventures in Anarchy." Although commentary-
oriented programming is sometimes a bit dry, Raunchy Rick proved
his production talent with high-quality comedy shows aired on an
irregular basis.

One deficiency that proved especially harmful to the popularity of Tangerine Radio was their Hallicrafters HT-32 transmitter.
Whether broadcasting with 100 watts on USB or 25 watts on AM on
41 meters, the audio was always a bit raspy and unclear. The few
programs that have been heard since the Spring of 1985 used clear
AM modulation, so they were probably relayed by another station.
Although it does not appear that the Tangerine Radio transmitter has
been in use for many years, station personnel are still in contact with
the hobby press and could easily return again in the future.

Since the decline of Tangerine Radio, a number of other
anarchist pirates were widely heard in the early to mid-1990s, including Anarchy One (Fig. 8-6), Radio Anarchy, the Voice of Anarchy, and
Defiance! 90. Anarchy One and Defiance! 90 featured the most

Fig. 8-6
*Anarchy One
featured
politics,
including
audio chunks
from the
Great Atlantic
Radio Con-
spiracy.*

serious political programming from these stations, with many commentaries. With more than 20 broadcasts in 1993, Anarchy One was also the most serious of these stations, in terms of activity, as it was heard with excellent signals across much of the western half of North America.

Radio Anarchy (Fig. 8-7) was aligned more with hardcore punk rock than overt political proselytizing. Since 1990, the self-proclaimed "enemy of the state" has been heard well, but a penchant for low-power transmitters has restricted the signal to the West Coast. The Voice of Anarchy is the oddball of the group; it has played a huge variety of music from swing to African pop to hardcore punk rock. In the case of the Voice of Anarchy, the anarchist ideals appear to be the personal, rarely expressed views of the operator, Leonard Longwire.

Another addition to the left-wing radio club, KNBS (Fig. 8-8), began broadcasting in early 1985 as a primarily one-issue shortwave station. Rather than taking an entire platform of issues like other political pirates, KNBS satisfied itself by only dealing with one: the decriminalization of marijuana. KNBS, a sort of phonogram for cannabis, was hosted by Phil Muzik, who formats his programming in a fashion similar to Tangerine Radio. In fact, Raunchy Rick has even appeared several times as a guest host on KNBS. Muzik editorialized between songs that support his statements. Occasionally, a few fake commercials, dealing with either marijuana or other liberal issues, were also thrown in.

Like Tangerine Radio, KNBS programs are very well-organized and -produced. So far, nearly every facet of marijuana has been covered to this point by KNBS—its history, uses, reasons for being banned by the United States government, actions being taken by

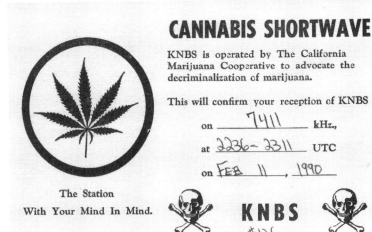

Fig. 8-8
KNBS was the first of the many pro-marijuana pirates that have broadcasted over the years.

CANNABIS SHORTWAVE

KNBS is operated by The California Marijuana Cooperative to advocate the decriminalization of marijuana.

This will confirm your reception of KNBS

on ____7411____ kHz.,

at 2236 - 2311 UTC

on FEB 11, 1990

The Station
With Your Mind In Mind.

KNBS
#126

those who support the decriminalization of it, *etc.* These editorials occurred regularly throughout the program and last several minutes. KNBS has been occasionally heard since 1985 with consistent programming. In the early 1990s, KNBS pulled off a one-shot broadcast from Canada; the station was then called CNBS.

Inspired by KNBS, other pro-marijuana stations, such as WEED and Radio Free Euphoria, were created. Neither of these stations took as serious a stand for marijuana, although both are almost entirely dedicated to broadcasting on the topic. WEED's programming consisted of a mixture of '70s hard rock and older punk rock, mixed with snips of audio from a variety of speeches. Radio Free Euphoria has spent the past ten years airing drug-related music, creating skits and parody songs, and making drug puns. See Chapter 3 for more station information.

As Radio Newyork International was attacking the Republican Party via WWCR in 1991 and 1992, another political pirate began developing. Radio DC started broadcasting in 1991 with some liberal political interviews. However, many of the Radio DC transmissions consisted only of an identification and "Don't Vote Republican" in either the voice or Morse code. Like its opposing station, WGOP, Radio DC disappeared soon after Bill Clinton was elected as President.

Gulf War Pirates

The Persian Gulf War inspired political programming from several stations, including Hope Radio, KUSA, the Voice of Oz, Radio USA, Radio Peace in Action (Germany), Saudi Sam, and Radio Boner International.

Interestingly, all of the stations were personable throughout the war even if they disagreed. KUSA said little about the war itself or the national policies during the war, but broadcast in support of the troops stationed in Saudi Arabia. The station began as the war started. The operator later stated in *The ACE*, "Everything was devoted to having support for the troops. Some of the broadcasts in some of the bands were to the Mideast." Saudi Sam was both a station and a person–Sam had been in the U. S. Army in Saudi Arabia and made several broadcasts in early 1992 concerning his stint.

Radio Boner International (Fig. 8-9) and Radio Peace in Action (both very political stations) were, like KUSA, created as a result of the Gulf War. Radio Boner International aired rock and pop music with commentaries and comments against the war. After only a few broadcasts, Radio Boner International disappeared. Unlike RBI, Radio Peace in Action was a serious political pirate that has survived the years following the Gulf War. RPiA was almost entirely different from all other pirates in Europe; almost no music was aired, just news, commentaries, and information. The station has outlived the war and grown to support and protest other topics using the same format.

Hope Radio, the Voice of Oz, and Radio USA were all regular stations that added political programming during the Gulf War. Hope Radio attacked both sides with protest commentaries against the U. S. and skits about "Sodomy" Hussein. The Voice of Oz carried several anti-Iraqi commentaries, including comments, such as "Saddam, you're out of time." Radio USA included some antiwar

Fig. 8-9 *A photo QSL from Radio Boner International, which described itself as a clandestine, not a pirate. Was this the dress code of their revolutionary army?*

commentaries, but mostly stuck with producing comedy skits that related to the situation. However, Radio USA frequently relayed anarchist programming from Defiance! 90 and the Great Atlantic Radio Conspiracy.

The wartime programming from these stations might have had some impact on the personal conflicts between pirate operators that arose about that time and the heated conflicts in the mid-1990s.

Canadian Politics

Throughout this book, the assumption is made that everything that is true in the United States also covers Canada. Part of this is because U. S. and Canadian cultures are very similar. Another is because shortwave signals travel so well that some of the stations assumed to be broadcasting from the United States are really coming from North of the border.

American and Canadian pirate identifications are kind of like gender representation•among cartoon characters: unless the character is wearing a dress and has eyelashes, it's assumed to be male. It seems that most listeners assume that all pirates are American, unless they specifically have a Canadian theme or callsign. Except for the FCC and Canadian DOC, no one really knows which side of the border these signals are emanating from.

The most political of any Canadian shortwave pirates is Johnny Canuck, which broadcasted regularly in the 41-meter band in late October and early November 1995. The station was semi-clandestine in nature and certainly serious. At the time, Quebec was considering breaking away from Canada. Johnny Canuck argued against the separatists and urged Canadians to vote no against the separation referendum. Canada remained intact throughout the vote and Johnny Canuck returned a few days later to announce that this was the last broadcast from the station unless Canada split apart.

The Politics of Pirating

Although shortwave and pirate radio listening might not appear to be an especially political topic, its range is so broad that the different listener clubs often establish guidelines to keep their organization controlled. What measures are taken against those who cheat on loggings? What land masses of the world are considered countries? Are pirates considered radio broadcasting stations or merely individuals who illegally operate transmitting equipment? All of these questions and many more are answered in one way or another by AM band DX and shortwave listener clubs.

In creating club policies, a political situation and potential controversies arise. During the late 1970s, one of the hotly debated questions concerning shortwave radio clubs dealt with whether or not pirates are actually stations. Those listeners who wanted pirate loggings banned from their newsletters claimed that the signals were produced merely by individuals committing illegal actions with improperly operated transmitters. On the other side, some DXers thought that since the transmissions were obviously broadcasts, the pirates must be considered broadcast stations. The subject might seem rather unimportant today, but many DXers at that time were not willing to see several pages of their newsletter "wasted" on pirates, and the free radio listeners clawed back to save their only sources of up-to-date information.

Needless to say, the listeners who enjoyed pirate radio considered them broadcasting stations, and those who disliked pirates did not. Some of those opposed to free radio scorned the music or unprofessional programming at that time and the rest felt that no one operating without a license deserved attention. Regardless, the conflicts no longer exist; with the formation of *The ACE*, most pro-pirate listeners have exited the international shortwave broadcast listening clubs. Today, the situation is rarely an issue (except for a few cases of censorship in the hobby), but just before the creation of *The ACE*, it culminated in a nasty duel between the Voice of Syncom and one or more shortwave listeners.

For example, the Voice of Syncom was attacked on the basis that it had been relaying European pirates in order to fool listeners into believing that it was a regular broadcast from those stations. The complainer listed the telephone number of the FCC Monitoring Watch Officer and encouraged everyone to call when they heard Syncom and any other station "imitating a European pirate." Unfortunately, he was unaware that Syncom offered one of the best DX services of any pirate at that time or ever. DJ Chuck Felcher often announced rare DX catches and upcoming pirate broadcasts, along with comedy skits and relays from stations across the world that always aired frequent "Syncom relay" identifications. After the magazine containing the complaints was sent out, Chuck Felcher furiously protested the false accusations in a later issue–seemingly unaware that the listener has later sent in an apology for the mistake. However, the havoc had already been wreaked, and Syncom rarely operated after that point. The listener, once an active reporter of pirates, never reported them to that shortwave bulletin again.

Another pirate with a less humorous nature from this era was WKKK/WKGB, which broadcast music and talk while its operators were in an outspoken, drunken state. One of the announcers claimed to be a particular columnist from the Ontario DX Association while telling racist jokes. This angered many radio listeners, especially the columnist named. By a letter sent to various DX clubs, he publicly disassociated himself with WKKK/WKGB, saying that he knew who was behind the broadcasts and he "never thought they could stoop so low." The station later toned down its style and continued for several months into late 1981 as WRNR.

Pirates in 1982 filled the air with stations responding to the criticism from some DXers that they were only "kids playing radio." Two pirates, ZKPR (Kids Playing Radio) and the Children's Radio Network, left little doubt as to how they stood on the issue. CRN used very young announcers and babies crying to make fun of critics, while ZKPR attacked the opposition with commentaries. Another station that operated several times in 1982 and 1983 was Radio Free C. M. Stanbury, a campaign against a well-known shortwave listener and magazine columnist.

Still other stations have reacted against pirates and the free radio listening community. During some shows, Dr. Why? from KMA frequently editorialized on how listening to KQRP was a boring waste of time. The complaints never caused an uproar because few listeners disputed the editorial, KQRP did not hear KMA, and whoever heard the KMA shows were probably just as bored with hearing put-downs of KQRP as they would have been if they were listening to that station. Radio Sine Wave once parodied the Voice of Tomorrow by airing confusing racist commentaries and the songs, "I Wouldn't Want To Be Like You," "You Sound Like You're Sick," and "The KKK Took My Baby Away." Later in the broadcast, it was announced that the program was coming from Radio Sine Wave, but some less careful listeners actually logged it as the Voice of Tomorrow.

One of the nastiest situations in pirate radio occurred when the Voice of the Night took to the airwaves in early 1992. The Voice of the Night apparently enjoyed the negative publicity from his unprofessional broadcasts. Not only did the operator attempt to annoy everyone listening by singing along with all of the music and provoking other stations, but he also jammed other pirates and even an amateur radio net! In general, infighting between the Voice of the Night (Fig. 8-10) and other stations made for unpleasant listening.

The politics of pirating reached a peak with the KGUN America/Radio USA situation. The operator of KGUN called a number of *ACE* columnists and threatened them and stations in the pirate radio scene with busts. Later, the station jammed Hope Radio, giving the name and telephone number of an individual who he said operated the station. Suddenly, in 1990, it was announced that WLAR, the Voice of Oz, and Hope Radio were raided by the FCC. KGUN was quick to brag that it had informed the FCC about these operators, and indeed the FCC mentioned that an informant had helped them.

During the

Fig. 8-10 *An anti-Voice of the Night QSL sheet from WREC in the early 1990s.*

Persian Gulf War, KGUN began jamming Radio USA, but it ended quickly after another station announced that "the jamming was coming from [name announced]." A few weeks later, fake KGUN verification letters with the same name and home address were received by a number of listeners. Several months later, the station changed its name to WJTA and it attacked several people in the pirate scene. When the FCC began a publicity campaign against the alleged operator of Radio USA in the summer of 1992, KGUN began calling itself "Radio USA," jammed other stations, and broadcast slanderous and offensive material, complete with listeners' home phone numbers. Even though the real Radio USA ignored the jammer and the hobby press rarely mentioned the station, the Fake Radio USA was still occasionally active at the time this book was being written. In fact, the situation has been skewed even further; in early 1995, several new

Fig. 8-11
This QSL card makes it clear that nothing in pirate radio is certain.

stations calling themselves "Fake Radio USA," "the Real Radio USA," "Radio Is Not Radio," (Fig. 8-11) and other names appeared on the bands. The end-result is confusing, and the old "you can't tell the players without a scorecard" statement is certainly accurate!

This situation cooled off in the mid to late 1990s, but never really ended. All of the fake stations that made fun of the fake station disappeared. Radio USA has rarely been active since 1997. However, the fake station still reappears about once every year or two, typically after 0300 UTC near Christmas or New Years with Nazi-related programming.

Several similar, but less noteworthy, mean-spirited "I'm going to expose you" type stations also appeared in the mid-1990s, but fortunately these, too, disappeared. CSHT attacked CSIC. Interstate 44 and KIRK, the Voice of the Ozarks, clearly pointed fingers at DXer Kirk Trummel. And another version of Let's Kill JTA Radio extolled the virtues of committing violent crimes against at least one *ACE* editor.

In Fall 2000, the scene became a bit ugly again when a feud broke out between Radio Bob (once one of the FRN moderators) and other participants on the Free Radio Network Vines (bulletin boards). Once banned from board, Bob felt that he had been wronged by Bill O. Rights from Radio Free Speech, and he sought revenge. During this time, several Radio Free Speech programs were broadcast in SSB (RFS only uses AM) with supposedly revealing information about the station. In November 2000, Bob assembled a broadcast known as "Fight For Free Radio," which was a vehicle to defame those who had banned him from the FRN. See Chapter 3 for more information on these stations.

These broadcasts ended in December 2000 and have thankfully remained off the air for 2001.

Conclusion

Political pirates and clandestines will broadcast in and for North America for as long as radio exists. Unlike the general pirate scene, which could disappear because it lives on publicity and enthusiasm for listening to unlicensed radio, political pirates are often removed from the hobby aspect and act on the enthusiasm for their ideals. For as long as radical political groups exist within this country, there will be those who decide that their particular group must have a radio voice. Because radical ideals have been followed since the creation of man, it is safe to say that political pirates and clandestines will stay around for as long as there are radios.

In addition, shortwave radio is particularly fascinating in times of war and civil unrest. You can hear programs from official government stations from all of the affected countries. And when those countries have pirates, the listening is unbeatable, such as in the United States during the Gulf War.

Spoofs and stations carrying the politics of free radio are another matter. Because no free radio politics can possibly exist without a free radio scene, the stronger the hobby is, the more politics will exist. Spoofing feeds from both the pirate radio and shortwave listening hobbies. Whenever many DXers are tirelessly striving to hear more and rarer broadcasters, some prankster will also be there transmitting fake programs.

In a perfect world, people wouldn't waste their time fighting, but it's not and pirate stations will surely knock heads in the future. In some ways, these little conflicts are interesting, but it's much less so, the closer you are to ground zero. For example, I've yet to hear entertaining programming from fighting pirates. And Internet sites like alt.radio.pirate and the FRN Vines fill up with mostly pointless flames.

Local and FM Pirates

A lot of the mystique of traditional pirate broadcasting has been the long distances that shortwave signals can travel. In typical conditions, the shortwave pirate signal that you are tuning in could be from well over 500 miles away and the output power could be as low as ten watts (slightly more power than a typical night light).

On the other hand, even when FM pirates run up to 100 watts, they probably won't be heard beyond a 20-mile radius (depending on antenna height, directionality, *etc.*). As a result, FM pirates are very localized. Because of the localization, few of these stations ever publicize their operations. If they did, the FCC would soon be visiting the small area over which the signal can be received and seeking out the source.

To further complicate the possibilities of publicizing an AM or FM pirate, the other licensed stations within these bands are particularly sensitive to unlicensed operations. These problems are at least somewhat justifiable; whenever a pirate broadcasts regularly on the AM or FM broadcast bands, that station is competing with the other stations for listeners. In one case, pirate FM station WKEY from Cleveland appeared in the Arbitron ratings for that city and was about to overtake the lowest-rated licensed broadcaster in the area. What a humiliation for a commercial station to fall behind a one-man hobby radio station using less than 50 watts!

Because of the pressure from the government and other commercial stations, few AM or FM pirates publicize their operations, and as a result, very little information has appeared in the hobby press. In the 1970s and early 1980s, most of these were located in the New York state area—especially near New York City. WRFI, Radio Free Ithaca, operated for months from Ithaca, New York, as a sort of college-style alternative station in the early 1980s. The station had a regular schedule with a number of staff members, and it was regularly listened to by thousands. Another group of stations from the New York City area combined forces to operate Radio Newyork International in the late 1980s and early 1990s. Some of these long-running and widely heard stations included KPRC, WHOT, and WJPL (see Chapters 2, 3, and 6). Dozens of local coverage pirates operated from the New York City area over the past decade, but none were as popular as these three.

Another small hotbed of local pirate radio activity occurred in central Indiana in the 1970s and 1980s. One of the largest of these was Radio Free Naptown, a commercial pirate with a regular staff that operated for approximately ten years. According to the former operator of Radio Free Insanity, the station had even advertised on a billboard! Radio Free Naptown ended in 1978, when one of its operators committed suicide. The best-known recent Indiana local pirate was Jolly Roger Radio, which also operated on shortwave with lengthy broadcasts for several weeks until it was busted by the FCC in 1980 (see Chapter 2). Jolly Roger Radio challenged the FCC for some time, then claimed to move to Ireland in search of better broadcasting venues. Ironically, the former operator of Jolly Roger Radio now owns several commercial radio stations in the region. The last well-publicized local pirate from the area was Radio Free Insanity from Indianapolis, which broadcast in the early- to mid-1980s with more than 100 watts of power. Radio Free Insanity was better known, however, for its shortwave broadcasts (for which it was busted in 1984) than for its FM transmissions.

Of course, many other low-power AM and FM pirates broadcast in the 1970s and 1980s (Fig. 9-1). For example, because I lived near Pittsburgh, Pennsylvania, for some time, I knew of several local pirates that broadcasted from this region in the 1980s.

However, they never received any media exposure, so they aren't included here. I'm sure that most other major cities have had pirates of these sorts over the past several decades, but the difference is the signal coverage, the population density within that signal area, and the publicity that the station received in the local media (which would have made the difference between few and many listeners). A few of the other better-known local pirates included the Black Rose from San Jose, California, WTPS from Milwaukee, Wisconsin, and WHDL from Boston, Massachusetts.

By now, you might have noticed that the AM pirates that helped shape pirate radio (listed in the earlier chapters) are not included here. Perhaps I'm splitting hairs, but for the definitions in

Fig. 9-1 *The dismantled homemade transmitter from WMJC, a low-powered early 1980s pirate.*

this book, the shortwave bands start at the top edge of the AM band (1600 kHz until 1994, 1700 kHz from 1995 on). Thus, the revolutionary AM pirates–such as WCPR, WFAT, Pirate Radio New England, KPRC, WHOT, WENJ, WKND, WJDI, etc.–are not included in this chapter. The difference is that these stations all operated on clear channels on the edge between the AM and shortwave bands and could be heard over large distances. For example, KPRC and WJDI were both heard from their home locations in New York to distances beyond Kansas. Also, most of these pirates crossed over into shortwave pirating. As a result, unlike the FM or "in band" AM pirates, these pirates were covered in the hobby press, received much publicity, were part of the shortwave pirate scene (to at least some extent), and were heard over vast distances.

FM Hobby Transmitter Technology

As stated in the last section, a number of FM pirate radio stations operated over the course of several decades. The 1990s, however, brought about a revolution of local radio pirates across North America. This revolution occurred as the result of several factors.

In the earlier decades, FM transmitters weren't particularly difficult to construct, but they had to be built from scratch (Fig. 9-2). Otherwise, the only avenues by which someone could reach the airwaves would be to purchase an old FM transmitter from a radio station or to buy an amateur 6-meter (50 to 54 MHz) or 2-meter (144 to 148 MHz) FM-modulated transmitter or transceiver and

Fig. 9-2 *One of the earliest FM pirate kits--a transmitter kit made in the United Kingdom in the 1970s.*

modify it to operate on the FM broadcast band. Unfortunately, this was difficult to do, often quite expensive, and amateur FM transmitters have restricted audio and are in mono. As a result, most of the FM pirate stations from before 1980 used old broadcast transmitters from commercial or educational radio stations,

This trend began to change in the mid-1980s, after some manufacturers of integrated circuits introduced chips for stereo generation in transmitters, such as the ever-popular BA1404. These chips became the heart of stereo "home broadcaster" kits. Part 15 of the FCCs regulations permit the legal use of low-power unlicensed transmitters. Part 15 covers such things as children's toy walkie-talkies, wireless microphones and intercoms, and cordless telephones.

Part 15 allows low-power devices to operate in the AM and FM broadcast bands, usually with a power equivalent to one-tenth of a watt, and the maximum coverage area of such transmitters is usually less than 100 yards under most circumstances. Most kit companies feel safe with building their transmitters (Fig. 9-3) to output between 20 and 100 milliwatts (mW). However, the FCC legal limit for power is rated in millivolts per meter at a particular distance from the transmitter. Very few people own equipment capable of measuring received radio signals, so the FCC has always been flexible toward

Fig. 9-3 *An inside-the-case view of a Panaxis transmitter kit.*

stations running 100 mW or less, but outputting an efficient signal.

Before long, dozens of different companies were offering their versions of these transmitters. The most famous of these kits, popular for either Part 15 legal broadcasting (or for use as an "exciter" for use with an external amplifier) is the Ramsey FM-10 kit, which has been replaced by the Ramsey FM-10A.

The available FM equipment received an extra boost from the car stereo market, which needed a method to connect the audio from CD changers to the car head unit. One of the easiest, most effective methods is to use a stereo FM transmitter on the changer and to merely tune the FM radio to the frequency that you are using. This influx of small, high-fidelity FM transmitters is also providing radio enthusiasts with an inexpensive source of FM exciters.

Good-sounding FM transmitters were the most difficult component to find for a complete FM pirate station. As stated earlier, it is possible to modify 6- or 2-meter transmitters and transceivers to operate on the FM band, but the audio is normally too restricted. With the FM stereo signal already produced by the tiny kit transmitter, only a slightly modified 6- or 2-meter FM amplifier is required to get the station operating. One FM pirate operator that I know built an amateur 2-meter amplifier and wound an extra turn of wire around each of the coils. Then, it could output between six to seven watts on the FM broadcast band. This amplifier is then used to drive a home-designed and -built amplifier that will output 25 watts on the FM

Fig. 9-4 *A well-made 25-watt FM amplifier (center) and a high-frequency wattmeter (right) sitting on top of a 1970s FM stereo tuner (bottom).*

broadcast band (Fig. 9-4). At this point, the coverage of this transmitter is unknown, but during one test, it covered a radius of seven miles. With inexpensive sources of equipment and relatively simple kits and modifications, it is a relatively simple matter for pirates to broadcast in the FM band in the 1990s.

By 2000, PCS Electronics, a small hobbyist broadcasting kit company from central Europe merged radio and computer technologies with the introduction of their Max-PC transmitter. This transmitter is a computer board that plugs directly into the slots on a PC's motherboard. The Max-PC transmitter includes control software so that all aspects of the operation can be changed or monitored via computer. Because the station is within the computer, all programming can be stored and played back via control software. Thus, the Max-PC transmitter is the ultimate "radio station in a box."

The 1980s Mass-Media FM Pirate Revolution
WTRA and Black Liberation Radio

As equipment became readily available for FM pirate broadcasting, it was inevitable that someone would push for a radio revolution. That someone was an unlikely trigger to the revolution—a blind man with a tiny half-watt voice on 96.5 MHz from Springfield, Illinois. WTRA ("Tenants Rights Association") began operations in November 1986, when DeWayne Readus felt that the black community in the city needed some leadership. The controversy and public attention began in 1989, after he aired viewpoints about police brutality in the Springfield area. Then, according to an article in the *Columbia Journalism Review*, the local police referred complaints about the station to the FCC. In April 1989, Readus received a visit from an FCC agent and five policemen, who ordered WTRA off the air. WTRA closed for a week but decided to commit an act of civil disobedience and returned during a press conference. Readus notified the Springfield police of his actions. They said that it was a federal case, so Readus went to the federal building and stated what he was doing. The end result was a $750 Notice of Apparent Liability from the FCC in March 1990. Readus refused to pay the fine and continued broadcasting.

In 1990, Readus changed his name to Mbanna Kantako and WTRA Radio became known as *Black Liberation Radio* on 107.1 MHz (Fig. 9-5). Otherwise, however, very little changed. The tiny station

Fig. 9-5 *Mbanna Kantako at the microphone of Human Rights Radio in Springfield, Illinois.* Human Rights Radio

continued to broadcast, now with a total of one watt output. Of course, Kantako continued to attack the local and national political structures. Various people associated with the station have been arrested on different charges, but Kantako asserts that it was because of their connection to the radio station. In fact, Kantako's nine-year-old son was even arrested and handcuffed allegedly for "getting into a shoving match" during a soccer game.

On November 5, 1994, Black Liberation Radio installed a 13-watt transmitter with a range of up to four miles away. At the time this book was written, it continued to be an inspiration to a number of pirate stations—especially those in the local "microbroadcasting" movement. Some major stations, including Free Radio Berkeley and Radio Free Detroit, publicly stated that they have been inspired by Black Liberation Radio. Although it has taken some time, several of the programs have been aired on shortwave via Solid Rock Radio.

After more than 12 years and thousands of days of consecutive broadcasting later, the station was still on the air. No fines had been paid, no equipment had been confiscated, and no one went to jail. Why didn't the FCC close down this pirate when they had made such a clamor over some shortwave pirates and even other FM pirates (such as Free Radio Berkeley and San Francisco Liberation Radio)? The FCC has said very little about the situation in Springfield; however, it appears that a number of external factors are influencing their decision to avoid confrontation.

In the second edition of *Pirate Radio Stations*, I speculated about the reasons why Black Liberation Radio was never closed by the FCC. One big problem for the FCC was that Kantako was an unemployed and blind black man living in the housing projects of a city with 100,000 people. Also, the broadcasts from the station were very political and controversial. If the FCC closed down the station and took the equipment, they knew that the broadcasts would quickly return via another transmitter. If this occurred, they would then be pressured to take a more forceful action against the station. However, a more forceful action could cause a number of other problems. Kantako would not pay any fines and it would be impossible for the FCC enforce a fine against someone who is unemployed.

I stated that in order to sentence Kantako to a jail term, the FCC would have to go to court. Because Kantako is both black and blind, a number of political organizations would quickly come to his defense. Additionally, how many jurors would want to throw a disabled man in jail for broadcasting low-powered community radio to his neighborhood? Chances are that the FCC would fight against several organizations with the power to fight a court case for years, and the FCC would stand a good chance of losing. If the FCC lost, it could set a precedent against them and give leeway to some forms of unlicensed broadcasting. The FCC wants to safeguard against court decisions, such as these, which would weaken their power in all respects—in politics, in enforcement, and in legislation.

If the station was closed and Kantako was jailed, he would become a martyr for black and/or disabled Americans, free speech, and pirate radio. His story would appear in even more magazines and then in television news programs. In the process, it would strain the already tense racial relations in the North—especially in Springfield. As far as radio is concerned, not only would other people continue his pirate radio work in Springfield, but he would inspire others throughout the country to begin operating low-powered FM broadcasters. For the FCC, this action would cause even greater problems than if they merely lost in court. Under these circumstances, racial violence could increase, and a large number of unlicensed broadcasters would take to the airwaves. One of them might eventually challenge the FCC and win in a federal court (maybe even the Supreme Court) at a level to set a precedent for other cases. In the end, the FCC could lose across the board just by trying to enforce one of its rules against a tiny radio station in a small city in the Midwest.

In the late 1990s, Kantako changed the station name to Human Rights Radio to better describe the scope of the issues covered in the broadcasts. In December 1998, Mike Townsend, a college professor and the main media contact for Mbanna Kantako since the WTRA days, felt sure that a raid from the FCC was imminent. Will Grey from the FCC hand-delivered a 10-day cease-and-desist order. As had happened nearly 10 years earlier, Kantako told Grey to leave. Townsend understandably felt that the renewed FCC interest in the station meant that a raid would occur at the end of the 10-day order—especially in light of the heightened enforcement against FM pirates in 1998. But the raid never occurred and Kantako continued broadcasting without a break.

Nearly 10 months passed and on the last day of September 1999, Grey arrived again with a fresh 10-day cease-and-desist order. Again, Grey was told to leave the premises and the 10 days again passed without incident. Human Rights Radio continued broadcasting regular programming throughout the end of the year and well into the next.

But in September 2000, a federal court issued a cease-and-desist order with more teeth. The FCC alleged that Human Rights

Radio was interfering with air traffic communications at the Springfield Airport between September 22 and 25. Will Grey stated that the operations posed a "safety-of-life issue" that required an "immediate response." One question raised by this incident relates to the FCC's role in the cause of interference. If interference was truly being caused by Human Rights Radio, then why didn't the FCC pull the station off the air on September 22 instead of waiting three extra days? If Human Rights Radio was truly causing interference to the airport, wouldn't the FCC be willfully and maliciously allowing dangerous interference to continue by not contacting Kantako immediately? On the other hand, after more than 4,000 continuous days of broadcasting, much of that time using the same equipment, it seems unlikely that it would suddenly begin throwing spurs into the aircraft band.

But, Kantako wasn't going to let some government agency and a court order end 12 years of broadcasting. So, he continued broadcasting. Finally, on November 30, 2000, eight members of a multi-jurisdictional task force plus a private contractor were sent to Kantako's apartment. The contractor first removed the antennas from the rooftop so that the raid would not be broadcast to the neighborhood, as had happened in the past. After the antennas were gone, the force went to the apartment and confiscated all broadcast-related equipment. Human Rights Radio was finally off the air.

Free Radio Berkeley

Broadcasting to thousands of people throughout the valley in which Berkeley, California, lies, Stephen Dunifer pieces together his scattered homemade equipment on a remote hilltop. Dunifer (Fig. 9-6) has openly broadcast in the area with this hodge-podge of equipment since 1993, from various hilltops, most of which required hiking in with a backpack station. Later, the station moved to safe houses in Berkeley and Oakland, California.

Like most well-known FM pirates, the initial broadcasting from Free Radio Berkeley did not catch the attention of the mass media nearly as much as their conflicts with the FCC. I heard my first report of the station in the *Ben Is Dead* fanzine (based in the Berkeley/San Francisco area). At that time, the station was not yet reported in the mass media; however, Free Radio Berkeley was already a high-profile station. According to this editor, the operator sometimes went to local underground rock shows and broadcast them live via a 10-watt FM transmitter that he had strapped to his back. An FM whip antenna was then plugged in to the top of the equipment pack.

Before long, Dunifer was appearing in a variety of mass

Fig. 9-6 *Stephen Dunifer of Free Radio Berkeley stripping some coaxial cable at his workbench.* Free Radio Berkeley

media outlets; he was featured on CNN, in the *New York Times* and the *San Jose Mercury News* papers, and in *Spin* magazine, among others. In addition to the station's rather eccentric methods of operation, Dunifer was also a persistent public relations machine. Dunifer, whose activist career began in the late 1960s, is a veteran political organizer; these experiences allowed him to know what resources were available and how to disseminate information quickly.

As a result, when the FCC handed Dunifer a $20,000 Notice of Apparent Liability, he immediately faxed a response to all local and some national media outlets. Before long, Free Radio Berkeley was featured on CNN and in the *New York Times*. The early, very positive articles and news features were a sharp blow against the FCC, considering that they were not prepared for the media blitz. At this point, the FCC still has not successfully fought off Free Radio Berkeley's public relations maneuvers or those of the National Lawyers Guild Committee on Democratic Communications.

As the glowing publicity was still fresh in the air, Free Radio Berkeley continued broadcasting and publicizing. These activities were boosted by such events as organized showings of *Pump Up the Volume* and other pirate/free speech-related movies, workshops on how to create your own "microbroadcaster," a regular station newsletter (*Reclaiming the Airwaves*) that has a circulation of 5000, and even a mini-catalog that features transmitter and amplifier kits in a variety of configurations for the AM and FM broadcasting bands.

In addition to news, schedules, and pirate hard-luck cases, *Reclaiming the Airwaves* also regularly features information about people who are being assisted by Free Radio Berkeley in installing their own pirate stations in various locations throughout the world. Obviously, the Federal Communications Commission felt pressured to do something; the station hadn't left the air, it was supplying others with equipment, it was telling thousands of people to fight the FCC and broadcast without a license, and it had ignored their $20,000 fine.

In the early stages of Free Radio Berkeley, Dunifer stated the station's purpose as the following (to the *San Jose Mercury News*): "We're trying to break the information stranglehold that's out there and show you—individual people—can have a voice by taking the airwaves back. If we are really going to have a democracy in this country, we need free and equal access to information—and that isn't happening. The catalyst (to start the station) was the orchestrated media campaign around the Persian Gulf War. That was the final straw. It was an MTV, Disneyland manipulation, a distortion of reality making it seem like a video game wiping 200,000 people off the earth."

Unlike the situation with WTRA/Black Liberation Radio, it was imperative for the FCC to close Free Radio Berkeley. The differences between the two situations were that Free Radio Berkeley was causing dozens more pirates to spring up and, from the FCC's perspectives, it was not dangerous to attempt to close down the station. In the situation with San Francisco Liberation Radio, the FCC called in for backup help from the local police department to apprehend the operator (even though he was quickly released). This scare tactic was not an acceptable option to be used against Free Radio Berkeley because the station was the key case for the National Lawyer's Guild's Committee for Democratic Broadcasting.

Although the FCC claims to have the authority to be able to enforce any measures that they threaten rule violators with, they were unable to close Free Radio Berkeley initially. After being unsuccessful at halting the broadcasts after over a year of threats, fines, and paperwork between lawyers, the FCC filed for a federal injunction to prevent Dunifer from broadcasting. In the Winter of 1995/ 1996, the FCC applied to the federal district court in San Francisco for an injunction to close Free Radio Berkeley. They were refused and FRB operated in the clear for two years. During this time, some FM pirates felt that the FCC's impotence in the Bay area was a nationwide phenomenon or that this somehow meant that the laws regarding unlicensed radio were void. Neither was true. Hundreds of pirates were closed by the FCC in 1997 and 1998. If anything, it appeared that their failure in Berkeley ignited a passion to close down all unlicensed FM stations.

Meanwhile, Free Radio Berkeley was an untouchable: operating in the open and popular in the community. In fact, in order to fall within the ordinances of the city, FRB applied to the Zoning Adjustments Board in 1998 so that their building would be technically allowed to house a radio station. Despite a 98-page opposition filed by the California Broadcasters Association, the board granted Free Radio Berkeley permission. FRB was truly an accepted part of the community, possibly even part of the establishment, in Berkeley.

Within days of the hometown FRB victory, Dunifer and the rest of the station faced a more difficult challenge from the out-of-town folks. The FCC managed to have the earlier decision overturned and were successful in getting an injunction against Free Radio Berkeley and Stephen Dunifer. Dunifer refused to give up and stated "Neither I nor the micropower broadcasting movement will be silenced by this injunction. It only encourages me to intensify my efforts in the creation of a truly democratic grass-roots broadcast media and facilitate the citizens of this country taking back what rightfully belongs to them—the airwaves." But the National Lawyer's Guild could not change the verdict and FRB was dead.

Of course, it's difficult to kill a pirate station with a staff ranging from 75 to 100. The only person condsidered to be violating FCC rules would be the transmitter operator, not the rest of the staff. FRB was gone, but pirate radio in Berkeley was merely evolving. In late November, a publicity-minded radio protest known as "Tree Radio Berkeley." For 11 straight days, TRB broadcast from a tall tree in Berkeley park. Two announcers/staffers were up the tree at any given time, along with a tape deck, transmitter, and antenna. Given the peculiar nature of the station and its catchy, play-on-words name, it was a hit with the commercial media.

After two years of silence and planning, large-scale pirate radio returned to Berkeley under the name of Berkeley Liberation Radio in August 1999. The latest pirate radio entry contains some former staffers from Free Radio Berkeley, but, as can be expected, new people are also involved. Like FRB, the new station broadcasts 24 hours per day on 104.1 MHz with more than 50 volunteers and a wide variety of programs. However, BLR takes a much lower-key approach to publicity than FRB did; rather than spearhead the national pirate radio movement, BLR seems to be focused on simply broadcasting.

Stephen Dunifer is still in the pirate radio business, promoting the cause. But, his website http://www.freeradio.org, is no longer filled with schedules for Free Radio Berkeley and information on station fundraisers. Now, Free Radio Berkeley is considered to be International Radio Action Training Education (IRATE), which is dedicated to educating people about operating radio stations. Sections of the website cover basic radio electronics, workshops on working with electronics, what to do when the FCC knocks, etc. In addition, a large section is a catalog of radio kits (but, of course, a legal disclaimer states that all equipment is strictly for educational purposes and may only be used on a dummy load).

Free Radio Berkeley is dead in name only.

Radio for the Fun of It

Not all of the FM pirates are as political motivated as either Black Liberation Radio or Free Radio Berkeley. Hundreds of stations are serious about playing a certain type of music, broadcasting community events, or designing their studio, transmitter, or antenna array. Others aren't a bit serious.

One of these pirates was Radio Addiction, which was a high school pirate that broadcast in 1998. The following story about the station was written by P, the driving force behind the operation. This story was originally published in the Winter 1998 issue of *Hobby Broadcasting*.

"Radio Addiction, 103.1FM, give us a shout, the request line is open, 452-0233, for whatever you need!!!" Words I miss saying. As the death of our station proves, not everybody is for free radio. Personally, I had been a shortwave pirate listener for a few years at the time, and was more than eager to put up a station. A friend and I got the idea and decided to give it a shot.

After checking many Web sites and armed with a copy of Andrew Yoder and Earl T. Gray's *Pirate Radio Operations*, the plan was put into action. The equipment was more easily accessible than we once thought. The first purchase was a Ramsey FM-100, a fully assembled 1-watt FM transmitter. We strung some wire up on the house and chose a frequency of 103.1 MHz because if we caused some interference, the stations were spaced far enough apart on the dial that we wouldn't "bleed" on them.

The signal was terrible and only got us a few blocks. So, we did the next best thing: got the equations, and put up a vertical dipole. This was decent, but we still didn't have enough range. We decided to completely rethink the idea, and to do it right this time.

After what seemed like endless hours of research, a 20-watt amplifier was found that could be driven by a 1-watt transmitter, and had a decent price. We now had an amp, but decided to go all the way and get a better antenna arrangement.

Some connections were made and the materials for a ground plane antenna and a SWR analyzer were acquired. I still can't believe we spent a whole day tuning the antenna for 103.1, but we did get the SWR's around 1:1.5, which was close enough after spending that long tuning the thing.

With all the transmitting equipment in place, many trips to Radio Shack for audio equipment, (we actually purchased two 6-disc CD changers from a station that was being taken down) and a extensive CD library we already had, it was time.

We were VERY shocked with the range. We had a good 10-mile radius (this sometimes depended on the weather) because we were located on one of the higher points of the city. But, after a few broadcasts, still no listeners that we knew of.

Fig. 9-7 *The newspaper ad for Radio Addiction.* Radio Addiction

OK, here is where we got insane with it, we decided to put an ad in the local newspaper and get a phone line (Fig. 9-7). This kind of helped in our favor at this point. Nobody questioned us about the phone or the ad stating a local radio station was to begin broadcasting within a week. Plus, the only "alternative" radio station in the area changed their format to "oldies" a week before our scheduled time to begin broadcasting.

As for the "main playas" or operators of the station, I called myself originally "Big Perm," but shortened it to "P," the other DJ was my friend "Mad Skillz." We had some technical help from Mr. Keith Richards (whose job, whenever he was in the studio, was to say "listen to the music, 'cause it's bangin'!"), and finally we had help from "The Captain" (as in Captain Morgan, Puerto Rican spiced rum).

The first two days of broadcasts were dead, so we made some stickers on the computer and passed them out at the high school during lunch. That did it. Slowly more calls came in, two or three on the first couple of days, (by the time it was over, the phone wouldn't stop ringing), and everyone LOVED the station. The broadcast schedule was 3 to 8 P.M. every weekday, we had regular listeners that would call in everyday to check up, and the word continually spread.

After about three weeks into the station's existence, it was like instantaneously, EVERYBODY was calling, the phone wouldn't stop. At times, as soon as we hung up on someone, another call would come in.

It was great. It was especially neat when we did our favorite line. Right before we would hang up with someone that we had on the air, we would say "now tell everybody, what's your station," their reply always being "103.1, Radio Addiction."

We still feel we did a good job where the music was concerned. We had a couple hundred CDs combined, and even though a few requests couldn't be completed, we could usually get what they wanted if they had a second request. The musical programming was mainly alternative, but there was also plenty of rap, R & B, reggae, ska, blues, and even a couple of local bands.

Around this time is when things went downhill, unfortunately. Well, some of the kids in the neighborhood decided to try and find us. We knew they were listeners, because they had called and we knew who they were. But we didn't expect to see them just walk in the back door. You see, the station was located in a basement, and we left the door open to cool it off because we would smoke down there. They said they were just walking around and heard us from the street. We figured that was BS; they probably saw an antenna on the roof, with an open door to a basement, and a few cars parked at the house and decided to investigate.

Well, there we were in the middle of broadcasting, and I

turn around. There are these three guys with grins on their faces. The only words I could get out were "Who are you, what are you doing, and who said you could come in?" Well they went crazy, being the fans that they were, so we figured, "Hey they already know our names," so we let them stay for a few minutes. After making them feel uncomfortable and no longer welcome there, we told them to keep their mouths shut, because "we will know if you leak a word" and all seemed to agree to this.

The following day, Wednesday, I decided to leave the station about a half an hour early, and on the way to my car, I looked up front and noticed a strange car sitting in front of the house. At first, I played it off as a neighbor, but noticed it had occupants and was running. Of course, my first thought was "the feds!," when they slowly drove by me standing beside my car, all three or four individuals shouted out of the car "Radio Addiction rules!" and drove off. It was then that I knew the station had to come to an end.

Thursday's broadcast went fine, with the exception that now people had spread the word of who we were and our location (which was not out of a blue, rusty 1979 Ford van, with shag carpet and gold rims, parked down by the river, like we claimed). That evening, we discussed our plans, which was to make Friday our "Free Radio Addiction" broadcast, all about our right, pirate radio, and the hows and whys that it is necessary. And we were to go on Saturday, with our entire crew (even the people that just helped get the station up and had never been on before) for a two-hour blowout and sendoff.

Friday came around, and with our free radio broadcast, we decided to start off the show with the theme from *Pump Up The Volume*, Concrete Blonde's version of "Everybody Knows" (It gave double meanings, seeing as to how everybody knew way too much about us, and it kind of gave a little tribute to free radio from Hollywood's perspective). After another song about freedom, we decided to do a little editorial, about what we thought of freedom and how we were expressing our rights, under the First Amendment, using the air.

Appropriately enough, we next played "Freedom" by Rage Against the Machine (unknown to us at the time, would be the final thing Radio Addiction would ever air). I went upstairs to take a leak, and on the way back, I saw something like a flash in the corner of my eye. I went to the window and saw two cars, and three men in suits walking to the front door. I ran so fast down the stairs I literally almost broke my neck, and yelled "SHUT THE THING OFF, RIGHT NOW!!!" (although now that I look back on it, I would have liked to at least have gotten on and said, "Well, the man is knocking on our door, with guns and badges showing, so we're going to sign off now, if you liked our station, you CAN start your own, 73s and fight for free radio!!!), but I didn't. In the heat of the moment, I just wanted to shut the thing off.

At 3:15 P.M. on Friday, October, 9th, 1998, Radio Addiction was forced to cease operations. That's when we heard the dreaded knock. So, we tried to be slick about it, disconnecting the amp and transmitter and strategically hiding them where we KNEW nobody would ever look. There were knocks at the back door, along with the doorbell ringing, but nobody ever announced themselves as the FCC, police officers, detectives, or anything. They just kept knocking.

After a quick prayer, I led the way upstairs. We peeked outside and they were kind of standing around in the front yard, we weren't going to open the door until cops pulled up. The assumption was made that they weren't going to just leave and to open up. Once we opened the door, the three men ran up, little did they know the screen door stayed locked. They introduced themselves as local detectives and showed proper ID and asked to come in.

They didn't have a search warrant, so we denied them entry to the house and said we would talk to them outside. I was told that because I was not a resident of the house I had to stay outside. The detective went in with my partner, Mad Skillz, and questioned him. They took pictures of the whole setup and questioned him. They were shown the transmitter and amp and they tried to figure out how the thing worked. Remember, these are small-city detectives, not FCC agents, this was the first time anything like this has ever happened around here.

When they finished checking the station, they talked to us outside a little. They asked us questions like, why we did it, what did we think we would gain, and if we knew we needed a license. We were then asked to come to police headquarters to write a statement. After being questioned on tape and writing a statement, which was pretty much everything said on tape, they told us they would get back with us and that the FCC would be down on Monday or Tuesday.

Monday and Tuesday passed, with no sign of the FCC, and Wednesday we got a call from one of the detectives telling us the Commonwealth Attorney was not going to press charges. We were also informed that the FCC had allegedly been in the city on the previous Friday (the date when we got busted) around 5 P.M., (which was two hours after we got caught) and didn't get in touch with anyone in the police department. Is it me, or does something not add up in this story?

We admit, that the station was kind of hard core. Yes, we did use vulgar language, and, yes, the show did have sexual content, but that is protected under the First Amendment, right? Surprisingly enough, we were told by the detectives that everyone they talked to seemed OK with having an underground station in the area, that wasn't the issue at all. But the issue was our content. We let things get out of hand and that is where we screwed up.

Thanks for everybody's time. Fight for free radio and don't let this corporate BS get any more out of hand that it already is. And if they don't let us get LPFM, tell everybody and their mothers to start a pirate station. They can't take us all down!

The LPFM Licensing Movement

In the United States, pirate and hobby radio has always been relegated to counter space on the other side of the fridge—it doesn't even make it to the back burners of the stove. For literally decades, such pirates as David Thomas and Allan Weiner fought to be licensed. But hobby broadcasting was just too much of a novelty for the general public to take seriously. Although most pirates were popular, as was media coverage of such stations, no one said, "there's something wrong that these stations aren't licensed," and initiated change. Instead, pirate radio stations were treated as one of those end-of-newscast fluff stories, like tap-dancing poodles or the family with the largest outdoor Christmas light display.

All this changed after the Telecommunications Act of 1996. Among other things, the act relaxed the restrictions on the number of stations that one corporation can own across the country (unlimited) and in a single market (eight). Before the act, one company could own no more than one station in a market and 28 nationwide. With the Telecommunications Act in place, radio stations around the country were quickly bought out by the major media corporations. By the late 1990s, the largest radio conglomerates artificially drove up the prices of radio stations. In just a few years, radio stations were bringing in offers many times their selling prices before the Telecommunications Act of 1996. With all of the stations going to the highest bidder, some companies suddenly had a near-monopoly in major cities. According to a Spring 2001 article on Salon.com, Clear Channel owned 1,200 different radio stations across the country—a far cry from the 28 limit of just five years ago.

Once a media corporation owns a group of stations in the same market, they can share (and reduce) resources (building, management, personnel, and equipment) and duplicate some programming. This allows the group stations to drop ad costs and offer packages of ads that run on all of their stations in the local area. These monopolistic efforts have financially devastated nearly all of the mom-and-pop stations that have dared not sell out to the corporations.

The end-result of the corporate buyouts included programming being run in parallel on two different stations, automation replacing live announcers, significantly reducing news departments, relying more upon syndicated programming, and reducing and simplifying the format types. Thus, fewer local musicians, entertainers, news, and issues are broadcast on commercial radio now than ever before.

The outcry against the effects upon commercial radio was loud. The loss of diversity fueled the community FM pirates that were appearing across the country. These stations, in turn, protested for the right to broadcast (Fig. 9-8). Pirate radio was transformed from being a weird, little frivolous hobby into a patriotic fight for one of our fundamental rights.

Fig. 9-8
Pete triDish flicks on the "free speech" switch of the mobile pirate transmitter at an LPFM protest at the Washington, DC headquarters of the FCC Mike Flugennock

To counter the negative effects on commercial radio, the FCC under William Kennard, agreed to listen to petitions for rulemaking concerning low-power, community broadcast stations in 1998. These proposals were for what was called LPFM (low-power FM radio). The first proposal for LPFM that received an FCC petition number was RM-9208, which was submitted by Don Schellhardt and Nick Leggett. The FCC requested that more suggestions for community radio be submitted. Soon followed RM-9242, submitted by Rodger Skinner, a displaced low-power TV owner; RM-9246, submitted by Harold McCombs, Jr. of Web SportsNet, Inc.; CDC, submitted by Stephen Dunifer and the National Lawyer's Guild; and CRC, submitted by Thomas Desmond, William Pfeiffer, Kent Peterson, and Bill Spry.

The proposals presented to the FCC represented a wide selection in service types. For example, RM-9208 offered commercial services with powers as high as 1000 watts, RM-9246 was intended for one-time and occasional radio sportscasts to regions around a stadium or arena; and RM-9208 was originally for a flea-powered (1 watt) commercial or noncommercial service (but the power was later increased). The FCC received comments on the different proposals, and the deadline was extended several times to allow more comments to be received.

Finally, a set of LPFM rules were accepted by the FCC, but they didn't resemble any particular petition. They established a noncommercial, two-tiered system (allowing for 10-watt and 100-watt stations) with an emphasis on community service. For example, stations that could air significant amounts of local programming would receive emphasis over those who could not. Also, the stations would be restricted concerning ownership. Among other points, only nonprofit organizations that own no other broadcast stations can own an LPFM and 75% of the board members must live within 10 miles of the antenna site.

For the most part, the FCC system ruled in favor of those who pushed the hardest for low-power broadcasting. However, two rulings almost entirely sliced the hobbyists and protesters out of the LPFM pie. First, anyone who pirated within the past two years was not eligible to be licensed. Second, the only applicants can be nonprofit organizations that have been in existence for at least two years. These two qualifications not only rule out the people behind the several hundred or thousand pirates that have broadcast recently, but they also slice out all of the people who would create a nonprofit radio club for the purpose of bringing interesting radio to a community.

Fig. 9-9 *Although WMOB had planned to take to the airwaves as a pirate (and even had these stickers printed), the organization instead rallied support for an LPFM station in the Baltimore area.*

Nonetheless, the petitions for rulemaking drew fire from the licensed commercial broadcasters, represented by the National Association of Broadcasters (NAB), and the licensed public broadcasters, represented by National Public Radio (NPR). From the beginning, the NAB produced anti-LPFM press kits for their member stations to use against local community radio organizations. In these early stages of the LPFM process, the FCC performed an engineering study on receivers to see if LPFM stations would cause harmful (or even noticeable) interference with presently licensed stations. The FCC's study found the interference requirements to be more than adequate for the FM band, using a variety of receivers. The NAB then performed their own receiver test to disprove the FCC results. According to an article by Pete triDish in the Summer 2000 issue of *Hobby Broadcasting*, "The standard of audio quality that the NAB used in its tests was so rigorous that over half of the radios tested failed to meet it— even when there was no interfering signal at all!"

If the three years of LPFM development were slow and painful at the hands of the FCC, then the months of Congressional anti-LPFM bills could be described as chaotic and reactionary. With the NAB pressuring Congress and National Public Radio lobbying Democrats to end LPFM, politicians clamored to create their own legislation, based on their own limited understanding of broadcasting and what information they had been fed by lobbyists. Ohio Republican Michael

Oxley put forth a bill to flat-out repeal LPFM forever, but it was worded a bit too harshly. Michigan Democrat John Dingle countered with a bill that allowed for the possibilities of LPFM, but eliminated most of the possibilities of such stations by allowing for much greater adjacent-channel protection for existing stations.

Arizona Republican John McCain switched sides on the issue and wound up pushing some legislation that essentially no one supported. McCain's bill allowed for LPFM stations, but gave the other stations the right to sue over interference; however, the low-power stations would not have been given the right to sue concerning similar interference from the previously licensed stations. The result would have been a nightmare for all but broadcast litigators.

In the end, Dingle's bill won. The final result of the great battle over LPFM is an essentially hamstrung ruling that will allow a few community broadcasters on the airwaves, but not the great outpouring that had been expected. Thus far, only about 250 applications have hurdled the restrictions and what percentage of applications will be converted into licensed, operating stations is anyone's guess (Fig. 9-9).

Current FM Pirate Outlook

At this point, it appears that the days of the large, highly publicized, confrontational pirates have mostly passed, at least for the time being. The publicity stunts and resultant mass media coverage have essentially disappeared. Likewise, the media-covered pirate webpages are either gone or haven't been updated in years. But all of the pirates haven't disappeared; it appears that the scene has reduced in size and is operating more clandestinely, as it had prior to the 1990s.

The following section is John Anderson's (about.com) outlook for FM pirate radio:

Things are definitely in flux. We originally saw the biggest boom in microradio in America following the Telecom Act of '96, passages, and the reality of industry consolidation set in. I think that as long as that law's still on the books, it'll continue to be a major factor spurring the creation of microstations.

And with the corpse of LPFM lingering around, there's defnitely been a backlash from the smell–new stations are going on the air like crazy. Many people who had hopes for LPFM and then got dissed by Congress are saying, "screw it"–and going on the air anyway. It seems to me, right now, that LPFM has only hurt the microradio movement.

Of course, from the FCC's perspective, LPFM serves a dual purpose; it allows the agency to say it's serving the public interest while giving it more credence to pursue a crackdown on unlicensed microbroadcasting. For FM pirates, LPFM has definitely been a 'carrot and stick' kind of issue. Some shut down in hopes of playing by the rules, but then the rules changed.

I think, in the long run, people will look back on LPFM and laugh, realizing it (in a big-picture sense) as the farce that it is.

In the interim, though, it's not clear just whether or not the FCC will jack up its jack-booted thugs. I think much of this has to do with the transition of power in Washington, as well as with pending plans for a massive agency restructuring.

For the pirates, it does seem like the nature of the game has changed. From what I'm seeing, most (if not all) of the new crop of micros are not like the "megastations" of the mid-to-late '90s, but are more along the lines of individual/small group-run operations who broadcast more sporadically and keep a lower profile.

The important thing to remember here is that microradio is just one part of a larger, still-growing movement for media democratization in America. Sure, microradio stations have been an invaluable source of alternative news and information, serving that niche quite nicely. But the more important act is the one of civil disobedience–by providing a concrete example that not only can seizing the airwaves be done, but it can be done by just about anyone.

Strategically speaking, the momentum is still with us: the FCC continues to underfund and underequip its 'radio police,' and they've already demonstrated that they're overwhelmed and can't shut them all down. By giving folks high hopes and then pulling the rug out from under them, the FCC (and Congress/NAB/NPR) may have given microradio a shot in the arm instead of dealing it a deadly blow.

Others involved in FM radio have mived feelings about its future. Some believe that the FCC's enforcement, the LPFM losses in the court system, ans the filling of open channels with unnecessary translator stations have ruined any hopes for hobbyist or community radio on FM. For these people, FM was a lost war, a territory controlled bu the powerful, bland enemy. These people have moved onto other outlets or medias: Internet radio, community-access television, zines and other print media, etc. As stated by Pirate Jim of the pirates WODD and WTRR, "Now with the FCC cracking down on unlicensed broadcasts and LPFM slowly making an appearance, no one is willing to take the chance anymore."

Some still are hanging on to dreams of FM broadcasting. These people are either processing applications for an LPFM station, waiting for a local LPFM station to begin broadcasting so that they can volunteer an airshift, or are still pirating. One of the surviving high-profile political FM pirates is Radio Free Santa Cruz, which broadcasts a mix of youthful anarchy and left-wing social activism 24 hours per day.

In the U. S., FM pirate radio is still popular with some of the communities of recent immigrants, particularly the Latinos and some other Carribbean groups. Florida cities are loaded with Spanish-language pirates, and these stations have proven to be very difficult for the FCC to remove from the air. Even if the pirate is closed for good, other new stations are always waiting to take its place. Spanish-language pirates have appeared in other cities across the U. S., but they don't dominate the illicit airwaves, not even in New York City. In addition to hobby FMers, New York is also famous for Haitian, Greek, Jewish, and Jamaican pirates.

It's clear that FM pirate radio has not and will not die, although it might be a long time until we can read about it in *Rolling Stone* or *Spin* again. Except for a few rare instances, they have returned to the fringes of radio society, instead of operating as 24/7 beacons on the dial.

For more information on community FM radio stations (both licensed and pirate), see *Rebels on the Air: An Alternative History of Radio in America* by Jesse Walker.

Tuning in to FM Pirates

FM local pirates and a handful of local "in band" AM pirates are scattered across every major city in the United States. In spite of the huge difference in publicity between the media-active shortwave pirates and the oft-forgotten local station, the local stations actually greatly outnumber those on shortwave.

Because of the greater number of local pirates than short-wave stations, you might think that they would be less difficult to hear. Actually, the opposite is true. The AM and FM pirates have much smaller ranges, usually no greater than a three-mile radius, and as a result, they are capable of covering fewer people than the shortwave pirates. These stations tend to congregate around the population centers, so if you live in rural America, chances are that you will never hear a local pirate.

On the other hand, if you live near or in a city, you might eventually hear a local pirate. As a group, the licensed FM stations with the lowest power are those in the so-called "educational" band (88 to 92 MHz). Because commercial stations are not permitted to operate in this region and the public and educational stations run with lower power, most of the clear frequencies on the FM dial are in this "educational" band.

If you are lucky, you might receive an advance notice that an FM pirate is operating in your vicinity. Such a notice would be a newspaper article about the station, a mention in one of the small newsletters on "microbroadcasters," an ad or flier in a local music paper or CD/record store (Fig. 9-10), or just a rumor from friends. Otherwise, your only recourse is to occasionally tune through the FM

Fig. 9-10 *A 1998 program guide from Vermont's Radio Free Brattleboro.*

RADIO FREE BRATTLEBORO 88.1 FM

OFFICIAL PROGRAM GUIDE

ISSUE 2 VOLUME 1 • NOVEMBER 1998 • 258-9879

radio free brattleboro (rfb)
STILL on the air!!

Young rfb deejays having a substance-free good time spinning discs behind the board at rfb's gleaming new studios on Flat Street.

We are a group of local teens and adults working together to provide non-commercial, independent radio programming to the Brattleboro area. RFB broadcasts daily at 88.1 fm and can be heard in town from the post office to Exit 1, and we are working on expanding our range. From 3 o'clock in the afternoon until after midnight, RFB plays the most diverse selection of music within a 50-mile radius.

radio (especially the 88- to 92-MHz portion) spectrum and look for new signals or peculiar-sounding stations. If you live in a city or are visiting another city, chances are that you will stumble across and suspect that a free-form college radio broadcaster is a real pirate. All you can do is listen for an extended period of time, catch an identification, and possibly check it in an FM broadcasting guide to see if the announced callsign is licensed.

The chances of stumbling across one of these pirates in much of North America are poor. It took me years to hear any FM pirates, aside from the "let me turn on my transmitter and show you how it sounds" types of stations. Even so, all of the FM pirates that I've heard have been located in Pittsburgh, where I have friends who sometimes know about the FM pirates around the city. In one case, I knew about the famous techno pirate Radio Carson (Fig. 9-11), which was audible across much of the city. With advance notice of the station available, I tuned for on several occasions while crossing the city and managed to hear it a number of times. In

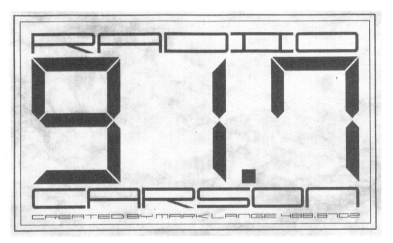

Fig. 9-11 One of the few FM pirates ever to cover a city from end to end, Radio Carson delivered techno music to the masses in Pittsburgh for 24/7 for more than a year.

the case of another pirate that I heard, a friend said, "sometimes a country music pirate appears on 92.3 MHz on Sunday evenings." I tried it and sure enough, there it was. The country music pirate was raided by the FCC soon after I heard them. Radio Carson lasted about a year longer before it was also evicted from the Pittsburgh airwaves. Since then, I've tuned around the FM band, but have yet to discover additional pirates.

Conclusion

FM pirate radio or "microbroadcasting" appears to be an important aspect in the future of pirate radio. As the FCC rulings and commercial station cost-cutting policies continue to erode the community service aspect of licensed AM and FM radio, more FM pirates will step into action to serve their neighborhoods. From a listener's perspective, it means that at least some aspects of broadcasting will be improving, even if the entire AM/FM scene is homogenizing. From a hobbyist's perspective, there will be many more difficult radio targets to monitor and more interesting stories to pass. From a reporter's perspective, it will mean more clashes with the FCC to determine what (if any) community broadcasting is permissible and exactly how much power the Commission has to license potential stations and punish offenders.

The FCC and Enforcement Actions

The *Federal Communications Commission (FCC)* is the administrative body that regulates radio and television broadcasting and transmitting in the United States. The FCC not only licenses radio and television stations for broadcasting, but its enforcement division also tracks down stations that are broadcasting without a license, are on the wrong frequency, or are causing interference.

The previous chapters of this book have covered pirate radio stations—hobby broadcasters that operate without licenses. Not surprisingly, they are always at odds with the FCC. As a result, the FCC's enforcement procedures are a key element in the pirate radio listening hobby. Although this chapter largely focuses on the historical aspects of the FCC's relationship with pirates, at the time this book was written there were proposals for sweeping changes in the FCC which could have a major impact on its ability to "police" pirate radio.

Procedures

The FCC consistently monitors various radio bands, seeking radio interference and unlicensed radio stations. Some of the frequently monitored areas are the standard pirate broadcasting frequencies (mentioned throughout this book), several of the most interference-prone amateur radio frequencies (such as 14313 kHz in the 20-meter amateur band), and frequencies above and below the Citizens Band (where a number of unlicensed radio operators transmit two-way communications to avoid interference within the band).

The FCC employs radio monitors at a handful of monitoring stations that are strategically placed around the population centers of the United States. When an offending signal is noted, the monitor enters the information into a nationwide communications network. From that point, the other monitoring stations also tune in to the signal and begin direction-finding ("DFing") the signal.

Each receiving station features high-tech receivers that are connected to a number of directional antennas. The equipment is interfaced with a computer system, which chooses the antenna from which the strongest signal is received. Each station reports the strength of the signal, and from which direction it is emanating, according to the computer's choice of antennas. This data is also entered into the computer; and, by comparing the signal strengths with the distances between each station, the approximate location of the signal can be determined to within 10 to 20 miles of its source.

The initial detection and long-range direction-finding stages are easy; the tough part is locating the transmitting equipment. The enforcement division of the FCC has several special mobile direction-finding vehicles to help locate the offending signals. Unlike the bright orange panel vans that were prominently displayed in the movie *Pump Up the Volume,* the FCC actually uses a number of discreet cars as mobile direction-finding (known as *MADF*) vehicles. The cars are generally late model American sedans, somewhat similar in make to unmarked police cruisers.

One of the major differences between a police car and an MADF vehicle is that the latter has no antennas. A radio car without antennas? Not quite, but the cars have no visible antennas.

Instead, the car's metal roof is replaced with a special identical fiberglass unit and the antenna is embedded within that enclosure. Furthermore, sleek solid-state equipment is placed in a console between the front seats. This equipment is capable of receiving virtually all of the electromagnetic spectrum and DFing it.

Usually, the FCC monitors will listen to unlicensed radio signals several times before they pursue the source. After the station has been adequately DFed at long range, several factors (such as the distance from the monitoring station to the transmission source, whether the FCC field agents are busy with another project, or the standard lengths of broadcasts from that particular broadcaster) help the FCC to determine whether or not they will seek the station.

In a worst-case scenario for the FCC, a station would operate sporadically (such as approximately once per month) for relatively brief periods of time (perhaps less than 30 minutes) on a variety of frequencies from various locations that were a six hour or longer drive from the closest monitoring station. On the other hand, the best-case scenario for the FCC occurs when a local station begins broadcasting long programs according to a certain schedule. As a result, nearly all of the stations that were listed as being raided by the FCC in Chapter 3 fit into the FCCs best-case scenario. Likewise, no FCC worst-case scenario pirate station has been known to have ever been closed down at the time this book was being written.

After the FCC has determined that a station is worth raiding, it will continue its monitoring operations. If a particular pattern is established, one or more field agents will be sent out to wait for the broadcasts to begin so that they can begin close-range DFing. Otherwise, if the station is close enough or if the broadcasts are lengthy, the agents are afforded the luxury of waiting on call until a broadcast begins, then driving to search for the secret transmitter.

By monitoring the signal and following its bearings, the agents merely drive until they receive the strongest relative field signal strengths. If the roads are clear and the transmitter site is in a suburban or rural area, this task can be relatively easy. The agents can also be tipped off by house lights late at night and large antenna arrays in the yard. Although neither of these tips are usable evidence, they do allow the agents to locate the area and talk with the residents of the house.

The standard FCC procedure is to show his or her FCC badge, ask to be allowed into the station, and talk with the responsible parties. After collecting evidence, positively identifying the individuals involved, and making technical checks on the equipment involved, the agent will leave. Approximately 30 days later, the operator of the station will receive a *Notice of Apparent Liability (NAL)*. The NAL normally issues a fine that must be paid within a certain period of time.

Ethics and the FCC

The previous section covered what normally happens with the FCC and a typical pirate radio station that has been caught. It's what is perceived in newspapers and books, and sometimes it occurs just this way in real life. Indeed, it is similar to the FCC raids that occurred up until the late 1980s.

However, times have changed. In the case of the raid against the Voice of Oz, the operator claimed that the FCC field agent posed as an insurance agent to gain access to the house. He also claimed that his $1000 fine was verbally reduced by the agent. After being traced by local government officials because of the outstanding fine, the operator pleaded his case over the telephone (not in administrative court), and the fine was completely dropped.

The operator of WKND was caught twice by the FCC in 1990. Each time he complied with the FCC field agent. The first fine was levied at $1000, but was later reduced to $100. In the second instance, no fine was levied, but the agent confiscated a trunk full of radio equipment and some listener reception reports. The alleged operator of Hope Radio met several state policemen and an FCC agent in his driveway one day after work. They questioned the man, but he denied everything. The man later received an NAL and a $1000 fine from the FCC. After some bargaining, the fine was dropped, but the man had his amateur radio license revoked.

At the end of 1991, the FCC installed a new fine structure that raised the standard fine for unlicensed broadcasting from $1000 to $8000. Armed with this new weapon, the FCC began handing out huge fines and undertaking major publicity campaigns when they did.

For example, one man accused of broadcasting programs from the Voice of Laryngitis for 13 straight hours was one of the first to receive the FCC's "full treatment." He was fined $10,000 and his name, name of workplace, and job position were released to a number of newspapers. Even though his case had not been decided in court, the FCC mailed press releases to many media outlets that definitely incriminated the man. In this case, many radio hobbyists believed the $10,000 fine to be excessive and the press releases—which were reprinted almost verbatim in several major newspapers and radio trade magazines—to be bordering on libel.

In one of the most publicized cases, a man who was accused of operating Radio USA was fined $17,500 in early 1992. In this example, the man was fined $7500 for not allowing the FCC agent into a house the supposed operator did not own (the agent was not carrying a search warrant and did not ask the actual property owners for access to the house). Also, even though the program featured a number of different announcers (including several women), the FCC insisted that this man was the voice on the program. As in the previous case, the FCC mailed incriminating press releases before the case had gone to court. Throughout the case, the FCC demanded payment for the fine even though it had not yet been settled in court and they had no confession and no proof (they had seen no transmitter, no antenna, etc.). The Radio USA personnel have continued their erratic style of broadcasting, and the NAL was thrown out.

These last two cases are interesting because both of the recipients of the enormous fines were highly visible in the pirate radio community. Both were published writers and had written on the subject of pirate radio. In one of these cases, an FCC agent mentioned that he had read some of the published articles of the person they were targeting, and, although this person faced a tremendous fine, it would be lowered if he "would talk." In both of these cases, it seems the FCC was attempting to send a message to other pirates and their listeners.

The FCC was pressured further when Stephen Dunifer of Free Radio Berkeley operated his 10-watt FM pirate in the open, as discussed in the last chapter. Not only was the station on regularly, but he informed the FCC and the local and national media of his presence. Dunifer was quickly handed an NAL for $20,000 in 1993, to which an attorney from the National Lawyer's Guild responded to. One year and a number of radio and magazine interviews later, Free Radio Berkeley was still broadcasting at an increased schedule. How the FCC will react to a federal court's refusal to grant an injunction against Dunifer (as described in Chapter 9) will certainly have a major impact on the FCC's enforcement efforts against pirates.

The cases since 1992 have brought up a number of questions about the FCC's integrity, intentions, and limitations. The fine structure that was implemented in late 1991 was challenged by one company, and was overturned by the Federal Court of Appeals in Washington, DC, in July 1994. (It is significant that the Washington court made this ruling, because it is generally considered that the Washington court of appeals is second only to the Supreme Court in its influence on other federal courts.) This finding made all of the FCC fines from late 1991 to mid-1994 questionable. If the structure by which a fine is levied is corrupt, has the entire case been nullified? At this writing, the FCC is still attempting to salvage these cases; it also promises to develop a new fine structure that would overcome the problems cited by the federal appeals court.

One of the most obvious and most pressing questions is, "Why is the FCC concerned about low-power shortwave pirate broadcasters that cause minimal interference?" I don't have the answer. Just one bad power-line transformer, one large, high-powered electric fence, or one "out of band" licensed shortwave station can cause more interference than all of the North American shortwave pirate broadcasters combined. To fine the pirate operators for potential interference and yet ignore the other, harmful forms of radio interference seems ludicrous. Although some hobbyists believe in a conspiracy theory about the FCC and the suppression of free speech, others believe that "the pirate problem" might be connected to the FCC's source of funding. Well-known shortwave author Harry Helms believes that this is the case:

> Although the U.S. constitution only specifies three branches of government—the legislative, the executive, and the judiciary—there has evolved a de facto fourth branch, namely the bureaucracy, of which the FCC is a part. Since it has no constitutional basis, the bureaucratic branch must continually come up with reasons for its existence. This is why the work of the federal agencies can never be done; those agencies have a desperate need for more "threats" that they can respond to. The EPA will always need more environmental problems, the DEA will need dopers and dealers, and the FCC needs pirates. Suppose there were no pirates and everyone obeyed the FCC rules? Whoops! Budget and personnel cutbacks aplenty at the FCC!

For the most part, I would have to agree with Mr. Helms' point on these matters. Although several inconsistencies in FCC policy have occurred over the past few years, they generally seem to be pushing the attitudes that radio pirates create harmful interference and that they can easily close down any unlicensed radio station. The truth of the matter is that pirates rarely cause any interference (much less than the shortwave stations that the FCC licenses to operate in the 7300- to 7550-kHz range) and that the FCC has been only moderately successful in closing down pirate stations.

When the FCC busted Radio Chaos International in Clark, New Jersey, the operator was fined $1000 and a large amount of equipment was confiscated for his operations on 7415 kHz. At the time, John Rahtes, engineer in charge at the FCC Philadelphia office, said that pirate stations, "not only cause interference on a local level, but on an international level. This country has treaties with other countries regarding allocation of frequencies. To interfere with that is against international law."

Although Mr. Rahtes and the FCC might believe that this ideal is worth upholding, the rest of the U. S. government obviously doesn't. Several months after Rahtes said this in respect to the operation on 7415 kHz, the Voice of America began operations on the same frequency from their Botswana relay station with 100,000 watts of transmitter power. Signals from this station were loud and clear throughout North America; in fact, it was far stronger than most pirates that had used 7415 kHz. Some radio hobbyists humorously referred to the Voice of America outlet as "Radio Free Botswana." In late October 1994, the situation was made worse when the FCC licensed shortwave broadcaster, KVOH, to operate on 7415 kHz. Today, Maine-based commercial broadcaster WBCQ has been licensed by the FCC to operate on 7415 kHz with 50,000 watts.

In the Autumn of 1994, the FCC licensed WWCR to operate with 100,000 watts on the frequency of 5065 kHz, five kHz above the top of the 60-meter broadcast band. This region is one of the tropical bands, identified as such because stations in North America and Europe are not permitted to operate that region. However, the FCC evidently felt that it was permissible to violate international treaties in this instance. Although this end of the band is not particularly crowded, WWCR now obliterates Radio CANDIP from Zaire, Ondas del Suroriente from Peru, and Sistema de Emissoras Progreso from Ecuador. Not only is the FCC now licensing stations that cause severe international interference, but they are sending a message to the rest of the world that they are willing to violate the spirit, if not the letter, of international treaties and occupy more frequencies adjacent to the tropical broadcast bands if they wish to do so.

The WWCR situation is interesting, in part because of the situation that occurred when its first transmitter went on the air. Pirate station XERK was busted in Donna, Texas, in October 1991 and fined $1000 for operating the station on 7435 kHz. After the station was closed, the FCC's Kingsville, Texas, field office sent a press release stating that the operator was fined "for operating an unlicensed pirate broadcast station on 7435. This frequency is assigned to the International Fixed Public Radio Service Band. Misuse of radio frequencies is a serious offense because of its potential for interfering with safety-of-life services such as aviation, law enforcement and marine." Less than two months later, WWCR was licensed by the FCC to operate on 7435 kHz with 100,000 watts, certainly more obliterating than XERK's mere 60 watts!

Legal Alternatives

If hobby broadcasting piques your interest, beware. There are presently no legal alternatives to pirate radio, although the Free Radio Berkeley case described in the last chapter might have created one. The lowest power permitted for unlicensed broadcasting on the AM or FM bands is 0.05 watts and 0.01 microwatts, respectively. The approximate maximum coverage radius for a transmitter of this type is 200 feet, perfect if your only audience is Mom, Dad, and your dog in the backyard. The minimum output power for a licensed AM or FM transmitter is 250 watts and 100 watts, respectively. Between application fees, attorney fees, licensing fees, and equipment, the cost of even one of the licensed low-power stations would easily be in excess of $100,000. If you think that shortwave might be an option, forget it. The minimum output power for a licensed shortwave broadcaster in the United States is 50,000 watts. And as I have heard before, if you don't have over $1,000,000 that can be tied up for several years, you won't have a shortwave station.

Still, the amateur radio broadcast bands are an inspiration to many that someday, the United States will allow amateur broadcast bands. Although some proposals have been put forth, few have gone past the pages of *The ACE.* Most of the previous proposals have pushed for low powered noncommercial licensed broadcasting in the 1600- to 1700-kHz section of the mediumwave band or in the 7300- to 7500-kHz 41-meter band, but one proposal pushed for part of 26 MHz to be used for hobby broadcasting. The problem with these frequency ranges are that the former is now part of the AM broadcast band in North America, and the other two are important internationally.

My favorite possibility is for the FCC to license noncommercial hobby broadcast stations to operate during the local daytime in the tropical broadcast bands. Because of the international treaties, U.S. radio stations are unable to use these frequencies. However, if low-powered stations (from less than one watt to 1000 watts) were licensed to broadcast during the daytime on these frequencies, no interference would be caused and this empty chunk of the radio spectrum could be utilized. By treaty, no North American stations are allowed to operate in these radio bands and the signals on these frequencies are much more limited in their range. As a result, the licensed signals on 120 meters (2300 to 2498 kHz) don't reach the United States until darkness arrives; even the signals on the highest tropical band, 60 meters (4750 to 4995 kHz), don't reach the United States until several hours before sunset. If the FCC would allow daytime noncommercial broadcasting here, the pirates would be satisfied, and normally unused frequencies could be utilized without causing any interference.

If these stations could then be licensed for less than $100 per year, the possibilities for engineering experiments and interesting free-form programming styles would be wonderful. The airwaves could be full of cutting-edge and mostly forgotten styles of music, information, and history on a huge variety of topics, and politics from a vast number of viewpoints.

The problem with a broadcasting system of this sort is that it would prove to be not only amateurish, but also volatile. Before you could say "political activism," some of these stations would be slapped with massive lawsuits from major corporations, who wouldn't want the truth or disinformation (depending on the situation and on your perspective) to be spread around. In order for these stations to stay alive, they would need to purchase expensive lawsuit insurance. This alone would close or prevent most of these stations from ever being able to hit the airwaves. Then, the whole amateur broadcast band would die as a result of what would appear to be lack of interest.

Because of these problems, I believe that the only manner in which noncommercial hobby radio stations can broadcast is if they do it without a license. I don't mean the all-out havoc that was created on the Citizens Band, but rather pirate radio—the same stuff that this entire book has covered. The difference between the present pirate radio scene and what I am envisioning is strictly a difference in enforcement. The FCC has "upped the ante" on an administrative violation that was considered to be minor just fifteen years ago. Most of the pirates that were caught in the 1960s and 1970s were only served up with warnings for first offenses. Even the second offenses only produced $500 fines. These measures seem to be more than appropriate for the level of the violation and they were effective in limiting the interference.

The fines of $8000 to $20,000 for hobby pirate broadcasting (until the fine structure was thrown out by a federal court) were neither appropriate nor effective. Not only are the fines higher than those for major felonies (such as drunk driving), but pirate broadcasting has been on the rise in North America since 1988.

The FCC should be providing leadership to minimize interference across the radio spectrum. On the whole, the pirate radio community has been very conscientious of clean technical operations and minimizing interference. In fact, over my years of listening, pirates have always been more courteous and conscientious than the licensed amateur radio operators. But rather than providing leadership, the FCC has taken a hard-line stance and frequently threatened and insulted pirate operators in the mass media. In one of the most quoted (by pirate operators) comments from the FCC, Serge Loginow from the FCC's New York City office described all pirate broadcasters as being "mildly disturbed individuals."

FCC Enforcement in the Late 1990s

After raiding a number of pirate operations in 1991 and 1992, the FCC became surprisingly quiet in 1993 and 1994. Over these two years, only WPIG and the unauthorized relay of WMXN were closed by the FCC. These actions were particularly noteworthy because WPIG aired numerous broadcasts while announcing the home address and telephone number of the station. The unauthorized WA4XN relay station broadcast for 24-hours per day for several weeks before it was finally closed. Because of the careless operations of these stations, it's no wonder that they were closed. Considering the circumstances behind both stations, the FCC did not publicize either raid. The only fined pirates that received national attention were the well-known California FM political pirates Free Radio Berkeley and San Francisco Liberation Radio.

San Francisco Liberation Radio was the focus of one of the strangest FCC raids against a pirate station. According to an article by the operator, he was stopped by an FCC agent on the way out of his house one day (not while broadcasting). The agent asked for identification and the man refused and left in his car. The agent then called for backup from the San Francisco police. The man was soon stopped in a roadblock by a number of police officers, who then handcuffed the man and waited for the FCC agent to arrive. According to San Francisco Liberation Radio, the police officers were disgusted to find that they had apprehended a radio pirate. He was quickly released and one of the officers even asked the man for the station schedule!

One of the most significant developments of FCC enforcement in the 1990s is the Free Radio Berkeley case. As stated earlier, the FCC recently failed to get a federal injunction against Stephen Dunifer to prevent him from broadcasting. Before failing in court, the FCC bypassed their own procedures to stop the station. Dunifer's lawyer complained that the FCC did not grant them their rights for an FCC hearing. However, it might be more important to note that the FCC evidently is, by enlisting the power of the federal court system, powerless. Otherwise, they would seize the equipment or enforce the fine. By placing the burden on the court, they were, in effect, showing that they were merely a go-between that can only recommend fines and station closures. These developments appear to be just the tip of important changes in FCC policy for the late 1990s and beyond.

In early 1995, it was clear the FCC was groping for a viable enforcement policy after its fine structure was struck down by the federal courts. An article in the March 1995 issue of *The ACE* described how the FCC is attempting to compel compliance with its rules without fines. Borrowing a technique used by such agencies as the Drug Enforcement Agency (DEA), the FCC showed up at the location of a suspected pirate station with a search warrant and a U. S. Marshall. The search warrant was for (in the words of the actual warrant) "equipment capable of being operated in the HF radio spectrum." Any equipment meeting that description at the location specified in the search warrant could be seized and recovered only through court action. In effect, the FCC must be sued to recover any seized equipment. Whether this technique will still be used by the FCC in the future is not known.

Even more sweeping changes may be in store for the FCC. In March 1995, FCC chairman Reed Hundt called his agency "horse and buggy regulators" and proposed, in effect, a massive downsizing of the FCC and shifting many of its functions to private industry and user groups. (For example, complaints against broadcasters would be handled by the National Association of Broadcasters.) One of Hundt's plans was to close all monitoring stations and consolidate all monitoring functions at the Laurel, Maryland FCC facility. A network of remotely controlled monitoring units would be installed and Laurel would be the "command center" of the network. A large number of existing FCC monitoring personnel would be re-assigned or dismissed under Hundt's plans; in fact, in early 1996, none of the existing FCC monitoring stations were being staffed from 10:00 P.M. to 6:00 A.M. local time!

Of course, with the volatile political climate of Washington, DC, where government agencies are headed by Presidential appointees, policies can change on a day-to-day basis.

For example, from about 1995 through much of 1997, the FCC was more flexible with unlicensed broadcasters than they had been in the past. Some stations had limited communications with the FCC and Free Radio Berkeley was fighting them in court. But on November 19, 1997, the FCC struck forcefully and made a statement in the Tampa area of Florida. At 6:00 A.M., a group of government agents (comprised of Federal Marshals, a SWAT team, local police, customs, and a self-identified CIA agent) led by the FCC raided the home of L. D. Brewer, operator of the Tampa Party Pirate.

According to a press release from Free Radio Berkeley concerning the situation, "With automatic weapons trained on them they were ordered to the floor where they were handcuffed face down with gun muzzles at their head. For the next 12 hours they were detained in their own home, not even allowed to go to the bathroom alone, while agents stripped their home of anything remotely related to radio transmission equipment. Police cordoned off the block around their home, the site of the micropower broadcasting station, and brought in a crane to dismantle his broadcasting tower."

"At the same time, other members of the same task force were busy shutting down two more micropower broadcasting stations in the Tampa area. One man, Lonnie Kobres, was taken to jail and not released until he signed papers agreeing not to take any action to recover equipment that had been seized. Radio X had its equipment seized as well with one of its members spending a brief time in jail."

On Tuesday, July 14, 1998, Lonnie Kobres, operator of Lutz Community Radio, was sentenced by a federal district judge. Found guilty of 14 counts of unlicensed broadcasting by a trial jury, Kobres was sentenced to three years probation, six months house arrest, and a $7,500 fine. That the case even went to criminal court is a dumb-founding novelty. The FCC was organized as an administrative organization only, void of any police power. Their due process had never included soliciting prosecution teams to curtail unlicensed activity. Lonnie Kobres of Lutz Community Radio knows better than anyone that times have changed.

In part because of the outpouring of support for Kobres from hobbyists, activists, and listeners alike, Judge Henry Adams reportedly refused to make an example of Kobres. The fact that this case is unlike any in the history of the judicial system would seem to indeed make Kobres the standard by which subsequent cases are measured. Kobres supporters are fully aware that the only way for Adams to not set a precedent would have been to declare the charges unconstitutional.

As it was, however, Kobres was facing a possible 3.5 million dollar fine and 28 years of imprisonment. Judge Adams' considerably kinder sentence was viewed as leniency by Federal Prosecutor Ronald Tenpas. According to an article by Bill Coats in the *St. Petersburg Times* dated July 15, 1998, Tenpas cautioned that the sentence would have "significant national ramifications beyond Mr. Kobres." Coats added that Tenpas had suggested a nine-month jail term and a fine of $10,000. The federal prosecutors are allegedly pleased, though, with the verdict and sentencing. The *St. Petersburg Times* also quoted Executive Assistant U. S. Attorney Monte Richardson. "This was the first criminal prosecution undertaken in this type of offense," said Richardson. "We think it had a significant deterrent effect."

Kobres contends that in the first written communication he received from the FCC, there was no indication that the Commission had been granted power to demand cessation of broadcasts by any judicial authority. "If the true intent of the FCC was to obey the law, it should have applied for a hearing to obtain an injunction signed by a judicial authority," he stated in a posting on his web page, ". . . be it known that all contact with the FCC after the 11/13/95 letter was in the form of dangerous, terrorist-style attacks. At no time has the Tampa FCC office demonstrated any willingness to deal with the issue except by means of violence and deadly force. Significantly, no representative of any branch of government has stepped forward to say that this method of dealing with an innocuous, community-based radio station is wrong!

Present and Future FCC Actions

FCC activity against pirate radio stations has dropped significantly in the first half of 2001, compared to the previous several years. John Anderson has been busy tracking FCC raids against pirate radio stations on his About.com web page, which is currently the most accurate database of such actions.

According to Anderson's list, by early July, only 11 stations were known to have been raided, warned, or fined through the mail. All of the stations were on FM, none were on AM or shortwave. Of course, other stations across the United States surely received some form of FCC action without receiving any media attention. However, it seems reasonable to assume that the ratio of raided known stations to raided unknown stations would be the same this year as in past years. Given this assumption, the number of FCC actions in 2001 is down to a projected 22 actions from 34 in 2000, 37 in 1999, and 73 in 1998.

Another interesting statistic is the number of known fines. In 2001, there have been three fines, with two occurring in 2000, none in 1999, and seven in 1998.

One oddity in FCC policy has been the differences in approaching FM and shortwave pirates. In many (not all) instances within the past few years, the FCC has only warned or confiscated equipment from FM stations. However, raiding FM pirates appears to be a top priority. On the other hand, shortwave pirates have been largely (if not completely) ignored since 1998. But the last shortwave pirates that were fined (between 1995 and 1998) typically received stiff fines in the thousands of dollars.

Conclusion

After the massive media attention for the radical FM pirates that were at the heart of the "microcasting" movement in the mid-1990s, it appears that the FCC changed its focus. Evidently, FM pirates were targeted while shortwave pirates have been "backburnered" for the time this book was written. Now that FM transmitter kits and higher-powered transmitters from overseas have become extremely common, it appears that the FCC will have its hands full with controlling the local pirate stations.

Of course, the FCC's priorities could and are likely to shift quickly at any given time. The recent years of shortwave pirate broadcasting without interference from the FCC are finite. Some authorities within the pirate radio scene believe that the FCC was operating conservatively after botching some of their investigations in the early 1990s. It was thought that rather than immediately jumping on the suspected pirate, the agents were sitting back and gathering plenty of evidence to be sure that their enforcement actions would "stick."

If these theories are accurate, pirate radio broadcasting will return to an upswing in the early to mid '00s, and several well-publicized FCC raids will also occur. Regardless, the best way to keep abreast of the FCC's activities is to check their official web page at http://www.fcc.gov. Another handy source is John Anderson's pirate radio site at http://pirateradio.about.com. For some interpretation of FCC policy as it relates to pirate stations, see *Hobby Broadcasting* magazine (http://www.hobbybroadcasting.com), subscribe to the *AMPB newsletter* (PMB 22, 2018 Shattuck Ave., Berkeley, CA 94704), or see the bulletin boards at the Free Radio Network (http://www.frn.net).

11 Verifying Pirates

or the most part, pursuing radio stations (licensed or not) over the airwaves is a solitary hobby. Unless you record every broadcast that you hear, there is no lasting "souvenir" of your experience. However, many hobbyists enjoy collecting lasting mementos of what they hear known as QSLs or verifications. These are cards or letters sent to the listener by the pirate radio station in response to a correct "reception report" sent to the station. Such reports let listeners give pirate operators information about their signals and programming, helping make the pirate listening experience more "interactive." Listeners enjoy collecting QSLs much like some people enjoy collecting baseball cards. QSLs from pirate stations—like those used as illustrations in this book—are often very creative!

About QSLs

To the listeners in the 1920s and 1930s, radio was a participant hobby rather than pure entertainment. Vaudeville comedians suddenly became radio stars, and the kid on the corner was soon the night announcer or engineer at a local radio station. Likewise, the stories in the print media just sat on the table; radio was alive and vibrant, never static. Before long, listeners were glued to their sets, laughing, learning, and even crying. Because there were so few stations even in metropolitan areas, listeners desiring new or different entertainment were forced to seek out new, and often distant, stations. Soon, looking for distant stations—what we call DXing today—became the main interest of many radio listeners. Listeners wrote letters to distant stations they heard, and stations answered with letters or printed postcards that confirmed the listener actually heard them. Listeners started to collected these cards or letters ("QSLs," from the radiotelegraph abbreviation for "I acknowledge and confirm"), and the hobby of radio listening and DXing was born.

Thus, radio was a booming hobby in the 1920s and 1930s. Even the mass-market magazines sometimes sponsored radio listening contests. Most AM stations eagerly gave out verifications, and the ECCO company even printed special stamps for nearly every station so listeners could collect stamps from every one they heard. Throughout the 1920s, newspapers even regularly included news articles about local DXers who heard a distant station (Fig. 11-1). This trend gradually gave

Fig. 11-1 *A 1930s QSL from licensed AM station WMBO. Not only did commercial stations offer QSLs then, but DXers were depicted as being typical members of the general public, sitting in front of the console radio.*

way to a less-personal commercialized radio that has dominated the post-World War II era. Today, few people have ever seen a verification card, let alone considered attempting to obtain one.

Verification cards are an integral portion of the free radio hobby to the average listener. Although the card does serve the purpose of verifying reception reports, it also provides a personal memento for remembering the station. Often QSLs, especially ones offered by pirate stations, provide a tangible, personal reflection of their broadcasts and individual personalities (Fig. 11-2). Although today's pirate stations are often on the cutting edge of societal and cultural trends (or just somewhere beyond, in left field), they are in some ways throwbacks to the early days of radio listening when finding new stations was an adventure and radio stations were extensions of

Fig. 11-2 *Okay, maybe KIPM offers a little too much of its own personality!*

their operators.

When you write to a station, make the report at least as valuable to the operator as a QSL would be to you. Although this might be difficult when writing to stations such as the Voice of America, the BBC, or the Voice of Russia, it is fairly easy to do with hobby broadcast stations. All it takes is a little time, honesty, and consideration to dish out a first-class report. To write a quality report, ask yourself what kind of report you would like to receive if you were the operator of that particular station.

Most of all, the report should be a help to the station by the information it contains. This is not breaking the law and collaborating with their broadcasting activities; it is simply presenting information in a friendly, personal, and factual manner.

All too often, listeners believe that every station must verify every correct report. However, this is unlicensed radio operated by individuals, with all of the costs coming from their own pockets. They have already provided the enjoyment of radio listening by adding another station to the DXer's list. However, if the station promises to verify all correct reports and does not, it damages their reputation.

Nevertheless, they do not have to do anything.

The issue of QSLing reports has caused a number of arguments in the pirate community—especially because stations that request reports usually ask for approximately $1 in postage. Stations such as these should verify if they request the postage; otherwise, they are being dishonest with their listeners. However, the fact remains that you cannot force a station to verify its reception reports. Trying to do so is quite tactless. Most pirates are quite independent; chances are that if you try to force an operator into QSLing, he will only become more stubborn.

Demanding QSLs is as tactless as making up fake reports for stations. Very occasionally, someone will take information from a radio bulletin and send it to stations that they have not actually heard with the hope of getting a QSL. Fortunately, these reports are easy to spot (especially if the station also receives the newsletter the information was taken from), and they are usually ignored. I have seen a number of faked reports over the years, but two of the best were from two radio enthusiasts in Pakistan. This pair reported WLIS on a day when they were not on, and the station has not yet been heard outside of North America. To top it off, the signals were reported as being quite good, and even though both listeners said that they enjoyed the program, they did not report any details!

Although friendly and honest reporting should always be a goal for the sake of the station, it also is productive for the listener. Rumors spread quickly about people with unkind or dishonest reporting styles. Eventually such people could become virtually blacklisted by both the stations and other listeners. In a hobby ruled by the honor system, everything is based on reputation.

Because of the risks that are taken by pirates on the airwaves, it is also best to make a few comments about the programming that you hear (if you are able to hear enough to comment). Programming comments are valued by all stations—the new stations want to know what people want to hear and the veterans want to know that their continued programming is of value to some listeners. These comments are becoming especially valued in the 1990s as the novelty of pirate radio in North America is wearing off a bit. Some stations, such as Radio Doomsday and Up Against the Wall Radio, even insisted on receiving comments on their programming or they wouldn't verify reports. Radio Doomsday has an excellent, long "why you did not receive a QSL" letter and information on what they want to see in a reception report. Although a number of listeners were offended when they received these letters, Radio Doomsday quickly sent off a

QSL when the listeners complied with the station requests.

Keeping A Logbook

The only way to write an accurate, helpful report is to copy the details of a program chronologically in a logbook. Detailed reports necessitate recording the exact time each part of the programming starts, determining accurately the frequency being used, the date, and a description of how well the signal was being received at the listener's location on the date of the broadcast.

A detailed, readable logbook is just as important as the amount of attention given to the actual DXing. Regardless of how well a broadcast might have been received, it will be difficult to write a convincing report if the details are not on paper. Although this aspect of listening takes on a sort of secretarial guise, it requires little effort to jot down a few details. In fact, if the listening hobby really entertains you, you should enjoy keeping a permanent history of your listening activity.

Often the logbook becomes a large resource instead of an impersonal "piece of history." It will become a tool that can be used many times to aid in listening and reporting. For example, maybe you heard a station last year with a telephone call-in format and accurately logged it. The next year, if you heard the station again but the call-in number was inaudible, you could find it in the logbook. Likewise, maybe only an announcer's name or a slogan was notable from a broadcast. If you had logged the station before, the latest broadcast could be tentatively identified from this information.

There is really no "best" way to keep a logbook. Each different book is a reflection of the person's individuality. Some listeners buy logbooks, some photocopy typewritten pages and make a logbook, while others merely use wire or ring-bound notebooks. Each type of book has its advantages and disadvantages. With professionally printed logs, the entries should be the neatest. Custom-made photocopies can accommodate any particular style or method of entering, but a blank notebook remains the cheapest and most flexible.

Regardless of which logbook style is used, it should be kept orderly. Don't hurriedly scratch the details down on an old piece of paper and make a mental note to recopy them at a later date; tomorrow stretches into next week and then next month. When you finally do return to recopy the loggings, you might not be able to read or decipher them, and they become useless.

Logbooks are most important for recording information about stations that you might want to catch in the future. If you have no intention of sending a report to a particular station, there is no hypothetical reason to be bound to a logbook. But with the erratic nature of pirate broadcasters, it is more crucial to write extensively about the infrequent operators than regular international broadcasters. An example is listening to European private stations. Back in the mid- to late 1980s when Radio Dublin International (on 6910 kHz) and Radio Caroline (on 6210 kHz) were audible nearly every evening in the United States, there was no need to take a detailed log of these stations every time they were heard. Furthermore, in a community with a local AM or FM pirate, it is not necessary to log them carefully during every broadcast if they go on the air frequently.

With frequently heard stations, such as some of the European privates or community service broadcasters, it is a good idea to make a short log entry. The time, stations, frequency, reception quality, and a few other details should be listed. These logs could be helpful with other DXing pirates; strong signals from the more powerful Europrivates indicate that conditions are improving to allow reception of low powered Europirates. Also, frequent logs of local pirates could determine a pattern in the way that a pirate operates.

You might want to keep all of this information in the same logbook or keep several to organize the material to personal specifications. Some listeners keep a simple logbook with all the stations they have heard and take detailed accounts of broadcasts only when they want to send for a QSL. Others just keep one logbook. I maintain a detailed logbook (Fig. 11-3) for all radio listening.

These days, many people might not even think of logging the stations on paper–they would rather use a computerized logging program. I sometimes use a computer to log reports. But I write much faster than I type and the computers that I have used have always caused buzzing interference across the radio. Unless I'm listening to a booming signal, I just turn off the computer and scratch the details in the logbook.

For some people, logging details on paper is just too inconvenient. Besides, they want to have the broadcast itself, not just information about it. Instead of logging the broadcasts, they record every program that they hear and keep a tape library organized by either station or date. For years, I didn't record the pirates that I heard, but over the past few months I have gained a greater appreciation for recording broadcasts. Still, as a logging method, it's much more convenient to track down information about a broadcast by glancing at the pages of a logbook than to listen to hours of tape.

If you have been keeping an organized logbook, writing a

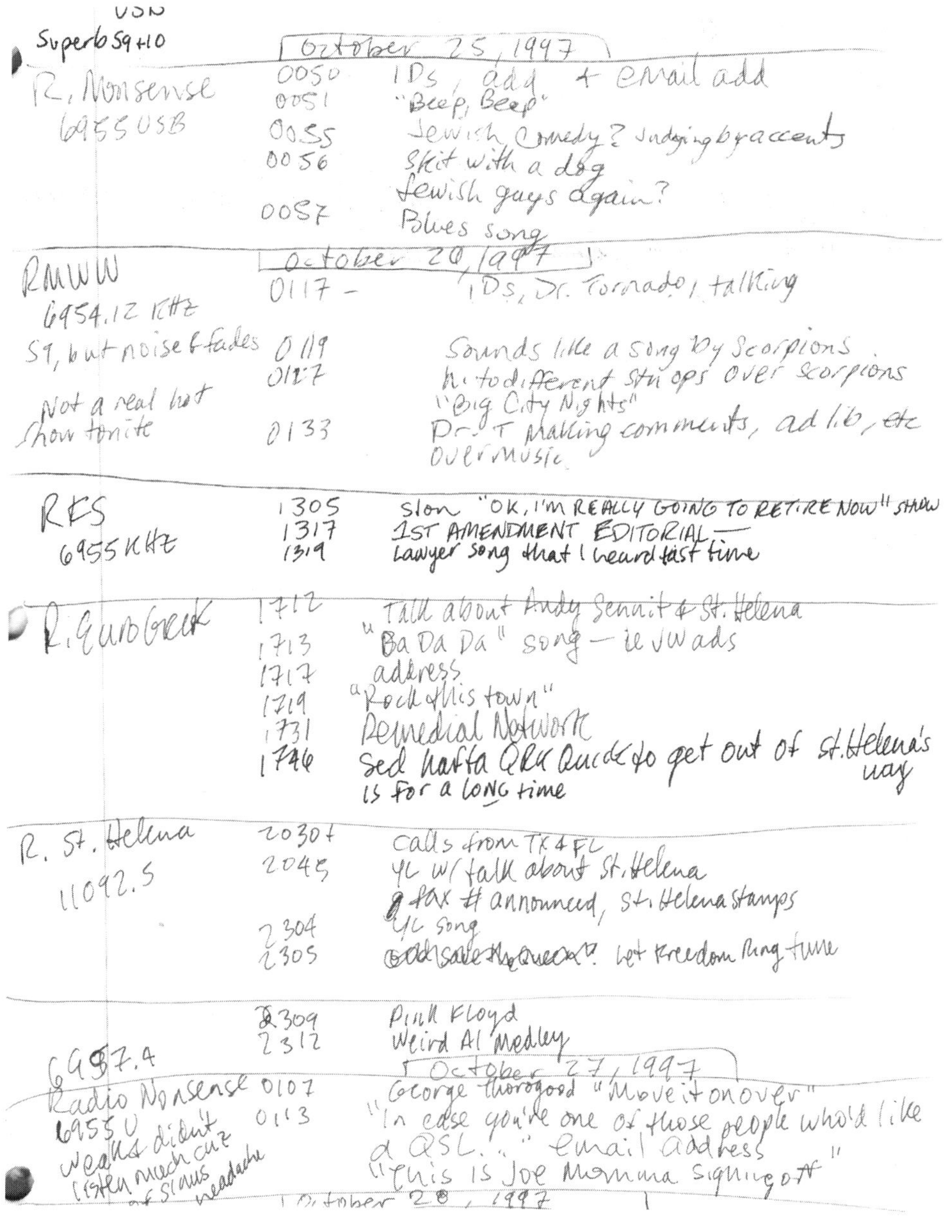

VSN
Superb S9+10

R. Nonsense
6955 USB

	October 25, 1997	
	0050	IDs, add + email add
	0051	"Beep, Beep"
	0055	Jewish comedy? Judging by accents
	0056	Skit with a dog
	0057	Jewish guys again? Blues song

RMWW
6954.12 KHz

S9, but noise & fades

Not a real hot
show tonite

	October 20 1997	
	0117 -	IDs, Dr. Tornado, talking
	0119	Sounds like a song by Scorpions
	0127	hi to different stations over Scorpions "Big City Nights"
	0133	Dr. T making comments, ad lib, etc over music

RFS
6955 KHz

	1305	Slon "OK, I'M REALLY GOING TO RETIRE NOW" show
	1317	1ST AMENDMENT EDITORIAL —
	1319	Lawyer song that I heard last time

R. Euro Greek

	1712	Talk about Andy Sennit & St. Helena
	1713	"Ba Da Da" song — ie JW ads
	1717	address
	1719	"Rock this town"
	1731	Remedial Network
	1746	Sed hafta QRK Quick to get out of St. Helena's way is for a LONG time

R. St. Helena
11092.5

	2030+	Calls from TR4FL
	2045	YL w/ talk about St. Helena
		fax # announced, St. Helena stamps
	2304	YL song
	2305	God save the queen? Let Freedom Ring tune

6997.4
Radio Nonsense
6955 U didn't
weake didn't
listen much cuz
of sinus headache

	2309	Pink Floyd
	2312	Weird Al medley
	October 27, 1997	
0107	George Thorogood "Move it on over"
0113	"In case you're one of those people who'd like a QSL..." email address "This is Joe Momma signing off"
October 28, 1997	

Fig. 11-3 *A page of my logbook from Autumn 1997.*

report to a station should require little more than merely copying the information onto another sheet of paper. In fact, I have received QSL cards by photocopying a page from my logbook and then writing a letter about the broadcast on the back. This is not a preferred method of writing stations, but if time is a factor, it is better than not sending a report at all. However, if this method is occasionally used, it is a must to have written legibly in the logbook. Just because you can read your own handwriting does not mean that the operator of a pirate station can.

Writing Reports

The first item to include in the report is the station name. This should be one of the simplest details to log, but many reports are sent to the wrong station or to the proper operator with an incorrect version of the name. The Voice of Syncom, named after a communications satellite, was listed as "Sin Com'" "Symcom," "Synton," "Symtom," and "Syncon" during its early shortwave broadcasts in the Spring of 1980, despite the fact that they frequently identified and spelled out their name over the air. Also, station names can be difficult to correctly log because of station spelling or other linguistics. For example, Radio Flattus (from the early 1990s) consistently spelled their name with two Ts even though the correct spelling has only one. And for two years, Psycho Radio (a current station) was known to listeners as "Sycko Radio" before announcing a new e-mail address. With static, fading, and interference from other stations, a proper identification can be tough, especially if the announcer mumbles or only delivers the callsign once or twice during the broadcast. But do make an effort to get the station name right.

Likewise, announcer names can be a problem to clearly identify. Often, listeners attempt to be more personal or show how much they know about the station by addressing the letter to one of the announcers. But do not do this if you are unsure of their names. It is not usually a serious problem, but an incorrect identity can only make the DXer appear foolish and demean his/her listening ability.

Next, list the exact frequency of the transmission heard. If the receiver has an analog frequency readout, report it to the nearest kilohertz or five kilohertz. Because few pirates ever use multi-transmitter broadcasts and parallel transmissions or even care if their programs are exactly on frequency, a close estimate should suffice if a digital receiver is not handy.

Just as the roots of a tree grip the ground and hold it upright in the soil, program details and their corresponding times comprise the basis of a reception report. The details should be arranged so that each piece of programming is identified or described, including the time of its start and end. For example, if the song "Break Down the Walls" by One*21 was played from 0346 to 0348 UTC, list the time first and then the song title and artist. Song titles and exact quotes from the announcers are the most verifiable details, so pay special attention to these items. Many listeners try to escape with reporting generalities such as "talk" or "music." It is important to be more exact than this.

Time can be especially perplexing to the novice pirate listener. Common questions include: Is the time to be listed in the time zone of the listener or the zone where the transmitter is believed to be located? If the time should be listed in UTC (the new term used to describe GMT), what if the operator is unfamiliar with this method? Usually just reporting in UTC is fine, but if the station's familiarity with the hobby is uncertain, list the time in both UTC and in local time. This should clarify any misunderstandings that the operator might have had from just one listed set of times.

Likewise, when the date is reported, use the same time zone that was implemented for reporting the times of the program segments. If only UTC is being used, just report the date in UTC; do not mix different time codes together without labeling the sets appropriately. If the reporting times are mixed without proper clarification, the operator might be confused and discard the report.

One particularly relevant section of the report to the station operator is how the signal sounded at your listening location. Because generally a small number of listeners frequently tune in free radio stations and different transmitter powers, locations, and antennas produce varied results, most operators would be delighted with a handful of reports with detailed descriptions of the reception in each location. Therefore, this section of the report requires careful attention. Although it will not make a great impact on receiving a QSL card, it does give the operator much-needed information.

Station personnel want their signal to be widely audible, but do not write that their station was coming in fine if it really was not. Sometimes listeners are inclined to believe that a great signal report will guarantee a QSL. This short-circuit logic usually fails and an operator might even become suspicious of a report with much better signals than usual for that particular part of the country. Also, the false information could confuse future technical operations for the station.

Unfortunately, an effective description of the reception of a broadcast could fill a page with type. To solve this problem, radio enthusiasts created a numerical rating system for the separate qualities of a transmission. While the system, known as "SINPO codes," should not fulfill the entire request for information on reception, it certainly does simplify the process. In addition to the SINPO code, a popular code for reporting broadcasts to radio bulletins is the "SIO code," a shorter, less complete version. Because this code is less complete, it is less helpful to the stations and less advisable for use when writing reception reports.

SINPO is an acronym for "strength, interference, noise, propagation fading, and overall merit." Each characteristic is rated on a scale from one to five. A SINPO of 55555 is perfect, with a distinct possibility that the broadcast can even be heard on a toaster; a 00000 SINPO is the type of reception possible when the electricity is off or when the radio needs serious repair. As might be expected, pirate broadcasts commonly range in reception quality (at least at my present location) from 11111 to 55555, with most landing somewhere near 33333.

This is not to say that the numbers should be or always are all the same in a SINPO code, in fact they rarely are. A particular broadcast with a certain time and frequency is much like a fingerprint, with no two being exactly alike. So a strong station with an equally strong RTTY (teletype) transmission on the same frequency could produce a SINPO of 51552, while a weak station on a clear day could pull a SINPO of 15552, for example. The numbers are the means by which the listener establishes how well the broadcast was heard in his area.

Strength and interference are not the only variables to greatly affect the overall merit. A station on a low frequency, such as 90 meters in the Summer months before sunset, usually contends with an unbearable amount of static (noise). Likewise, tuning in a pirate on or below 41 meters near midday is often barely worthwhile because of the heavy fading (propagation).

After the SINPO is reported, describe how the broadcast was received. The operator might not fully know the SINPO code, so a description can help identify exactly what is meant by the numbers (usually listeners vary slightly with their standards). A simple statement that might follow a reported SINPO of 32333 might be, "The signal, noise, and propagation were all fair, although there was some interference from a station on an adjacent frequency (list the station name and frequency, if known)." You might describe the type and height above ground. Other things that might be included here are the listener's exact location using longitude and latitude lines, the height above sea level, or the weather for the area during the reception of the broadcast. Whenever I am low in details and/or have the time to write a longer report, I include information about the weather. I'm not saying that anyone could use the weather details, but if nothing else, maybe it will be interesting to them.

If I have a chance, I like to send something along with the report, although I do it less often than I should. Some DXers include trading cards and others include a postcard from their local area. Ed Kusalik in Coaldale, Alberta, Canada, sends along a professionally printed SWL cards with a photo of him in his shack. It's an easier way for the station operator to make a personal connection with you (Fig. 11-4)

Next, try to mention something about the format, audio, etc., of the station later in the report. When writing the operators, remember to praise their station in some way, and if appropriate, provide polite constructive criticism. Sarcastic comments might anger the operator, so if you want to help the station, do not take this approach. Even if the station has little to praise, try to compliment something. Then suggest ways they could improve their signals and/or programming. Because most operators are eager to find out ways to better satisfy their listeners, polite advice is usually well-received. Several stations have greatly improved their audio transmitter stability or programming because of listener response. See Fig. 11-5 for a sample report.

Fig. 11-4
Long-time DXer Ed Kusalik sends this photo card along with reports to pirate and licensed broadcasters.
Ed Kusalik

Short wave DX'er since 1963 Call Sign: VE6DX1AJ
Amateur Radio Operator since 1992 Call Sign: VE6EFK

Receiving Equipment: JVC NRD-525 (modified) Drake R8A
Member: North American Short wave Association since 1969
Ontario DX Association (Founding Member) since 1975
Numero Uno since 1987 (Now Explorer)
A*C*E* since 1990;Free Radio Weekly and Short wave Pirates

March 4, 2001

Howdy!

I am quite pleased to report reception of Radio Shadowman. If this report is correct, could you please verify it with a QSL card?

\\\
Radio Shadowman   I heard Rod Stewart "Maggie" at 0704
3/4/01   Slade "Cum on Feel the Noize" at 0708
6300   Man announcer that I couldn't understand because of FMing?
SINPO: 23332
0655-0712 UTC   For details on my reception of your station, please listen to
the cassette tape that I made of your broadcast
///

I listened to the program on a Kenwood R-5000 communications receiver connected to a 500-800 foot (approximately) longwire antenna. My home location is too noisy to hear Europirates, so I found a nearby baseball field with surrounded fields. I parked my car close to a pavillion and ran the antenna from the car across the baseball field, and through the fields. Almost all of the antenna was lying on the ground.

I placed the cassette deck, R-5000, and a small lamp on the rear deck of my 1986 VW Golf diesel. It was about 40-degree F, so I really wasn't too cold. This was definitely the warmest evening I've had yet for DXing Euros! However, a snow storm has just started and we're expected to receive at least 12 inches today!

Last night when I went back to string up the antenna, I policeman watched me drive into the baseball field area. I assumed that he would have come back later (fortunately, he didn't). I wasn't doing anything wrong, but I thought that he might not understand what I was doing and make me leave. When I started to string the antenna, I noticed a skunk about 20 feet in front of me! I slowly backed around and tried to avoid him! I was imagining getting sprayed by a skunk, then getting in trouble with a policeman! Fortunately, neither happened.

Your signal was really strange. I think the transmitter was FMing, but I'm not sure if it was FMing or if maybe a variable-frequency utility was right on top of your signal. Either way, something that I was listening to was quickly varying in frequency and creating a severe "warble" on the audio.

BTW, I had a great time DXing for Europirates over Christmas and New Year's Eve. In the past two months, I have heard: AL Int'l, RECH/VOTN, UK Radio, Radio Torenvalk, Radio Borderhunter, Laser Hot Hits, Radio Nova Int'l, Radio Bluestar, Radio Eastside, Union Radio, Radio Black Arrow, Radio Foxfire, Radio Marabu, Radio Black Power, FRSH, XTC, Radio Perfekt, Radio Wonderful, Delta Lima Radio, Astoria Radio, Tower Radio, Mike Radio, Radio Shadowman, Wreckin Radio Int'l, & West Coast Radio. I heard 10 Europirates on 2/25 alone, although only two were new for me.

The town where I live is named Mont Alto and it is about a 15-minute drive from where I work and receive my mail in Blue Ridge Summit. Mont Alto is in the south central portion of the state of Pennsylvania. Pennsylvania is in the northeastern portion of the United States. In the summer, the temperatures are normally in the 80s Farenheit, and in the winter, they are typically in the 20s or 30s Farenheit (although a few years ago, the temperature dropped to -32 degrees Farenheit one very cold night!). The area around Mont Alto is agricultural. The area is filled with farms that raise cows, corn, soy beans, etc. Washington, DC is about 80 to 100 miles south of us.

Fig. 11-5 *The first page of a reception report that was sent to Dutch pirate Radio Shadowman. The second page contains a few paragraphs about myself, the area where I live, and this book.*

Fig. 11-6 *An International Reply Coupon, which is redeemable in dozens of countries for cash. IRCs are available at post offices in the United States, Canada, and Europe and are frequently used in place of return postage when writing to radio stations in other countries.*

Maildrops

Nearly every hobby pirate in the United States and Europe uses maildrops to handle their mail. The station announces the maildrop's address, and the maildrop forwards letters to the operators. The stations' replies are then either immediately mailed out or they are sent back to the maildrop to be postmarked by that post office. With this method, it is impossible to trace the location of a station through the mail. Also, it is completely legal for the listener, maildrop, and station to communicate with this method.

Unfortunately, maildrop security does have its costs. Although the postage for a single letter is not a great deal of money, pirates request their listeners send three unused first-class stamps with all reports that ask for a reply. Some might call pirates "cheap" for not paying everyone's postage, but this practice could end up costing some stations hundreds of dollars per year. These extra stamps are used to cover the postage needed for delivery between the listener, the maildrop, and the station. Pirates in Europe ask for two International Reply Coupons (IRCs), which are available for a small fee at most post offices. (However, I have encountered workers in many small post offices that did not even know what IRCs were. If problems occur while you are searching for IRCs, go to a larger post office for help.)

Sometimes maildrops are blamed for negligence by listeners when they do not receive QSL cards. One rule to keep in mind is that maildrops are not affiliated with stations past the point of forwarding their letters. If a QSL has not arrived, then it is the station's fault or perhaps the listener has not waited long enough for it. On the average, a reply to a reception report arrives within a month or two, although some have taken a year or longer. Regardless, do not blame the maildrops for QSL problems. Chances are that they have had nothing to do with anything that could have happened to a long-lost QSL.

Inside A Maildrop

"P.O. Box 1, Belfast, NY 14711" is one of the most recognizable addresses in free radio and even in the shortwave hobby as a whole. Thousands of letters have flooded the box since it opened in 1989 after moving East from Hawaii.

Many other maildrop boxes have operated since forwarding became popular in the late 1970s, but most of these were intended solely for one or two stations. Thus, when the stations were closed down or faded away, so did the maildrops. In the early 1980s, several "powerhouse" maildrops opened to handle mail for many different stations. One was located in Arcata, California (this maildrop later moved to Hilo, Hawaii), two were closed by people who lost interest in pirate radio, and one from Washington, DC, only handled mail for some of the more mysterious broadcasters.

Considering the nature of unlicensed broadcasting, it might appear that operating a box is dangerous. Indeed, any time that I go to a shortwave listening convention, someone asks me why anyone would ever operate a drop box "because they're sure to get busted." However, John Arthur, proprietor of the Belfast maildrop, says this notion is entirely fictional. Contrary to popular opinion, running a maildrop for pirates is neither illegal nor dangerous. Now in his third decade of forwarding listener letters, he has never had a direct encounter with the FCC.

Even without the legal disputes, many shortwave hobbyists feel that because the maildrops are "siding" with the unlicensed broadcaster by handling their mail, they are unethical. But the major maildrops claim to side more with DXers rather than with the pirates. "Because I'm a DXer, I like to know what is on the air," says Arthur. "And if it's there, I'd like to send a report to get a QSL. It is one way to not only get QSLs for myself, but to help others, too."

Where Do All of Those Stamps Go?

Nearly every pirate station that requests reception reports in North America asks for three first-class stamps. You might wonder if you are being ripped off or if the station operator is trying to get buzzed by licking stamp adhesive. The explanation is simple: postmark hiding.

Fig. 11-7 *A yellow, red, and black RFM QSL that's more notable for the stats (1625 kHz AM on New Year's Day) than for the artistry of the QSL.*

Most maildrops open the letters that they receive, take one of the stamps, "packet up" the reception reports, and use the stamps to mail the packet to the station operator. At this point, two stamps remain. The operator reads the reports, fills out the QSL cards, addresses the envelopes, and uses one stamp for each letter. All of these responses are bundled up and the last stamp (per letter) is used to send the QSLs back to the maildrop. The maildrop receives the package, opens it, and drops the letters off at the post office. The end-result is that a station can receive and send mail without having the actual location divulged.

As a result, if you send fewer than three stamps along with a letter, it might be trashed. It depends on the generosity of the maildrop and the station in question.

QSL Returns

Some stations actually encourage listeners to send in reports of multiple receptions by offering many varieties of verifications. In the past, RFM (Fig. 11-7), Voice of Laryngitis (Fig. 11-8), Radio Doomsday, Radio Airplane, CSIC, WLIS, and Radio USA (Fig. 11-9) all have made extended efforts in this direction. Although the many stations usually produced photocopied QSLs, the Voice of Laryngitis used a variety of humorous black and white photographs in the early 1980s.

Current North American stations with different QSLs include WLIS, Radio Azteca, WHYP, and KIPM. KIPM QSLs are amazingly crisp, colorful, strange 8.5"-x-11" inkjet images printed on photo-quality

Fig. 11-8 *Fun photo QSLs like this one only added to the mystique of the Voice of Laryngitis in the mid-1980s.*

paper. Because the KIPM imagery is sometimes grotesque or disturbing, the station's QSLs might not be everyone's favorite, but on the basis of colors, size, creativity, and paper quality, they are probably the best (overall) ever sent out by a pirate.

Many listeners enjoy writing several times to stations, but it is important to try to make the report timely. Old reports are of less value than those of recent reception to the station because antennas and transmitters are often changed or reconstructed. Thus, a report of a transmission from a now-defunct transmitter or antenna will not give the operators any new information or new ideas on improving the station. Also, if the report is late, the operators might suspect that the reception details could have been copied from radio newsletters or someone else's logbook. Because of these reasons, the chance of receiving a QSL lessens. However, if you want more QSLs from a station and/or would like to get in touch with the operators again, using old reports is fine, especially if they were friendly in the past.

Few stations offer many different types of verifications, but some do have professionally printed cards. Contrary to the excellent

Fig. 11-9 *A 10th Anniversary Radio USA photo QSL, showing some of the designs used over the years.*

production of their European counterparts, American stations usually send out homemade or photocopied QSLs. However, the verifications have improved significantly on this side of the pond. Some stations (such as Voice of Juliet, CSIC, He-Man Radio, Hope Radio, and KNBS) have *tee*-shirts for sale, KNBS has pins, Radio Airplane and KMCR have printed stickers, and (for very special listeners) RFM had watches made while WVOL had hats printed. And for scavenger-hunt winners at the SWL Winterfest, WPN has given out full-size station wall clocks.

On the other hand, The Voice of the Phoenix, a station heard just several times on shortwave, announced that it would only verify reports from listeners who reported the station in *The ACE* and live outside the United States. Another antagonizing station, the Voice of Epileptic Catfish, gave out one fake address in "Ankara, Czechoslovakia" and another using pieces from two real maildrop addresses. Although these stations are almost certainly run by radio enthusiasts playing jokes on their peers, it can become upsetting to realize that they will never reply to reports.

One strange reversal of this technique was used by UNID, a station that popped up on shortwave several times toward the end of 1984 and beginning of 1985. Main announcer John Anon aired some comedy skits including a fake newscast and later said (while laughing) that no QSL cards would be issued. Unexpectedly, everyone who reported UNID in *The ACE* began receiving QSLs for the broadcasts—

five months later. It just goes to show that pirate rhetoric cannot be trusted; even when it looks as if they are trying to be nasty, they might be arranging for a big surprise.

Conclusion

Reporting broadcasts and receiving verifications from any station, especially a pirate, is a lot of fun. With free radio as a whole being the most unpredictable outlet on AM, FM, or shortwave, their verifying processes range from friendly to reclusive. Verifications from pirates offer a personal and humorous look at radio that exists nowhere else. With just a little time and effort, a listener can manage a detailed logbook, write quality reception reports, and have excellent results receiving QSL cards.

Equipment and Information

The equipment necessary to listen to pirates is often confused. After writing several magazine articles about pirate radio listening, I have received a number of questions along the lines of: "Can I hear pirates on my boom box?" . . . "Can I hear pirates on my Dad's ham radio?" . . . "Can I hear pirates on a Radio Shack shortwave radio?"

The most basic requirement for listening to pirates is a dedicated shortwave radio. By "dedicated," I mean that its prime frequency coverage is shortwave (although most shortwave radios also cover the AM band and some of the newer models even cover FM or the air band). These radios are often called shortwave radios or communications receivers in their advertising, packaging, or even on their front panels. Beware of the old AM/FM radios that also included "SW1, SW2," and/or "PS" bands. With one of these radios, you'll be lucky if you can hear any of the enormous government shortwave broadcasters, let alone any pirates.

Carpenters can accomplish little construction without their tools; likewise, to even casually listen to pirates, it is important to have the necessary tools. It should be common sense, but a surprising number of shortwave listeners expect to hear as many pirates as veteran DXers even though their equipment, experience, and information cannot compare to those of the experts.

To hear the maximum number of pirates (and radio stations), it is very important to have no less than an "average" quality general coverage shortwave receiver, an antenna, at least one source in up-to-date pirate information, pirate DXing friends, and a great deal of dedication. You can skimp on any of these categories, but doing so will greatly affect the number of stations you hear and the amount of enjoyment you get from the hobby.

Receivers and Antennas

Dozens of books have been written about picking out suitable receivers and building better antennas. A list of these is contained in this book's bibliography. Because even entire books can never seem to completely cover either of these subjects, I can't possibly do more than lightly cover them here. Receivers are another prime topic, and the receivers currently on the market vary enormously in quality.

DXing is a personal hobby and each listener has his or her own taste in receivers. Many receivers completely outclass others, but the bottom line depends on what features the listener enjoys the most. For example, a nostalgiaist might want a huge tube receiver because it is an old model. Likewise, someone easily impressed by new technology might select a small, light model, or one with lots of gadgets: filters, memories, and flashing lights. Sharp selectivity is certainly important, and so is good audio, but those two qualities are somewhat mutually exclusive.

Buying a receiver can be an expensive task, so as a novice you should prepare yourself before picking one out. There is no excuse for a hastily chosen receiver. Reviews of receivers are featured in such magazines as *Monitoring Times* and *Popular Communications* as well as annual publications like the *World Radio-TV Handbook* and *Passport to World Band Radio* (Fig. 12-1). Even if these sources are not

Fig. 12-1 Passport *and the* WRTH *feature plenty of receiver reviews.*

readily available, adequate receiver information can be obtained from other shortwave listeners or maybe even radio equipment dealers. The sales pitch from equipment dealers needs to be sorted out, but, after talking to a few dealers, you should be able to find a receiver that meets your needs.

While searching through the reviews, recommendations, and sales pitches, several factors must be considered, such as selectivity, image rejection, size, workmanship, audio, availability, ease of operation, and price. Each should be carefully considered with respect to the maximum utility for your needs. Sensitivity, selectivity, and image rejection all are aspects of performance and are especially important to the "hardcore" DXer. Here are definitions of those and other important factors to consider when selecting a receiver:

- *Sensitivity* is the threshold at which a radio signal is audible on a receiver. An example of poor sensitivity is an old car radio that will not pick up any stations outside of the immediate vicinity.

- *Selectivity* is the capability of a receiver to tune in separate stations on nearby frequencies. On a receiver with poor selectivity, a weak station near a strong one would be obliterated; on a good receiver, both would be audible separately.

- *Image rejection* systems produce a radio spectrum with signals occupying the frequencies that they are transmitted on. A radio without quality image rejection, however, allows duplicates of the actual signals ("images") to appear on many other frequencies.

- *SSB (single sideband)* is a particular mode of transmitting audio via radio mentioned in earlier chapters. A normal AM signal consists of two sidebands and a carrier. All the "intelligence" in an AM signal is contained in the sidebands, and the two sidebands are identical to each other. A single sideband transmission consists of just one sideband. SSB is popular with amateurs and pirates because the method of transmitting is much more efficient and higher output powers are much more easily attained. However, a sideband needs a carrier signal before it can produce intelligible audio. Many receivers come equipped with a circuit known as a *beat frequency oscillator (BFO)* to produce a replacement carrier; such radios have "SSB, USB, LSB," or "BFO" selection positions on their front panel. However, some inexpensive portable receivers are not capable of receiving an SSB signal that you can understand. Considering that most North American pirates broadcast in SSB, you must choose a radio with SSB reception (or a BFO) if you want to hear many pirates.

- *Size* is not always a good indicator of receiver performance.

Receivers vary in size from 100-pound military surplus heavyweights to portables that fit easily into your pocket. When considering the size of a receiver, think about how often you might be taking it on trips or moving it to other locations. Small portables are new technologically, but often manufacturers compromise on performance and cost compared to the large receivers.

- *Workmanship* is the quality and construction of the material used in the receiver; will it withstand years of use? Most receivers are constructed well but many new microprocessor-controlled digital models are fragile compared to the large mechanical equipment of the past.

- *Audio quality* is important to DXers as well as program listeners; it does little good to be able to hear a weak signal if you can't understand it! This is an area where older tube radios really shine compared to some modern receivers. There's nothing like finding a strong, good-quality AM pirate signal and selecting the filter position wide on a classic vacuum tube communications receiver like the Hammarlund HQ-180. On the other hand, most contemporary small portables have very poor, tinny audio (Fig. 12-2). Even some modern communications receivers that are otherwise excellent performers suffer from poor audio quality; the Icom R-71A and Japan Radio NRD-525 models were notorious for their audio failings. On the other hand, such contemporary communications receivers as the Drake R8A and Lowe HF-150 and HF-225 models have outstanding audio quality by any standard.

- *Availability* refers to the cost and abundance of parts in case the receiver breaks down. Finding tubes for an older, used radio can

Fig. 12-2 *One of the few portables noted for its excellent audio: the tough-to-find Magnavox D-2935 from the early 1980s.*

be a lengthy and expensive process. Some high-performance receivers made before 1965 are rather inexpensive, but replacing parts if trouble arises is not always easy (or even possible!). When looking at equipment of this nature, it is best to ask the owner or dealer about limited warranties, find local shops where it could be fixed if necessary, and inquire where replacement tubes can be bought.

- *Ease of operation* applies more to novice listeners than to DXers. Few receivers on the market are so difficult to use that a brief examination won't help demonstrate the way to operate them correctly. However, if the novice has had problems using a stereo system, obtaining detailed information about operating a particular receiver might be wise. On the other hand, technical enthusiasts sometimes like their receivers more complicated than necessary with more buttons, knobs, and readouts to play with.

The most important consideration for the average DXer or program listener is the price. Few shortwave hobbyists of any type have the resources to dish out several thousands of dollars for a professional-quality Watkins-Johnson receiver. Therefore, you have to determine which options comprise the best receiver you can afford. This is where the controversy enters the scene. What is the best affordable receiver? Each buyer must decide for him or herself.

Personal Receiver Picks

I have used a number of different receivers over the years, and for lower budget listening, I have a few choices for new receivers in different price ranges. Here are my favorites:

- For less than $200, I like the Grundig YB-400. It is very solid and small, the key pad is easy to use, the sensitivity is very good, it won't easily overload (when images of other strong radio signals are heard all over the radio bands) with an external antenna, and it has SSB detection. The audio is tinny, but for even under $200, I have heard a number of pirates on this receiver.

- For less than $400, I pick the Sony ICF-2010. I have used this receiver for a number of years and it is still known as "the Cadillac of portables." The sensitivity is great, it's well-built, and it has digital tuning readout to 100-Hz intervals. Over the years, I have heard hundreds of North American pirates and over a dozen Europirates with this receiver.

- For less than $600, I like the Grundig Satellit 800 (Fig. 12-3)/ Drake SW-8 (Drake phased out the SW-8 and designed the Satellit 800

Fig. 12-3 *The semi-portable Grundig Satellit 800 is one of the top-rated receivers for the price.*

for Grundig, based on the circuitry of the SW-8). This receiver is a strange adaptation of a desktop receiver for portable use. The 800 features excellent audio, great sensitivity, and incredible image rejection (it can be used with massive antennas without overloading). The SW8 was extremely well-built, with a metal cabinet, massive rubber feet, and a solid plastic front panel; the Satellit 800 uses a lesser-grade cabinet. Unlike any other portables that I have used, the Drake is excellent in the AM band–perfect for tuning in AM pirates. With the SW8, I heard some Europirates that were not heard by anyone else in North America, not bad for a portable.

- For $1000, I would pick the Drake R8 (including the R8A and R8B versions), although all of the receivers in the $600 to $1000 price bracket offer excellent performance for listening to pirates. The R8 is the "bigger brother" of the SW8. It has all of the features of the SW8 (except for the ruggedization for traveling) and a few extra filters, an analog signal strength meter, synchronous AM detection, an input port for control via computer, and much more. You can't get much better for hardcore DXer or pleasant program listening than the R8.

- As a runner-up, I do like my Kenwood R-5000 (Fig. 12-4). It's a great DX radio and the audio is very good. However, the radio only came with two filters (2.3 and 6 kHz) in the five filter positions. To Kenwood, filters are options, which seems akin to offering different gears on the manual transmission of a car as options ("First and fourth are standard, but you'll have to pay extra for second, third, and fifth.").

"Budget" radios can provide reliable service, but the old "you get what you pay for" axiom applies. You must be a careful buyer. My original listening set-up only consisted of a Yaesu FRG-7 receiver and a 40-foot piece of "hook-up" wire strung out the window. Yet I heard

Fig. 12-4 *My Kenwood R-5000 in my temporary radio room--a bedroom in the middle of remodeling.*

approximately 100 pirate stations over the late 1980s. The FRG-7 receiver was made in the late 1970s as an inexpensive table model. Today, this receiver can commonly be found in good condition for as low as $75 (although eBay has been driving the price into the $200-$300 range as of late). Other decent-quality portables are breaking through on the used market for reasonable prices as well, but buyers should be wary of radios that are cheap in all characteristics, not just price.

Over the past 10 years, I have picked up a bunch of old tube receivers from the 1930s through the 1960s. These radios have been an excellent value in the cost/performance ratio, and frankly, I think that they're a lot more fun! However, these radios have often had technical problems and I have either had to do some minor repairs or take them somewhere to be repaired. I now have better receiving equipment than I did 10 years ago, and I can hear a few more stations with better reception. However, I can only hear a few more stations than I could have with the FRG-7 (mostly distant stations, such as those from Europe). The point is that many pirates deliver relatively solid signals into North America; although you will notice a difference in performance between a used $150 receiver and a new $1000 receiver, you probably will not hear that many more pirates on the expensive receiver. Evaluate your needs, your available funds, and shop carefully!

Antennas

Antennas for shortwave listening are much less controversial. Although an elaborate antenna works better than a random wire, a simple longwire or dipole should be adequate for most pirate listening, especially if you're located in the eastern half of North America where pirate signals are numerous and strong.

Simple longwires or dipoles are inexpensive, costing as little as, $4 for wire and insulators for a longwire or a random-wire antenna. However, you will need some more expensive coaxial cable from a dipole antenna to your receiver. Random-wire or dipole antennas bought from companies usually run at least $70 and often perform no better than a homemade antenna. Small "active" antennas often cost over $100 and generally perform no better than a random wire. Because of the cost considerations, active antennas are recommended by most experts only if the listener lives in an apartment building or some other area without sufficient land on which to erect antennas. Many books about antenna construction include plans for the listener with limited space; such antennas can be built for much less than the cost of a manufactured active antenna.

However, it is true that you'll hear more pirate stations with a better antenna. Many DXers find ham radio antennas for the 40-meter (7000 to 7300 kHz) ham band to be solid performers when chasing pirates operating just above and below 40 meters. East Coast DXers often use specialized longwire and beverage antennas to dredge up weak European signals from the murky noise floor.

Finding antenna plans to suit your needs is not an exasperating task. Antenna plans are often included in shortwave/ham radio books and magazines at local bookstores and libraries. Both books and plans can be obtained from radio-supply and mail-order companies. After finding these plans, just build one that best fits your ideal for cost, size, and ease of construction.

Media Guides

A subscription to a regularly published radio bulletin is necessary for the listener to keep abreast of the pirate scene. Ever-changing addresses, frequencies, and patterns of operation are often confusing even to those with up-to-date information. Someone without a subscription to regular newsletters containing pirate information will be lost—and generally unsuccessful—in listening to and verifying free radio stations.

Several types of regular radio guides containing a variety of monitoring information are widely available for the pirate listener. Monthly magazines such as *Popular Communications* and *Monitoring Times* (Fig. 12-5) offer a professional approach and stories about a

Fig. 12-5 *On the newsstands:* Monitoring Times *and* Popular Communications.

variety of communications topics. Monthly and weekly radio bulletins or newsletters contain few articles; they generally contain news and loggings about radio listening. The last type is the pirate-only newsletter, published monthly or weekly with some articles, but they mostly contain raw information and loggings.

Monthly magazines greatly supplement other sources of information with "after the fact" articles, equipment reviews, and construction tips. However, a communications magazine alone does not provide the necessary up-to-date information that is useful to the pirate listener. The articles and columns in these magazines are usually at least several months old; in that time, a pirate could get busted, change frequencies, become extensively active, etc.

A new approach to radio magazines was taken by *Hobby Broadcasting*, which started publishing in 1998. *HB* is a quarterly magazine that focuses on feature articles (historic features, interviews, construction projects, *etc.*), rather than on loggings and other bits of timely information that can quickly become dated. *HB* covers pirate radio, part-15 transmitting, LPFM and FM "microcasting," carrier-current radio, offshore broadcasting, syndicated radio, *etc.* A number of the topics in this book (such as WCPR/WFAT/WHOT/RFNY, the Voice of the Voyager, LPFM, Tampa FM pirate raids, WDAB, *etc.*) are included in much greater depth in *Hobby Broadcasting* (Fig. 12-

Fig. 12-6 *Hobby Broadcasting magazine, with the baseball parody cover.*

6). As an extra benefit for all readers, at the end of this book is a coupon for $1 off a subscription to *HB*. I really enjoy *Hobby Broadcasting*, but I do have my biases—I publish it!

Regular newsletters are becoming a relic of the past. Because of the increased cost of printing and postage combined with the competition from the Internet, hardcopy logsheets and newsletters are becoming extinct. Club newsletters have been around since the 1920s, but how can they compete with online and e-mail newsletters, bulletin boards, and chat rooms? They can't and nearly all of the shortwave newsletters and bulletins have either folded or are seeing rapidly declining membership numbers.

Two old general-interest shortwave-listening bulletins still exist and do publish information on pirate radio. *NASWA* (North American Shortwave Association) and *ODXA* (Ontario DX Association) have been publishing monthly bulletins for decades, and it appears that both groups are successfully weathering the e-publishing storm. To many shortwave listeners these days, particularly those interested in pirate radio, NASWA is best known for hosting the SWL Winterfest near Philadelphia every year. The NASWA club bulletin runs approximately 100 pages of shortwave broadcast loggings, schedules, columns, and QSL reports. Most of the material applies to licensed shortwave broadcast stations, but Chris Lobdell (editor emeritus of *Free Radio Weekly*) writes a column about pirate radio. For more information, see http://www.anarc.org/naswa or write to: NASWA, 45 Wildflower Rd., Levittown, PA 19057.

The Ontario DX Association also publishes a monthly bulletin, but they include pirate radio loggings in with the standard shortwave broadcast loggings. The successful survival strategy for ODXA has been to change from an A4 standard newsletter format to a full-size magazine-style publication. The magazine looks more professional now and is distributed throughout Canada. For more information, see http://www.odxa.on.ca or write to: ODXA, Box 161, Stn. A, Willowdale, ON N2N 5S8 Canada

CIDX (Canadian International DX club) publishes *The Messenger*. Richard Sim writes the "Captain's Log" column, which features pirate radio loggings and more. For more information, see http://www.anarc.org/cidx or write to: CIDX, P.O. Box 67063, Lemoyne, St. Lambert, QU J4R 2T8 Canada.

Information on the various North American radio clubs is available from their umbrella organization, the Association of North American Radio Clubs (ANARC). All ANARC member clubs are trustworthy, but you should make sure that the one you pick best suits your needs and regularly carries pirate information. The current mailing address for the Association of North American Radio Clubs is: ANARC, 2216 Burkey Dr., Wyomissing, PA 19610
For a current list of ANARC member clubs, send a large self-addressed stamped envelope to the address.

Pirate-only newsletters and bulletins focus on the hobby and thus provide the best information on the free radio scene. The only pirate newsletter of this genre in North America today is *The ACE*, published by the Association of Clandestine radio Enthusiasts. In addition to printing loggings from pirate listeners, *The ACE* contains many other columns, articles, and news clippings. European pirate news, microcasting, the latest clandestine radio events, covert communications, a biyearly comprehensive guide to station addresses, and other features are included regularly.

There is a strong tradition of publishing pirate radio newsletters in Europe. Since the late 1980s, I've subscribed to or traded copies of newsletters with the following: *Pirate Chat, Free Radio News from Ireland, FRS Goes DX, Offshore Echo's, Radiotelex, Pirate Connection, Free-Radio-News, R.W. News, Free Radio News Sheet, Wavelength, Activity Magazine, Free Radio Sheet, Free-DX, Airspec News, Weekly Report*, and *Free Radio Desaster* (Fig. 12-7).

Fig. 12-7 *A bunch of different Europirate newsletters from the 1980s and 1990s.*

European pirate newsletters are helpful for hunting stations from that continent. Today, only a few free radio newsletters exist throughout Europe; most of those from the 1980s and early 1990s have been eliminated by the Internet. The only real newsletter that's currently active in Europe is *Pirate Connection*, a high-quality radio bulletin that publishes sporadically. *Hobby Broadcasting* also includes Europirate information, but in lesser quantities. We are just going through a little drought in pirate periodical publishing and I'm sure that more European newsletters will begin in the near future.

Computer Information

The world has changed significantly since the last edition of this book. No longer are computer communications only for the rich or technically determined. Now, with literally tens of millions of people using the Internet in the United States alone, the best sources of pirate radio loggings and test broadcasts are online. This section covers online information sources, but not Internet audio (both streaming and archived files)

Not all pirate information sources are newsletters published by listening clubs. Several of the best sources of pirate information could be accessed via telephone lines. The three major systems were all newslines that operated via answering machine. Participants would call in, listen to the opening message, and leave a message at the end if they had any information to share. From time to time, the moderator would then update the messages, based on the information that was recorded on the machine. Two of these from the late 1980s were regional newslines from Maryland and New Jersey with information on all types of radio DXing. The Newsline for Free Radio operated for approximately a year in the mid-1990s and offered an immense amount of pirate information. The newslines were all rendered obsolete by the Internet.

The Internet is an incredible source of pirate information, which is an enormous change since the previous two editions of this book. The first edition predated the open availability of the Internet. In the second edition, the Internet was a relatively new technology, experienced mostly by tech heads and college students. In those days, the Fidonet (an old international computer service carried by hundreds of computer BBSes) was a much better source of shortwave and pirate radio information than the Internet. Today, literally hundreds of different web pages include pirate radio in some capacity and the increases in bandwidth and improvements in audio compression allow for old shows to be archived and live programming to be streamed.

My favorite location for posting and reading information from others is the Grapevines section of the Free Radio Network (Fig. 12-8): http://www.frn.net/vines. This area contains separate boards

Fig. 12-8 *The Free Radio Network (www.frn.net) contains some of the best resources for the pirate radio listener.*

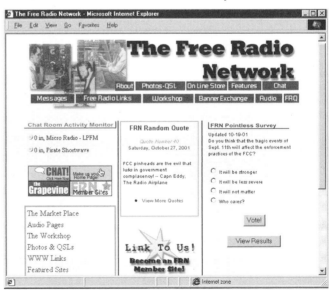

for shortwave and LPFM, technical talk, test broadcast announcements, and loggings for *The ACE*. Literally thousands of posts are on the FRN board and it's a terrific way to follow pirate radio. In fact, some pirates even request loggings on the Vines for QSLs. For shortwave fans, alt.radio.pirate on the Usenet would place a distant second. Very little shortwave information is here. For years, many of the postings were flames or involved FM pirating. It has improved over the past year or two, mostly because of the influx of Europeans. It can be a handy source of information for LPFM, but it still rarely contains anything concerning shortwave.

Aside from the "raw" logs, the best bet for fast information concerning North American pirates is *Free Radio Weekly,* a weekly online loggings sheet. The only catch with *FRW* is that you must contribute logs in order to receive copies of the newsletter. To participate, contact frw@frn.net.

For chat rooms, again, my favorite is the chat on the Free Radio Network. Frequently, a number of different pirate listeners (and sometimes even pirates) check into the chat rooms to "talk" about what's on the air, etc. I have spent essentially no time in the FRN chat since starting work on this edition of *Pirate Radio Stations*, but I'm looking forward to checking back in when I have a chance.

The selection of Europirate resources is a bit more varied. For immediate information, I think that the SW-Pirates listserv is best because some committed EuroDXers post loggings and the subscribers appear to be primarily European. This listserv is moderated by Alfred of Alfa Lima International and you can read more about it at: http://groups.yahoo.com/group/SWpirates (Fig. 12-9)

SRS News and the weekly Dr. Tim newsletter are the best newsletters from Europe. *SRS News* is published in Sweden, but the information covers all of Western Europe. To see the latest issue of SRS News, go to: http://www.srs.pp.se/news.html. Dr. Tim is a pirate station from Germany, but his newsletter covers information from all of the Europirates. This is probably my favorite European e-newsletter because it covers so much background information about pirates. For years, most European pirate newsletters only covered loggings and very little information about programming. The major problem with Dr. Tim's newsletter is that it is published only in German. I use one of the free online language translators to translate it into English. There's enough shortwave jargon and slang in the newsletter that some sentences really don't make any sense, but I can probably understand about 80% of the content, which is still very helpful. Check the Dr. Tim site at: http://www.doctortim-news.de.vu/

Fig. 12-9 *The home/signup page (groups.yahoo.com/group/SWpirates)* *for the SW-Pirates listserv.*

DX Friends

Don't overlook in any hobby (radio listening or otherwise) the friends and associations you can make. Free radio listening, while fascinating in itself, cannot be sustained for lengthy periods of time under solitary conditions. Many pirate DXers that were involved in the hobby for varying lengths of time dropped out because of the lack of friends interested in shortwave.

It is really not important to meet other listeners in person (although doing so is fun). Some of my best friends in the hobby over the past few years have been people I have never met. We write a lot of letters and talk on the telephone, but the distance between us is too great for us to meet each other. As the years have passed and as more get-togethers have occurred, I have been fortunate enough to meet some of these friends.

Friends in the hobby can improve your DXing considerably, depending on how many stations each listener hears and how well everyone communicates. For example, an article appeared in *FRENDX* (the former name of the NASWA bulletin) a number of years ago entitled "What Communal DX?" that detailed one listener's personal experiences with his friends. The author of the article lived within an hour (driving time) of ten other shortwave broadcast DXers. The group often staged overnight DX sessions where everyone brought their receivers to a particular house and hunted stations all night. The result of the DX session was a greatly increased number of stations and countries heard for each listener..

More importantly, the author of that article enjoyed conversing about a wide variety of subjects and listening to shortwave radio with his friends. Although the group mainly concerned themselves with legal shortwave broadcasting, the same rules apply to pirate radio listening because the latter is a subset of the former (Fig. 12-10).

Fig. 12-10 Dave Valko (left) pages through a QSL album while Jim Kay (right) listens to Dave's Collins R-388 at a small DX session.

Friends improve each other's listening results and longevity in the hobby. Not everyone has the time to hunt stations on shortwave during every hour of the day, so listeners often call others when a pirate is broadcasting. A procedure that was often used in the 1980s to notify other listeners of free radio activity was the "pirate alert." In order to save time and money, whenever a DXer logs a pirate, he/she called•other hobbyist friends and hung up the phone after it rang once. This notified listener then had to find the station, but that usually was not too difficult because most pirates normally use only a few frequency ranges. This method of notification might not give any details about the broadcast, but it does save on long-distance charges.

Often, pirate listening enthusiasts jump into the hobby as teenage renegade shortwave listeners. Legal shortwave DXing is not as exciting as the thrill of hearing someone, who might also be a teen, breaking the broadcasting codes of this country. Being friends with other listeners counteracts low points in pirate activity and personal enthusiasm that might encourage a hobbyist to drop away from free radio listening.

Radio Gatherings

Following the concept of "DX friends" is that of radio gatherings. Unlike classic car or science fiction enthusiasts, there aren't many opportunities for radio hobbyists to visit conventions. Aside from the personal get-togethers or DXpeditions with a few listeners, only a few shortwave broadcast or pirate-related gatherings occur every year.

The major event in the United States and Canada is the SWL Winterfest, organized by NASWA. This very low-key (no local advertising) event is held annually in Kulpsville, Pennsylvania (near Philadelphia) and draws about 200 radio enthusiasts from several countries. The Winterfest features forums on a variety of shortwave-listening topics, including pirate radio (Fig. 12-11). A number of different pirate operators have attended the Winterfest over the years, so it's fascinating to see who will drop in for the weekend. I've attended the past 14 SWL Winterfests and I've greatly enjoyed each one. For more information, see http://www.trsc.com/winterfest.html.

The ODXA also sponsors a DX camp regularly in Ontario. These

Fig. 12-11
A room of listeners awaits the beginning of the pirate radio forum at the SWL Winterfest.

events are much less formal than the Winterfest and are more actively radio related. Instead of just talking about radio, participants spend their time in a cabin, tuning the spectrum, searching for tough radio catches. At every DX camp, the participants log pirate stations, so tuning in pirates isn't discouraged. However, the DX camps would probably be much more enjoyable for pirate listeners who also have an interest in licensed shortwave broadcast listening.

European pirate gatherings have been growing and attracting more attention at the end of the 1990s and the early part of this decade. All of the major pirate radio events are insider-type events— something that's fun to read about on someone's web page, but you would probably only consider attending if you are very serious about pirate radio.

Two of the most popular pirate radio gatherings in Europe are held in The Netherlands in July and in Germany in August. The Dutch meeting is a relatively new one, first being held in Baarle-Nassau last year (Fig. 12-12). This year, more than 40 pirate enthusiasts found their way into the woodsy campsite for a weekend of pirate radio. The Merchweiler DX camp has been an annual event for considerably more years, dating back to the early 1990s. It's also better-known and attended than the Dutch upstart.

The largest shortwave gathering is the EDXC, which is held at a different location in Europe every year. The EDXC is very similar to the SWL Winterfest, with forums about international shortwave broadcasting. However, the EDXC does not emphasize pirate radio information and most European pirate hobbyists seem to avoid it. Thus, the EDXC is not one of the best choices for pirate enthusiasts.

Aside from these events, several offshore radio gatherings are sure to occur every year. Most seem to commemorate a particular

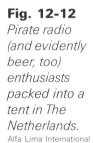

Fig. 12-12
Pirate radio (and evidently beer, too) enthusiasts packed into a tent in The Netherlands.
Alfa Lima International

station or event, and are joined by several offshore radio disk jockeys. Although the offshore stations would be included as a part of the general pirate radio topic, there is very little overlap between the hobbyists at an offshore event and at one of the pirate events.

Conclusion

True enjoyment from free radio listening requires a decent receiver, up-to-date information, and friends within the hobby. Unlike collectors of antique automobiles, rare records, coins, or other expensive items, the pirate radio listener is not motivated by monetary gain. Because the hobby is generally greed-free and based on the aesthetic value of listening to rare and interesting unlicensed broadcasters, it is driven by the flow of information and enthusiasm surrounding it. This is quite a bit different from most hobbies, considering that the equipment and nominal objects involved do not make or break a good DXer.

Computers and Radio

I f you stay in touch with the radio community—attend hamfests, read radio magazines, talk with other listeners, etc.—you will often hear the opinion that "computers are killing radio." Many of the old-timers voicing this thought will note that amateur radio was thriving in the 1950s and 1960s, but now the kids would rather play video games than experiment with the radio.

There is a kernel of truth to these statements. Fifty years ago, radio was the only game in town. If you wanted to be involved in electronics or communications, you became interested in radio. The fact that World War II was in full swing and tens of thousands of Army, Navy, Air Force, and Marine recruits from around the world were trained to operate two-way radio equipment added to this interest and the domination of the hobby. Upon entering civilian life in the late 1940s, these people automatically took to amateur radio as their primary hobby. And, in the 1950s, if you were young and bright, you were probably either interested in amateur radio or model railroading.

Competition for the youth of the world began in the late 1950s—both television and rock music became widely accepted. Then, more kids began to sit around and watch TV or listen to records than listen to the radio. The CB radio craze of the late 1970s attracted more shortwave listeners and amateur radio operators, but that fad only lasted a few years. With the 1980s began the so-called MTV generation and the personal computer revolution. And in the 1990s, the great social upheaval of the Internet occurred, leaving many people saying, "Will there be a place for radio in the 21st century?"

As the focus of youth education and culture changed to computers and rock music, radio as a hobby has been left behind. In the 1980s and 1990s, I have rarely seen amateur radio or the shortwave-listening hobby represented as a regular service to the general public.

Very occasionally, an amateur radio club will have a display in a mall or will be featured in a newspaper article (often something to the effect of "Local man networks with thousands worldwide"). Unfortunately, the shortwave broadcasters don't advertise; even though the Voice of America is a massive radio network, it is virtually unknown in North America.

So, with little attention and sudden competition from the Internet and DBS satellite TV, is there any room for shortwave? I certainly believe that shortwave broadcasting and two-way operations will continue into the 21st century. Why?

1. Shortwave radios can be inexpensive, but Windows-compatible computers aren't.

2. Internet connections are expensive—typically a minimum of $15 per month, yet shortwave radios operate for free.

3. Internet connections are complicated and are very difficult to protect. Shortwave radio transmitting antennas are easy to install and the radio signal can't be tampered with as it travels through the sky.

4. The less that the shortwave spectrum is used by the general public, the more it will be used by those who want their communications to be kept secret.

Because of these factors, I believe that shortwave radio will continue to be used regularly—even if some of the large shortwave stations and small shortwave networks in developing nations diminish. Shortwave will continue to be used by powerful nations who want to influence people around the world; by people in locations that are difficult to access (such as the mountains of Peru), by organizations who want their communications to be relatively secure (utility stations), and by those who want to reach niche audiences (such as pirate stations and those who buy airtime on commercial shortwave stations).

Clearly, shortwave radio can't compete as a regular broadcasting service in developed nations, where people have access to hundreds of AM and FM radio stations, TV and cable channels, newspapers, and Internet broadcasters and web pages. But for the niche market in the Western nations, shortwave radio is tough to beat. With shortwave, very little power can cover many listeners over thousands of miles. Even if the density of listeners per square mile is mighty thin, a large overall audience can still be attained. And if the word gets out to a particular niche market that shortwave is the place to be, those people will buy radios and listen.

A perfect example of this niche-market situation is pirate radio, where thousands of shortwave receivers have been bought either primarily or solely to listen for these underground radio stations. A good example of this phenomena was displayed on the Allan Handelman show, a syndicated FM rock radio interview program. Handelman bought a shortwave radio years ago with the sole purpose of listening to pirates, and during one broadcast, described how pirate radio is exciting and fun to listen to, but "everything else on shortwave is boring."

With such publicity, it will be fascinating to see if the shortwave broadcast bands remain a forgotten radio ghetto or if a sort of coolness factor will develop and encourage others to broadcast. Will a wealth of pirates soon be on the bands? Will the pirates inspire more people to purchase airtime on time-for-hire licensed commercial shortwave outlets? Will more commercial stations vie for the right to broadcast on shortwave? In short, will the shortwave neighborhood be thoroughly gentrified, leaving little space for the pirates to broadcast? It's doubtful, but the theories are interesting nonetheless.

Pirate Radio into the Digital Era

Aside from pirate radio, the best source of hobby broadcasting is on the Internet, where several large organizations are streaming audio from numerous individual sources. The two largest of these are Shoutcast.com and Live365.com, which feature literally hundreds of different audio feeds. The beauty of these sites is that anyone can post up a stream. There are very few restrictions on what material can be streamed and none on who can be doing it. The cost is nothing, although because of bandwidth restrictions, most of these stations can only serve a maximum of 10 to 100 listeners at any given time.

Already, some pirate radio has been appearing on the Internet. Check Live365.com at any given time and at least a few channels will be broadcasting old recordings of European offshore pirates from the 1960s and 1970s. Also, the popular old-time pirate, the Voice of Laryngitis is now broadcasting their material via Live365.com. Does this move signify a wholesale move of shortwave pirates to Netcasting? John Anderson, editor of the pirate radio channel on About.com doesn't think so, "I kind of chuckle when people say, 'What about Netcasting?' Netcasting has its place, but in no way will it replace terrestrial radio any time soon. If anything, I think the threat is more from radio going digital itself—some of the business models proposed include things like turning stations into 'wireless ISPs' and the like.

Programming, in that sense, may become secondary—to the betrayal of the essence of what radio has always been. That scares me."

Blackbeard from Jolly Roger International, an early 1990s shortwave pirate, echoes these sentiments: "Netcasting is not going to become a problem to pirate radio until we have high-speed wireless Internet in the hands of most people. Even then it comes down to programming. If you have good programming, people will listen no matter how they receive it. When that time comes, I guess we will have to sell the 'JR-Eye in the sky.'" And Blackbeard isn't at all opposed to Netcasting; in fact, he embraces it. Jolly Roger International shows are also streamed via Live365.com, 24 hours per day, and this stream is linked to their webpage. But like many of the shortwave pirates, Blackbeard views Netcasting as only a promotional tool—a great way to give people a taste of the programming, but not a replacement for the radio broadcasts.

Frederick Moe, host of Seldom Heard Radio, which is broadcast via commercial shortwave station WRMI in Miami, Florida, sums up the problems with streaming audio, "The main contrast that comes to mind when thinking about radio and Internet/computers is that pirate radio is broadcasting, whereas the Internet is narrowcasting. Anyone with a portable shortwave radio and an antenna can receive shortwave pirate broadcasts. Not everyone has the hardware to listen to Internet broadcasts, and even if they did, there are so many competing 'Internet stations' and audio streams that you could spend years tracking them all down and sampling them."

The worst problem that I've noted with the multitude of streaming audio stations in 2000 and 2001 is the lack of creative or diverse programming. Almost none of these outlets are "real" broadcast stations. Most are simply audio jukeboxes that exemplify the worst aspects of commercial radio and computer automation. Instead of having a real personality, most transmit a random "shuffle play" of a few hours of digital audio files. Most of these digital jukeboxes don't have announcing segments, promotions, or even IDs. It's just a rotation of songs. Although it's nice to hear a different selection of songs than what I can hear on commercial radio, I don't find most of the Netcasting stations to be particularly engaging. In fact, most are so devoid of personality that it's tough for me to think of the name of a single station!

But Shoutcast.com and Live365.com have some fascinating potential uses aside from the typical Netcasting. Pete Costello has connected the audio of a 900-MHz headset unit to the PC. By using Live365.com, he can stream his radio's audio onto the Internet. Combining this with chatting on IRC to shortwave listeners around the world, he can allow other people on the internet to hear his radio in real-time. With this use, not only can Costello give people without shortwave radios a taste of current pirate radio, but he can also provide pirate operators with a real-time proof of reception and signal quality.

Streaming audio could also be used to feed live programming to a shortwave transmitter. A pirate station could set up a transmitter in a remote location (even on the other side of the Earth) and accept audio for broadcasting. For example, if someone had rented a cabin or apartment several states away, the transmitter and a computer could be placed there. Then, they could program the computer to connect to the stream at a particular time every day. The transmitter could either be turned on remotely via telephone or regularly with a timer. The advantage of such a system would be that people could broadcast openly and freely without breaking any rules; only the owner/programmer of the arrangement would be subject to FCC actions. And if locations were moved frequently and the broadcasts were kept short, such actions could take some time.

Other Uses for the Internet and Computers

Magazines and science books from the 1950s through the 1970s that dared to predict the future dreamed of such things as personal helicopters, underwater cities, space centers, and wall-sized televisions. We are technologically so far behind most of the predictions that life seems much more like 1955 than the predicted 21st century. However, the science futurists of the past totally missed the computer telecommunications revolution of the 1990s. Communications have advanced far beyond what was expected. The closest that I've found is that one children's science fiction book predicted that in 2000, people would be able to go to a post office, have their letter keyed into a system that would send it via satellite, and it would be retyped out at the destination post office. This was called "electronic mail." There was no way to predict the web and all that it would provide by 2001.

Pirates, who by definition, make creative use of radio technology, also find ways to maximize the effectiveness of the Internet. The first area of improvement for pirate radio is just simple communications. Today's pirates can e-mail advance notices of broadcasts to selected listeners (or post the details on a bulletin board or alt.radio.pirate, if they are less discrete) and have a mid-sized audience even if an odd time or nonstandard frequency are being used. This is an amazing advancement from the 1980s, when pirate fans would have to either call each other with the details or leave one-ring telephone "pirate alerts" so that everyone would know that a pirate was on the air. I think it's been at least eight years since anyone has given me a pirate alert.

Where the Internet has really improved communications is among the different continents. Information from international scenes was very difficult to come by in the 1980s and early 1990s. Back then, the only international communications existed between Europe and North America, and even those were limited. In the late 1980s, I received telephone calls from Radio Silverbird (Holland), Radio Pirana International (Europe), Weekend Music Radio (Scotland), KIWI Radio (New Zealand), and several others, asking me to spread the news about upcoming test broadcasts. By contrast, I called Bill Lewis from Live Wire Radio (England) once or twice around Christmas to help me identify some of the weak Europirates that I heard. Unless I wanted to wait a few more weeks to receive a Europirate newsletter in the mail, I'd have no idea what I had heard. Today, the information is so much more available, cheaper (than calling Europe or South America on the telephone), and you don't necessarily have to be in the loop.

It's also interesting how Internet communications have improved the general flow of incidental information. For example, KMUD is a low-power West Coast pirate and I'm on the East Coast. Although, I've seen the test announcements from KMUD, I've never heard the station. However, despite never hearing KMUD, I've e-mailed the guys from the station a few times, I've read about some of their experiences, and I've seen some photos of their transmitter, antenna, and broadcast location. When KMUD was broadcasting back in 1987, I knew nothing about the station (aside from the fact that they were on the air), but with the advent of the Internet, not only am I familiar with the station, but I've even made contact with the operators.

Alfa Lima International in Holland uses the Internet to determine whether he should be broadcasting to North America. Some people have shortwave receivers connected to the Internet, and they can even be tuned remotely. Alfred turns his transmitter on to 15070 kHz and starts broadcasting. Then, he goes on to the Internet and goes to the Web page of an online receiver in New York and remotely tunes it to 15070 kHz. If he can hear himself with a good signal via that receiver, he knows that ALI will be audible all over

North America, and he continues broadcasting. If not, he can just shut down.

Hunting and Gathering

Thanks to fan web pages, you can find just about any type of information online. Thus, pirates can find material for broadcasting, whether it's ideas, material to read or perform, or audio bits for airplay. For example, I heard an acoustic, folky pirate radio song on KRMI and a few days later, I noticed that the songs were on the About.com pirate radio site. Evidently, the operator of KRMI saw it on the About.com site, downloaded it, and plugged it into his radio program. Captain Disturbio of WVDA does the same thing. "Do you want some rare or 'different' musical selections? BINGO! Just download a peer-to-peer sharing network and you can have MP3s galore."

The only problem with downloading material, like anything, is that it can be used a bit mindlessly and uncreatively. Some listeners have complained about how pirates have been downloading fake ads from Internet comedy and radio sites, rather than creating their own material from scratch. But of course, the stations that download everything are the same ones that would not have created their own material in the 1980s or early 1990s, so the Internet is not robbing us of creative, homegrown programming.

Captain Disturbio takes the idea of downloading material one step further. "Yet another reason is clip-art for QSLs, etc. There aren't a lot of clip-art images that come pre-loaded from Microsoft or Macintosh on your everyday computer. With the Internet, you can download appropriate art for your specialized QSL card (that is if you use a computer to construct a QSL)." Indeed, the quality of the QSL cards has greatly improved since the 1970s and 1980s. You just don't see the crudely hand-drawn cards, like those from Voice of the Voyager, WCPR, or KQRP anymore. And the full-color QSLs from KIPM have taken pirate QSLs to another level.

Studio in a Box

Aside from the Internet, computers have also revolutionized radio in terms of actually producing the programs. Just a few years ago, producing a radio program involved lots of audio equipment. For example, a nice pirate studio from the 1980s would have had one or two audio mixers, at least one turntable, at least one cassette deck, at least one CD player, a microphone, an open-reel tape deck, some audio effects (reverb, flanger, etc.), and several cart machines. New, you could expect this equipment to cost at least $700. Of course, such a system would still be excellent in the 21st century, but it doesn't have to be so expensive or space consuming.

Blackbeard from Jolly Roger International says that the only savings aren't space and money. "What used to take days to create now takes hours. You no longer need cart machines, a radio console, CD players. All you need is a computer with audio editing software and a transmitter. You can even use a laptop computer as a portable studio."

Programs such as Goldwave and Cool Edit are like turning your computer into a mini-broadcasting studio. These programs can mix several audio sources at set audio levels. They can also edit bits much faster than splicing open-reel tape with a razor blade. Best of all, the chances for permanent errors are minimal, thanks to the mighty Undo command in Windows.

In addition, these programs also offer the ability to add special effects to the audio. Some of the effects include flanger, reverb and delay, audio level alteration, distortion, and backwards audio. Such effects can really dress up a radio show and, if used properly, can make an amateur show sound more professional and well-crafted than many commercial stations.

Computers for Listening

Computers don't just improve life for the pirate station; they can be directly used to make listening easier and better. Computer professional Pete Costello said, "I have my AOR AR3030 connected to my PC via an RS232C serial data cable connection. I also have written my own radio control program that interfaces the AR3030 radio to my PC allowing me to control many of the operations remotely. This along with a wireless 900-MHz headphone lets me work on the computer (my job) or work on my computer (personal home and life management) or play on the computer, while tuning the band. My control program allows me to be alerted when scheduled broadcasts should be tuned in. When publications in print or computer-based tickle-my-fancy, I input the data as an alert. I use this sometimes to catch scheduled pirate broadcasts when they exist."

Costello also said, "I have also used IRC, Internet Relay Chat, using programs like PIRCH (my favorite) or mIRC to access an IRC server network to 'talk' to groups of pirate radio listeners in real-time one line at-a-time. I have used this technique to verify a West Coast USA pirate playing Grateful Dead music that I was hearing faintly here in New Jersey with a bunch of listeners in California."

Even better than IRC is the chat room on the Free Radio Network site. Just about any time a shortwave pirate is broadcasting, you can expect that someone will be in the chat room to talk about the show. I've gone on the chat room a number of times to check what station was on the air, and I've gone on many other times to let people know what I was hearing. In the most interesting chats, several people are listening to the broadcast and the station operator is also chatting. It gives you a chance to hear where different program elements were acquired, request songs, and comment—all in real time. It's fascinating to receive such comments as, "I just rotated the beam antenna 10 degrees to the South," or "The tubes are getting kind of hot, so I think I'll shut down the transmitter now."

Pete Costello also noted that he knows some other radio enthusiasts who use computer audio capturing to record and then edit broadcasts. Often this means cleaning up distortions, noises, and changes in volume so that a clearer, cleaner audio is achieved. Although I haven't used audio editing programs for this purpose frequently, I have recorded some segments digitally and then cut out static crashes, etc.

Some pirate DXers, particularly the North Americans who listen for the Europirates, find that it's especially handy to either e-mail these brief files to the pirates in question or place them on a personal webpage so that the stations (and other interested parties) can listen almost immediately. Best of all, not every pirate has "DXer's ears" and can pick the details out of a weak and static-filled recording. A little audio cleanup can improve matters significantly.

In general, the best programs for removing noise from audio files would be the same programs that pirates use for audio editing. Most audio editing programs have the ability to filter out different audio frequencies and also to edit out segments. Again, Cool Edit and Gold Wave are two favorites for shortwave listeners.

List of Pirate Radio Audio on the Internet

The following list includes plenty of pirate radio audio that you can find on the Internet. Some of the links are to old shows that can be downloaded or played back on a media player (such as RealAudio player or Winamp). Other links are for live streaming media from FM pirates. Still others are off-air recordings from DXers or older pirate stations.

If you have ever searched for information on the Internet, you will realize that the medium is extremely volatile. Web pages disappear long before the search engines can make the changes. The problem is compounded by pirates, whose existence is tenuous at best. Even if you backordered this book right off the press, expect that some of these links will be dead. However, some of these will last for years, so I think that it's worth including them in the book for that reason.

87X
http://www.elastik.com/87x
Tampa community station that's live on Live365.com on the weekends

BBMS
http://ourworld.compuserve.com/homepages/bbms4ozone/
RealAudio clips from UK shortwave pirate BBMS.

Classic Rock Radio
http://www.geocities.com/classicrockradio
Snips of audio as heard by listeners from around the world

The Irish Era
http://dxarchive.blackpool.ac.uk/rahome.htm
RealAudio clips from dozens of Irish community pirates from the early 1980s.

Dutch pirates audio archive
http://dxarchive.blackpool.ac.uk/radutch.htm
RealAudio clips from several 1970s Dutch pirates

DX archive
http://dxarchive.blackpool.ac.uk/rasw.htm
RealAudio clips from European shortwave pirates of the 1970s and early 1980s

Earthradio
http://www.earthradio.co.uk/
Radio Caroline audio, offshore jingles, etc.

EGMC
http://www.EGRN937FM.com
Classic rock from Colorado in RealAudio and on 93.7 MHz

The European Shortwave Pirates
http://www.fly.to/piraterad
A number of off-air clips of Europirates

Flux FM
http://www.fluxfm.nl/fluxjingle.htm
Jingles from this Dutch FM pirate

FRDM
http://www.geocities.com/krvlfm
Streaming audio from Naked Truth Radio, Tucson

Free Radio Clips
http://members.tripod.com/radioclips/
Tons of clips of UK FM pirates

Free Radio Network
http://www.frn.net
Programs in the True Speech format from dozens of different shortwave pirates

Free Radio Santa Cruz
http://members.cruzio.com/frsc
All-volunteer community radio on 96.3 MHz

Free Wave FM
http://www.deejay.demon.nl/freewave.html
Several audio formats, 24/7 from this Dutch pirate on 96.4 MHz

Free Waves
http://www.alpcom.it/hamradio/freewaves/suoni.htm
WAV and RealAudio jingles from Italian pirates

F.U.C.C.
http://www.sleepbot.com/fucc
Pearl Jam operated this station from Seattle on 89.1 MHz

Guerrilla Love Radio
http://www.chifreeradio.org
Streaming audio from a Chicago FM pirate

Jolly Roger International
http://www.geocities.com/radiojollyroger/
Site links to live365.com streaming audio of their old programs from the early 1990s

Beerus Maximus

"I find it amusing that the "Internet" is predicted to change everything as we know it and cause the death of shortwave, ham radio, social interaction and societal morals (among other things, on a long list). But that, I think, is a simplistic view that borders on irrationality. It happens every few decades when the pundits and marketing people usher in some "paradigm shift" in the way we live our lives.

But when you actually link everything together, very little has changed. Radio didn't kill the telephone, and television didn't kill radio. Likewise, I don't think computers and the internet will steal life from pirate radio or shortwave listening in general. 1/3 of the world's population lives without electricty completely. They *can* slap some batteries in a battered old shortwave, but they *can't* hear that new cultural interval signal "You Have Mail!" And that's not likely to change soon, or perhaps ever.

A radio is a one time investment that returns years of reward whereas telephone and electricty is a subscription service that the bottom rungs of our humanity will never enjoy. Really, 100 years after their introduction, 1/3 of our world's population lives without electricty or telephone. Amazing when you think about it, and astonishingly sobering as well.

The Internet will become a

Jolly Roger Radio International (JRRI)
http://listen.to/jrri/
A large RealAudio file from this Irish shortwave pirate

K-2000
http://www.geocities.com/Broadway/
Stage/2370/pgms.htm
RealAudio segments from their programs in the mid-1990s

K-Art
http://members.tripod.com/kart_fm
A 40-minute archive show from this FM pirate

Laser Hot Hits
http://scorpius.spaceports.com/
~laserhot
24/7 on the Internet and short-wave from the UK

Lick 106.7
http://www.angelfire.com/nt/
realvideo/menu.html
Some audio from their first show

Liquid Radio
http://www.frn.net/wwrb
Saturday nights, live on the web, from Minneapolis on 102.3 MHz

LOZ
http://www.home.zonnet.nl/
zwetsloot44/lozsite2.html
Streaming audio from an FM pirate on 104.7 in Zoeterwoude, The Netherlands

North Side Network
http://free.freespeech.org/nsn
Streaming hip-hop FM radio from the Midwest

Offshore '98
http://www.offshore-radio.de/
98offshore.htm
RealAudio from the world's first hobby offshore pirate

(Beerus Maximus continued)

corporate outlet, like Walmart or Home Depot. You'll find it on every American corner, and you can go in and buy anything you could possibly imagine, but ultimately it will be owned and controlled by corporate entities. Already we're starting to see the inevitable concentration of broadwidth, much like in the waning days of the '40s and '50s with radio and a little later with television. Once this happens, the great novelty will wear off and it will just be another $60 a month utility bill that helps you do mindless things.

Of course some of those things will actually be beneficial to pirate radio (and shortwave, and ham). I mean, I can pick a DX-60B off of eBay any day of the week, if I so desire, whereas five years ago, I might have had to do a few months worth of hamfest hunting. Rather than waiting for *The ACE* to arrive every month, I can get my radio jollies on the FRN, anytime. The Internet removes some constraints, makes information retrieval easier, but it has no personality. It will not hurt pirate radio any more than it will hurt golf, hunting, wrenching on your car, or collecting stamps. It'll make these things easier, if any-thing."

Offshore Radio Guide
http://www.offshore-radio.de
Huge offshore pirate site with some audio

Ozone Radio
http://ourworld.compuserve.com/homepages/bbms4ozone/
RealAudio clips from Irish short-wave pirate Ozone Radio

Pirate Radio Hall of Fame
http://www.offshoreradio.co.uk
An amazing amount of offshore pirate data and audio

Pirates Cove
http://www.wireless.org.uk/pirate
Offshore radio clips

Radiate 88
http://www.889fm.org
Cedar Rapids alternative radio, live and on 88.9 MHz

Radio 11 Onweizen
http://listen.to/11onwiezen
MP3s of past programs from this Dutch FM pirate

Radio 101
http://www.radio101.de
Some jingles from this German CB & FM pirate

Radio Alpen Adria
http://www.alpcom.it/hamradio/freewaves/alpe.htm
Just a jingle from this Italian shortwave pirate

Radio Blackbeard
http://stationss.www7.50megs.com/page8.html
live365 audio from this UK pirate

Radio Bluestar
http://home.wanadoo.nl/bluestarradio/live.htm
live programming 24/7 from this powerful Dutchie

Owsley of Up Against the Wall Radio

UATWR, Friday Radio, and a host of one shots owe their existence to computers. Not only did we do the first digital QSLs, but the entire productions were thanks to the computer.

I came into radio from a computer background. I didn't have cart machine, mixers, and several turntables, but I had a Soundblaster and some software. It was crude in those days, but Doomsday and Radio Airplane quickly converted to using their machines to make shows after seeing how easy it was done. This was several years before the legit stations were running their signal with a computer and MP3s, which is so common today.

The early use of 6955 was popularized by a cadre of stations all being generated by computers. Suddenly, you could almost count on hearing something if you just sat on 6955 during a weekend.

I like to think that the small resurgence of pirate radio during those days was partially due to the quantity (as well as the quality) of programming being aired. Without computer "studios" based on Soundblasters, it would never have happened.

Radio Caroline
http://www.radiocaroline.co.uk/
*New programs from this former
offshore pirate*

Radio Europe
http://www.alpcom.it/hamradio/
freewaves/europe.htm
*Jingles from this Italian relay
station*

Radio East Coast Holland
http://listen.to/eastcoastholland
*Jingles from one of the world's
most powerful pirates*

Austin Resistance Radio
http://www.austinresistance.net
*A clip of Free Radio Austin
getting raided*

Radio Free Clare
http://www.radiofreeclare.com
*Michigan Patriot radio on 88.3
MHz and live on the web*

Radio Free London
http://www.geocities.com/
SunsetStrip/Villa/2375/broadcast.htm
*Official site with hours and hours
of audio!*

Radio Free London audio archive
http://dxarchive.blackpool.ac.uk/
rflra.htm
*A fan site with RealAudio clips
dating back to 1968!*

Radio History
http://home.hetnet.nl/~oldies45/
index.html
*Several Live365.com streams of
'60s offshore pirates*

Radio Marabu
http://www.radio-marabu.de
*Many hours of audio from this
German alternative station!*

Radio Mariquita
http://www.alpcom.it/hamradio/

Bill O. Rights of Radio Free Speech

I can only address this from an op's
viewpoint. I'm in a remote area of the
country and have to go into town, to
the public library to get access, but
have found the Internet to be very
helpful in a couple of respects. One is
viewing propagation charts in real
time. I can see if it'll be a waste of
time to transmit on any given night,
since my signal has to go a long way to
reach anyone. Second is reception
reports. Once I've been on the air, I
can go to the Logs section and see
within a few hours who heard me, and
based on old QSL reports, roughly
where they are, and what the signal
strength was.

I echo what others have said
about streaming and netstations. Nice
term Narrowcasting and you could
spend weeks looking for a station only
to have it drop out on you after
listening for a few seconds. I don't see
netcasting as any threat to pirate
radio for a lot of the same reasons
stated by others. Nothing beats the
thrill of sitting in the shack, trying to
pull that signal out of the noise.

The FRN has played a key role
in helping the pirate community stay
in touch with listeners and Ops and we
should all thank John Cruzan for what
he's done. So in that respect, the
Internet has helped bring a very
diverse and alienated group of people
together, to exhange ideas and views
over a huge geographic area but has
done little if any, to dilute the magic
of pirate radio.

freewaves/mara.htm
A jingle or two from this Italian shortwave pirate

Radio Mistero Ghost Planet
http://www.alpcom.it/hamradio/
freewaves/rmgp.htm
Two jingles from this Italian shortwave pirate

Radio Northlight
http://www.free-radio.f2s.com/
northlight
Many off-air recordings of this German shortwave pirate plus Live365.com streams

Radio One Austin
http://www.radio1austin.com/
Austin pirate with streaming on Shoutcast and Live365.com

Radio Scotland Janim Archive
http://www24.brinkster.com/
scotland242
Archive of RealAudio from offshore pirate Radio Scotland

Radio Silver
http://www.alpcom.it/hamradio/
freewaves/silver.htm
Several jingles from this Italian shortwave pirate

Radio Strike
http://listen.to/radiostrike
Some jingles from an Italian shortwave pirate

Radio Viviana
http://www.alpcom.it/hamradio/
freewaves/viviana.htm
A jingle from an Italian shortwave pirate

Radio XRP
http://www.radioxrp.com
Live365.com nightly, and simulcast in Philadelphia on FM

Captain Blackbeard of Jolly Roger International

I think computers offer a great enhancement to Pirate Radio. As you may remember, Jolly Roger International only accepted signal reports over the old ANARC BBS at first. It was a great way to get reports very quickly. DXer's now have a way to let fellow DXer's know when a Pirate Radio station is on the air via E-mail. Pirate Radio Web sites.

The biggest improvement computers have made is radio production. What use to take days to create now takes hours. You no longer need cart machines, a radio console, and CD players. All you need is a computer with audio editing software and a transmitter. You can even use a laptop computer as a portable studio.

As far as netcasting, we have embraced it as a promotional tool for not only us but for the free radio movement in general. We have placed most of the old Jolly Roger International programs on Live365, which netcasts 24 hours a day and is linked to our web site. Netcasting makes a great studio-to-transmitter link that has worldwide coverage. Great way to feed multi-transmitter sites just like the big boys.

Netcasting is not going to become a problem to pirate radio until we have high-speed wireless Internet in the hands of most people. Even then it comes down to programming. If you have good programming, people will listen no matter how they receive it. When that time comes, I guess•we will have to sell the JR-Eye in the sky . . .

RNI Janim Archive
http://nl.internations.net/rni/
index.html
*Archive of RealAudio from offshore
pirate Radio Nordsee International*

Rock-It Radio
http:// www.palmsradio.com/
main.html
*An online rockabilly station
frequently relayed by pirates*

San Francisco Liberation Radio
http://www.liberationradio.net
*7 hours/7 days per week on 93.7
MHz and web streaming*

Solid Rock Radio
http://www.solidrockradio.net
*This SW pirate turned netscaster
might again return to SW*

Station Sierra Sierra
http://stationss.www7.50megs.com/
page8.html
*On Live365.com and RealAudio clips
from different Euro shortwavers*

Transatlantic Radio
http://users.bart.nl/~trans/sound.html
*Audio clips from this Dutch
shortwave pirate*

UK AM audio archive
http://dxarchive.blackpool.ac.uk/
lbp.htm
*RealAudio clips from 1970s UK
land-based pirates.*

Voice of Laryngitis
http://www.azstarnet.com/~dwahl/
index.html
*Audio from the old comedy
programs of this classic SW pirate*

Voice of Peace Janim Archive
http://www.zyfect.com/users/
vopsounds
*Archive of RealAudio from offshore
pirate Voice of Peace*

John Anderson of http://pirateradio.about.com

Computers, IMHO, have definitely helped pirate radio as a whole. Others have mentioned the production and communication benefits computers have afforded everyone—although I think the communication improvement definitely has been the revolutionary one.

The ability to impart "scene reports," as they were, with much more speed and ease is one thing—but the ability to collaborate with like-minded folks is quite another.

I kind of chuckle when people say, "what about netcasting?" Netcasting has its place—but in no way will it replace terrestrial radio anytime soon, for many of the reasons outlined in posts above. If anything, I think the threat is more from radio going digital itself—some of the buisness models proposed include things like turning stations into "wireless ISPs" and the like. Programming, in that sense, may become secondary—to the betrayal of the essence of what radio has always been. That scares me.

On the other hand, instead of using netcasting to replace radio, why not use netcasting to make some good radio? On the microradio side of things, one station I know of is experimenting with this very concept—their first incarnation was as a website where anyone could upload any mp3 they wanted, and it would automatically get dropped into the playlist

WBET
http://sites.netscape.net/wbet983
A commercial Illinois FM pirate on 98.3 MHz

Weekend Music Radio audio archive
http://dxarchive.blackpool.ac.uk/wmraudio.htm
RealAudio clips from Scotland's long-time shortwave pirate.

White Lake Radio
http://thelaser.webprovider.com
live365.com audio from a Michigan LPFMer

Wizard's Free Radio Recordings
http://www.magicspell.free-online.co.uk/recordings.html
Off-air clips of UK shortwave pirates

WRCR
http://www.fortunecity.se/hultsfred/bowiebacken/135/audio.html
A program from this UK shortwave pirate

Wreckin' Radio International
http://website.lineone.net/~12256
On-line programs in the Destiny format

John Anderson (continued)

rotation.

The next incarnation will actually involve a piece of software you can download—kind of like an mp3 player—where you'll build a playlist and then set it in the background. The station's server will query the list when there's an open spot programming-wise, and your computer will stream it to the station, which then rebroadcasts it. They call it "public access pirate radio" and I think it's pretty cool.

Computers and net connections have, without a doubt, been more of a boon than a bane to free radio broadcasters and enthusiasts. For the forseeable future, I think it'll stay that way.

Conclusion

Computers are seriously developing as an important tool for both audio and communications. However, if anything, the increased use of computers is making homemade radio seem more appealing, rather than rendering it obsolete. Frederick Moe of Seldom Heard Radio said, "For me, pirate radio is a true alternative to the direction that the rest of the culture is travelling, for better or worse. Computers can help with program content and audio mixing, however I will always prefer the crackling of radio signals appearing out of the aether." With the aid of somputers, it appears that pirate radio on shortwave will only grow in popularity.

Update

My major difficulty writing this book was that the pirate radio scene changes too fast. As I wrote the book, new stations came on the air, old stations left or reactivated, and stations were raided. Because so much had changed from the second edition, large portions were rewritten. By the time I had finished large sections, some of those portions had already become dated. This especially occurred in Chapter 3, which became a nightmare of writing, rewriting, and editing. Work on the assembly line was passing by me. I received warnings from McGraw-Hill that I was falling too far behind, so I just cut my losses and decided to stop revising (for the most part) and place new events in an update at the end of the book.

While I was writing and assembling the book, I received some telephone calls requesting that I be interviewed on radio programs. It's always a blast talking about pirate radio, and I sent along audio CDs of pirate clips, so that other people could get a taste of the fun. The first interview was in late August on the Allan Handelman Show, a syndicated FM rock talk show in the Southeastern U. S. I've been interviewed several times in the past there and it's always a blast to talk about pirates with Allan. The next interview was really big time, the Art Bell Show, which is syndicated on more than 500 radio stations in the U. S. He-Man Radio even broadcast during the show. I could hear He-Man in Pennsylvania, and Art even tuned him in (while on the air) on his in-studio shortwave receiver.

The day of the Art Bell Show wound up being more significant than I would have imagined. It was broadcast on the morning of September 11, 2001. Several hours later, I was at work exhausted after only getting one hour of sleep overnight. Reports trickled in about planes crashing into the World Trade Center and American life changed.

The events of September 11th (and the subsequent anthrax contaminations) are both terrible and tragic. However, they have been personalized at great length by the mass media, so I won't comment on them further, except in the context of pirate radio.

Such a tragic event could either drive pirates off the air or to the airwaves. After a few days of silence, a number of pirates returned to the air. Uncharacteristically serious, WHYP broadcast a patriotic 9/11 program. Take It Easy Radio did the same. Overall, few pirates aired special programs about the event, but many included patriotic comments or special clips for the survivors. The only dissenting voice was United Patriot Radio, which some listeners complained had an "I told you so" kind of attitude about the events.

By October, shortwave pirate radio was more-or-less back to normal in North America. In fact, with the improved propagation, it was probably better than it had been all year. By now, anthrax had been spread around New York and Washington, and B-52s had bombed Afghanistan many days in a row.

The programming on pirates apparently was not affected, but pirate radio was greatly affected nonetheless. Suddenly, data transmissions sprung up all through the traditional pirate frequencies. The 6940- to 6950-kHz area was blown out by a SITOR station and RTTY has been heard on 6955. Data of this sort had not been heard on these frequencies in literally years of listening every day to 6955 and 6950 kHz. These are evidently both military in nature and the pirates scrambled to get out of the way.

Although some pirates are still being heard on 6950 and 6955, mostly at off-times, most are moving down between 6900 and 6935 kHz. One note about this frequency change is that the SITOR station moved down a few kilohertz at the end of October (to avoid the pirates?) , leaving 6950 clear. So, 6950 might still remain on the active frequencies, but it's too early to know if this was a permanent move for the SITOR station or if it will be wiping out 6950 kHz again next week.

Really, the frequency moves are beneficial because it leaves frequencies available for more pirates to broadcast at the same time. And this situation was taken advantage of in late October when numerous stations were on the air while KIPM was marathoning on 6900, 6925, 6928, and 6935 kHz.

KIPM has been absolutely amazing. The science fiction stories are excellently written and read–truly a highlight of the current pirate radio scene. With the right promotion, I could imagine people buying shortwave radios specifically to listen to KIPM's horror-laden science fiction. I don't know whether Max writes all of the radio plays himself, whether he plagiarizes from other sci-fi writers, or whether he reads other people's short stories, but the programming is seamless. It's a perfect example of the creativity that is lacking in commercial radio. In addition to the program quality, KIPM is sometimes on the air all night with tremendous signals, such as October 29, 2001, when DXers in Australia tuned in to good signals for more than four straight hours!

One of the stations on during the KIPM pre-Halloween marathon was Paragon Radio, which was overlooked in Chapter 3. Paragon broadcast through the year with a format of blues music and random comments. However, the operator didn't really seem too interested in the music. He talked overtop of it about different topics—it was almost like filler between his thoughts. With comments like, "you have to accept me for what I am, completely unacceptable," "I've just discovered the truth, and can't understand why everyone isn't eager to hear about it," and "I'm just waiting to tell you I have nothing to say," maybe the operator should have named the station "One-Liner Radio."

United Patriot Radio

One of the most important radio events of recent years was end of United Patriot Radio. I expected the end to be near for UPR, but the outcome was nothing like what I suspected in Chapter 9! Steve Anderson was never (at least not yet) raided by the FCC for unlicensed broadcasting. However, he made a big splash in Middlesboro, Kentucky, news on October 16, 2001. On that date, Anderson was pulled over by Bell County Deputy Sheriff Scott Elder for not having tail lights on his pickup truck. According to Elder, Anderson was initially angry, but then calmed down . . . until he asked for registration papers for the guns in Anderson's truck. Then, Anderson reportedly recited bylaws from the Kentucky State Militia and drove away. According to the *Middlesboro Daily News*:

> "I then started the pursuit and he turned across the median and tried to ram the cruiser, two or three times, then he started southbound in the Northbound lane of traffic and started firing his rifle at oncoming traffic," Elder explained. Anderson then crossed the median again and in a straight away, just before Sam's Mountain, Anderson stopped his vehicle in the inside lane. "I stopped about 50 yards behind him and he stepped out of his truck and opened fire and I returned fire and that's when he went up on the hill (Sam's Mountain)," he said. During this pursuit Elder was assisted by Pineville Police officer Greg Hendrickson. Anderson reportedly struck Hendrickson's police cruiser also.

According to the reports, Elder's police cruiser was "disabled" by the gunfire. In fact, deputy sheriff John Hoskins said "He tried to cut it in half." After escaping, Anderson's pickup was found in the nearby mountains. Except for several guns and pipe bombs, the truck was empty. The National Guard brought in a helicopter with the ability to locate heat-producing sources. Assisting in the manhunt were Middlesboro Police, Kentucky Department of Fish and Wildlife, the Bell County Volunteer Fire Department, ATF, and FBI. Two weeks later, Anderson still had not been located.

Anderson was, obviously, off the air, but United Patriot Radio lived on in the hearts of pirate radio fans. Radio Bingo's "United Patriot Militia Bingo" series continued almost nightly with updated programs concerning Anderson's flight. In one program it was claimed, while doing a Steve Anderson impersonation, that Anderson stole all of the Bingo winnings and the broken-down bingo machine. Later episodes featured Anderson being spotted by various pirate listeners. These broadcasts were mostly near 6925 kHz.

Europe

Although no North American pirates were raided by the FCC while this book was in production, several Europeans were. The first was Radio Northlight (Nordlicht), which was raided in Germany on September 29, 2001 by two German post and telegraph agents (German FCC equivalent) and two police officers. The station had not caused interference and only a few pieces of equipment were confiscated. Northlight had been heard the weekend before in North America.

The other station raided was one of the better-heard stations in North America, Mike Radio. Hours after the raid, Mike e-mailed the following message:

DURING THE LAST TRANSMISSION ON 48 MB 28 OKTOBER 2001 TIME AROUD 0.00, MIKE RADIO INT.IS RAIDED

4 POLICE CARS ONE OF THE RCD AND ONE VAN, FOR THE TRANSMITTERS

ALL TRANSMITTERS AND AUDIO IS TAKEN WITH THEM

IN TOTAL 9 MEN CAME TO THE STUDIO!!!

I WILL COLLECT STAMPS..................

mike

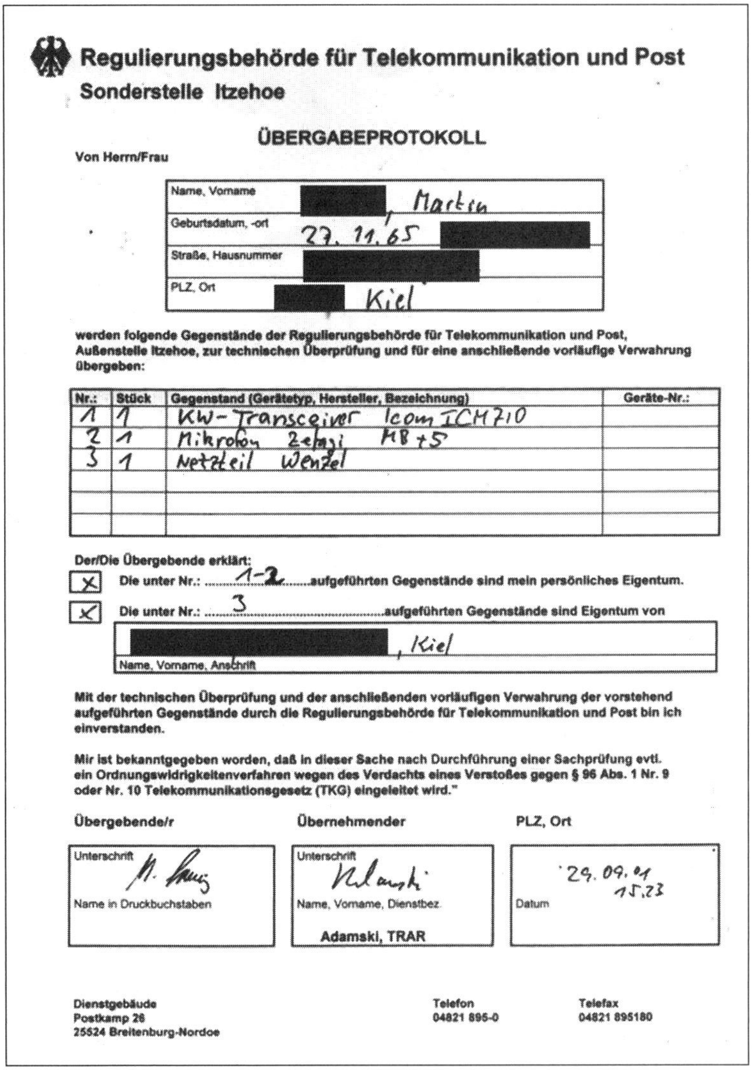

Fig. U-1 *A copy of the German PTT's version of a Notice of Apparent Liability against Radio Northlight from September 2001.*

This raid is significant for Mike because he had several very large (and presumably costly) Rohde & Schwartz German military surplus transmitters (pictured on a QSL in Chapter 7). Several days later, Mike's web page noted that the station would soon be returning from another location with much more power. When Mike Radio was raided, there was some question concerning interference because he was operating on 48 meters (home to most Europirates and also many licensed maritime stations), rather than his scheduled standard

Sunday morning frequency of 9290 kHz. Some other Europirates began digging through the frequency manuals, looking for clear frequencies in 48 meters that are not licensed to governments for communications.

Now, a brief explanation about the government communications on shortwave. Although a frequency might be licensed to a particular service, these frequencies are not always used. A case in point is 6950 and 6955 kHz, which are licensed, but had not even been tested upon for literally years. However, since September 11th, these frequencies have been alive with data. 6200 to 6300 kHz is the home for European pirates, but these frequencies are rarely used for SSB communications or data transmissions, particularly during Sunday mornings. So, although these frequencies are full of licensee stations, they contain very little activity. Thus, most pirates would not consider themselves to be negligent for broadcasting on an unused frequency-even if it is licensed to another station. If the station decides to reactivate in five years, they will concede the frequency.

Now, some pirates are taking preventative measures to move out of the way of licensed stations that might decide to transmit in the near future. On 48 meters, the clearest range of frequencies (in terms of licensees) is 6300 to 6320 kHz. Otherwise, it's impossible to predict whether the bulk of the Europirates will remain on 48 meters or if they will move to clear frequencies in 41 or 31 meters . . . or go somewhere else altogether.

Other stations on the air

Halloween always brings out some of the more interesting stations and 2001 was no exception. Probably the most fascinating was the New Voices of the Purple Pumpkin, which covered some of the history of the original Voice of the Purple Pumpkin from 1970. Some of the information broadcast matched things that I've heard in the past from people "who might know," so I tend to believe that this is either a return of the original station or a broadcast by someone who knows the original crew. Regardless, it did not appear to be one of the many VoPP one-offs that pirates in the past had pulled off.

Take It Easy Radio returned in the Fall (as noted previously in this section) and made several well-heard Halloween broadcasts on 10/28 and 10/31. Radio USA returned with a repeat of a Halloween program from the mid-1990s. WHYP aired another installment of the Brownyard family Halloween program, digging out scary songs from the past, including "The Zeller Mash." After fighting a battle with a data station on 6945 kHz, Z-100 announced that they would be testing on the unusual frequency of 13555 kHz. WAIR and the Purple Nucleus of Creation were new stations relayed by KIPM.

I've talked with some people who have felt that this was the best Halloween for pirate activity that they had ever experienced. Because I've been listening for quite a few years, I wouldn't consider this the best Halloween, but it would probably place in my top five. That said, it appears that we're in the middle of a great era for pirate radio listening.

With a great crop of pirates on the air and as fears of anthrax and terrorism abound, this is a great season to stay at home and listen to the shortwave.

Conclusion

Thus, the contacts made within the hobby and information received from other sources are of great importance. Information breeds enthusiasm, which in turn breeds a desire for hobbyists to begin broadcasting and pirates to operate more often. This increased activity breeds more information. However, this is not to say that by actively listening for pirates that you will cause others to broadcast illegally. This book is intended to aid the DXer and radio hobbyist in receiving signals and verification cards from unlicensed broadcasters. Nothing more is intended; all methods of pirating and biographical details from the various stations were added to help the radio hobbyist to better understand different broadcasting methods and odd occurrences. The reader is not encouraged in any way to operate radio transmitting equipment against the rules of the Federal Communications Commission.

However, unlicensed broadcasters do exist and cannot be ignored. This book is dedicated to those that will (or do) enjoy listening to pirate radio.

Happy listening!

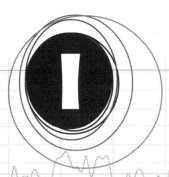

Index

CD Track List

This CD is a compilation of snips of audio from about as many different stations as I could squeeze onto the CD. As a result, the audio for many of these stations is poor, and some of the clips aren't particularly funny or dramatic. But there's a lot here and if you're a fan of pirate radio, you'll probably enjoy this . . . it might even bring back a few memories.

Many of the stations on this CD have disappeared without a trace. Only a few of the stations from the 1980s still exist, and many from the 1990s are gone, too. Even if you knew some of these operators personally, tracking down audio from these stations could be a daunting task. For example, the operators of the Voice of the Voyager were easy to locate, but they have absolutely no recordings of their hundreds of hours on the air.

I believe that the volatile nature of pirate radio makes these recordings all the more essential. For all I know, if I wouldn't release these CDs, the audio record of some of the stations contained herein might become extinct.

1. **Union Radio (The Netherlands)** Sign off, 6210 kHz, January 21, 2001.
2. WABE FM pirate from Houston, PA. This clip was from the summer--sometime between 1972 and 1974.
3. **Radio Nova International (The Netherlands)** Clip of Frank Carson DJing in Dutch, from 2001.
4. **WDX** 1620 kHz, January 2, 1984. A recording from one of the last New York City area pirates.
5. **KPRC** 1616 kHz, January 21, 1984. Pirate Joe signing off the air with their signature tune, "Goodnight Irene."
6. **The Pirate Blaster** 1616 kHz SSB, January 9, 1984. Jamming KPRC with occasional comments about the programming.
7. **RBCN** 6955 kHz, October 1, 2000. Radio Bob describing putting $17,000 into recording this show, so the listeners had better write good reports!
8. **Kentucky State Militia Radio (KSMR)** 3260 kHz, 0300 UTC, March 17, 2001. With Steve Anderson. Unfortunately, his amplifier had a nasty hum during this show. Later, this station became United Patriot Radio.
9. **United Patriot Militia Bingo** 6925 kHz, 0130 UTC, October 21, 2001. Radio Bingo parodies the United Patriot Radio situation after Steve Anderson is

wanted and on the run.

10. *Voice of the Lake Superior Circle Route Radio Network* 6955 kHz USB, 0245 UTC, July 23, 2000. J. Spencer and Jimmy Hix.

11. *East Coast Beer Drinker (ECBD)* 6955 kHz USB, 0425 UTC, September 3, 2000. Distortion and feedback galore!

12. *Radio Torenvalk (The Netherlands)* Studio ID from Spring 2001.

13. *WPN* Via Radio Free Speech, 6240 kHz, 1430 UTC, November 26, 2000.

14. *WKND* "We're Kanine Dog" Studio sign-on, November 1994.

15. *KMUD* October 13, 2001. A slogan or two and a Morse code ID.

16. *Voice of the Angry Bastard* 6950 kHz, 2240 UTC, December 15, 2000. A quick ID amidst the adjacent-channel SSB interference.

17. *WHYP* Christmas program 2000.

18. *Paragon Radio* 6950 kHz USB, October 28, 2001. Quick sign off.

19. *Z-100* 6945 kHz USB, October 28, 2001. A quick ID.

20. *WSKY (Whiskey Radio)* 7415kHz, December 25, 1991. A segment of their Christmas special.

21. *Jolly Roger International* 7415 kHz, January 1, 1992. A brief ID from their New Year's special.

22. *Sycko Radio* Some bits from around Halloween 2001.

23. *Radio First Termer* FM pirate from South Vietnam, operated by American G.I.s, guesstimated date of 1972.

24. *Voice of the Epileptic Catfish* This novelty pirate appears approximately once every five years, for only brief periods. This ID is from 1984.

25. *Radio Newyork International* Testing on July 28, 1987 with the crew from WHOT.

26. *Radio Blandengue* South American pirate, studio clip, 2001.

27. *Voice of the Voyager* 6220 kHz, 0300 UTC, November 5, 1978. Interval signal and sign-on announcement.

28. ***Radio Free London*** Mark Ashton DJing on a Wednesday night FM program in 1995.

29. ***Radio Free Speech*** From 1996, one of Bill O.'s little broadcast-ending editorials.

30. ***K-2000*** An ad for that new DXing movie . . . and a top-10 list from the Bob Rock show. I believe this program is from 1995.

31. ***Radio Free Naptown*** Bassy and a bit distorted, this is a listener recording of this Indianapolis FM pirate from 1978.

32-33. ***Radio North Coast International (RNCI)*** ID and ad for the RNCI Scum Sheet from May 26, 1987.

34. ***Voice of Laryngitis*** The evil doctor and Gomez lose control of the vilest weapon ever known to man! January 1, 1984.

35. ***Radio Airplane*** Time to call The Radio Airplane, October 2, 1993.

36. ***Radio Garbanzo*** Fearless Fred wants you to dine at the International House of Chili. From November 1989.

37. ***Free Radio Santa Cruz*** One of the few major FM pirates from the mid-1990s currently still on the air. 96.3 MHz on February 18, 1996.

38. ***WENJ*** The last major upper AM band pirate from the East. Promo from

August 1988.

39. **Hope Radio International** MJ makes a final desperate plea to gain some leverage against the FCC. It doesn't work. Hope Radio is raided, but MJ only loses his amateur radio license, not his freedom.

40. **9X2V** "The Voice of 1932" mysteriously appeared across the dimensions of time and space 60 years later.

41. **WR International** Studio recording, from Sunday, March 18, 2001.

42. **Radio USA** An ID and one of the long-running, Joe Bob's series of ads from an oldie. July 2, 1991.

43. **KNBS** Phil Muzik discussing the possibilities for marijuana, on one of the longest-running, U. S. pirates. July 1985.

44. **WREC** An ID from program #3 from the early 1990s. WREC was one of the most active stations of the 1990s.

45. **WSLH** 1605 kHz, December 21, 1970. An ancient clip from this AM pirate, believed to have direct connections to the Voice of the Purple Pumpkin.

46. **Voice of the Purple Pumpkin** 7335 kHz, July 28, 1970. The earliest-known recording of a North American shortwave pirate . . . and one of the most copied stations. Unfortunately, it's also a terrible-quality recording, (right channel only).

47. **Voice of the Purple Pumpkin** 7410 kHz, October 25, 1982, 0511 UTC. Is this really the same station that was on 12 years earlier?

48. **Radio Morania** A 1969 shortwave program that has been aired by pirates for decades. Not an easy station to QSL!

49. **Munchkin Radio** Just a brief ID from the Spring, 1983 program. One of the best-known "mystery pirates" of the 1980s.

50. **WCPU, Silicon Valley Radio** The first all-computerized pirate station, from 1984. Voice processing has advanced a bit since then!

51. **Radio Confusion** One of the best-known U. S. pirates of the early 1980s. I believe this program was from 1981 or 1982.

52. **Christian Rock Radio** A brief section of interview with Mbanna Kantako of Black Liberation Radio, discussing, his viewpoints on racism. From January 1992.

53. **WRPD, Warped Radio** 7414.4 kHz, 2200 UTC, January 25, 1992. A snippet from their first show.

54. **WAM** A New York City top-of-the-AM band pirate from the early 1980s. This reocrding is from Halloween.

55. **WRKS** A New York City FM pirate being relayed on AM in the early 1980s.

56. **Voice of Revolutionary Vinco** A classic shortwave DX parody pirate from somewhere between 1979 and 1981.

57. **Pirate Radio New England (PRN)** A great off-air recording, from 3:43 A.M. of this classic 1620-kHz phone-in station. This recording was from sometime between 1979 and 1983.

58. **KVHF** Some clips, showing off the professional sound of the station. These clips are from 1980.

59. **Radio Wolf International** September 15, 1990 The Radio Animal, Harry, and Sparky talking about *Pump Up the Volume*.

60. **Voice of Free New Jersey** 1615 kHz. From sometime in the 1970s or early 1980s.

61. **KDOR** 830 kHz, September 11, 1978. One of the early community pirates, K-DOOR broadcast for several years from Los Angeles.

62. **WFUN** 1632 kHz from the early 1980s. Maybe "W-GRUFF" or "W-SURLY" would have been more appropriate. Needless to say, after listening to the accent, this one was from the New York City area.

63. **Take It Easy Radio** 6950 kHz, October 28, 2001. The end of a Halloween bit, raving about bunny rabbits and Al Fansome.

64. **Radio Caroline** Clip from the Summer of 1968 when Caroline was hoping to become legally recognized by the UK government.

65. **Radio North Sea International (RNI)** Long clips from when the *Mebo II* was firebombed and the crew was about to abandon ship. The following clip is the news from the next day, announcing the damage.

About the Author

Andrew Yoder discovered a 40-year-old military-surplus Hallicrafters SX-28A in a friend's basement in 1981 at the age of 13. Since then, pirate radio has become the focus of his short-wave-listening hobby. In 1987 and 1988, he wrote the first edition of *Pirate Radio Stations*. In 1989, Andrew started the radio logsheet *Pirate Pages (PiPa)*, which he published until 1998 to make way for *Hobby Broadcasting* magazine, which has been published ever since. His magazine articles have appeared in *Electronics Now*, *Popular Electronics*, *Popular Communications*, *Radio!*, etc. He has appeared as a guest on the Art Bell Show, the Voice of America, Radio Netherlands, Allan Handelman Show, WBZ, KOA, WYSP, and many other stations. Of course, he has been a Pittsburgh Pirates fan for nearly his entire life.